The Minerals, Metals & Materials Series

Corby Anderson · Graeme Goodall ·
Sumedh Gostu · Dean Gregurek · Mari Lundström ·
Christina Meskers · Stuart Nicol · Esa Peuraniemi ·
Fiseha Tesfaye · Prabhat K. Tripathy · Shijie Wang ·
Yuanbo Zhang
Editors

Ni-Co 2021: The 5th International Symposium on Nickel and Cobalt

TMS

Springer

Editors
See next page

ISSN 2367-1181 ISSN 2367-1696 (electronic)
The Minerals, Metals & Materials Series
ISBN 978-3-030-65646-1 ISBN 978-3-030-65647-8 (eBook)
https://doi.org/10.1007/978-3-030-65647-8

Cover illustration: Top left: From Chapter 7 "Selective Sulfidation and Electrowinning of Nickel and Cobalt for Lithium Ion Battery Recycling", Casper Stinn et al., Figure 2: Lithium, nickel, manganese, and cobalt were selectivity sulfidized from synthetic nickel-manganese-cobalt oxide cathode material. Optical microscopy (a), SEM microscopy (b), and EDS mapping (c, d) reveal that upon selective sulfidation, a cobalt-rich sulfide (c, purple, Co0.67Ni0.33S), nickel-rich sulfide (c, pink, Ni0.75Co0.25S), and manganese oxysulfide (c, yellow, MnO0.2S0.8) phases formed. These phases coalesced to sizes between 100 μm and 1 mm, suggesting that nickel, manganese, and cobalt may be liberated via comminution and physical separation. Minimal inclusion between manganese and nickel-cobalt is observed. While lithium is indiscernible on the EDS maps, XRD reveals lithium to exist as a sulfate, suggesting oxygen-rich regions in (c, d) lacking Ni, Mn, or Co may correspond to lithium sulfate. Due to differences in aqueous solubilities between sulfides, oxysulfides, and sulfates, lithium may be potentially recovered from sulfidized metals via leaching. https://doi.org/10.1007/978-3-030-65647-8_7.

This Springer imprint is published by the registered company Springer Nature Switzerland AG
The registered company address is: Gewerbestrasse 11, 6330 Cham, Switzerland

Editors
Corby Anderson
Colorado School of Mines
Golden, CO, USA

Sumedh Gostu
Air-Liquide
Newark, DE, USA

Mari Lundström
Aalto University
Espoo, Finland

Stuart Nicol
Gopher Resource
Brisbane, QLD, Australia

Fiseha Tesfaye
Åbo Akademi University
Turku, Finland

Shijie Wang
Rio Tinto Kennecott Utah Copper Corp
Riverton, UT, USA

Graeme Goodall
XPS - Glencore
Falconbridge, ON, Canada

Dean Gregurek
RHI Magnesita
Leoben, Austria

Christina Meskers
Helmholtz Institute for Resource Technology
Frieberg, Germany

Esa Peuraniemi
Boliden Harjavalta
Harjavalta, Finland

Prabhat K. Tripathy
Batelle Energy Alliance (Idaho National Laboratory)
Idaho Falls, ID, USA

Yuanbo Zhang
Central South University
Changsha, Hunan, China

Preface

As when I was a co-organizer of Ni-Co 2013, the Markets, Mineral Processing, Extractive Metallurgy, Recycling and Physical Metallurgy of Nickel and Cobalt have been a focus of activity for both TMS EPD and MetSoc of CIM. As such, these two metals have evolved significantly since ancient times. Then, they were both named after unidentifiable demonic spirits who prevented the production of other metals. As such, Kobold was initially designated for the goblin-like Cobalt impurity. Old Nick, the chief spirit of evil, was designated for the Nickel found in copper ores. Now, these metals are better understood and utilized and form an essential part of our society's needs. We currently deem them Critical Materials as they are key metals for batteries and modern materials.

After 2005, TMS and MetSoc decided to collaborate on future programming in this area. Hence Ni-Co 2021 represents a continuation of this mutual effort. And given the sudden global challenge of COVID-19 in early 2020, the organization and completion of this conference became exponentially difficult.

So, it is both a tribute and a dedication to the members of the Ni-Co 2021 Organizing Committee, the contributing authors, and the TMS staff that this set of proceedings has actually been produced given the immense challenges that were suddenly thrust upon humanity. I personally commend them all as this has undoubtedly been the most challenging conference that I have ever been part of organizing in my over 40 years in the industry. Suffice it to say, shear dogged determination has won the day with the production of the Ni-Co 2021 proceedings as it will eventually in defeating and controlling COVID-19.

As a special note, I would like to personally thank both of my early mentors and longtime friends Mr. Jim Gulyas and Dr. Gus Van Weert for introducing me to the applied Extractive Metallurgy of Nickel and Cobalt. Jim was a highly skilled former Sherritt Gordon engineer who was instrumental in the adoption and application of their famed ammoniacal based hydrometallurgical technologies worldwide. Unfortunately, he has recently passed away. Gus is a former Falconbridge engineer and still practicing technologist who was instrumental in the development and application of their famed chloride based hydrometallurgical process that is still being practiced on an industrial scale in Norway.

In summary, we hope you find this proceedings volume a useful addition and reference to the continually growing understanding of Nickel and Cobalt.

Eur Ing Dr. Corby Anderson
PE CEng QP FIMMM FIChemE
Ni-Co 2021 Lead Organizer
Harrison Western Professor
Mining Engineering Department
and George S. Ansell Department
of Metallurgical and Materials Engineering
Colorado School of Mines
Golden, CO, USA

Contents

Part IV Pyrometallurgy

Abstracts from Presentations not Represented by Manuscripts

Plenary

EPD Distinguished Lecture: Ferronickel—Thermodynamics, Chemistry, and Economics

Rodney Trevor Jones

Ferronickel is used for producing stainless steel and other ferrous alloys. Therefore, the smelting of nickel lateritic ore to produce ferronickel is of great importance to the stainless steel industry. A novel simple thermodynamic model allows one to map the relationships between the Fe:Ni ratio in the ore and the FeNi grade and Ni recovery. The chemical composition of the lateritic feed material, especially the SiO_2/MgO ratio, strongly affects the composition of the slag, and how the furnace is operated. The choice of the FeNi product grade is affected primarily by economic and marketing considerations. A blend of thermodynamics, chemistry, furnace design and behavior, and economics allows good choices to be made around a particular project. The talk will also touch on a little of the history of laterite smelting and will provide a brief overview of world FeNi production.

Batteries

Processing of Leach Solution of Spent Li-ion Batteries for Generating Pure Nickel Sulfate and Cobalt Sulfate Salts via Solvent Extraction Route

Harshit Mahandra and Ahmad Ghahreman

This study pertains to synergistic extraction and recovery of nickel and cobalt from processed leach liquor containing 5.22 g/L Ni, 5.85 g/L Co, 4.12 g/L Mn and 1.64 g/L Li using organophosphorus extraction system. Dilute sulfuric acid is used as a stripping agent for both metal ions. Nickel and cobalt are recovered as pure (>99%) metal sulfates via evaporative crystallization of their respective stripped phases. The economic viability of the process has been confirmed by the recyclability of the extraction system up to five extraction cycles. The manganese and lithium are recovered as oxide and carbonate, respectively using precipitation methods. The recovered products were characterized by XRD and FE-SEM techniques. The study can be extended from bench to plant scale for industrial applications to provide economic and environmental benefits.

Generalization of the Dissolution Mechanisms of Battery Materials and Impact on Their Recycling

Emmanuel Billy, Marion Joulié, Adrien Boulineau, Eric De-Vito, and Richard Laucournet

Much attention has been focused on hydrometallurgical routes to recover valuable metals from spent Li-ion batteries. A lot of works has demonstrated that this method is an effective approach toward the recovery of a large panel of metals constituting LIBs and particularly the positive electrode material. Despite the large number of studies on hydrometallurgical processes for LIBs recycling, the phenomenon taking place during positive electrode material dissolution remains unknown. This work is dedicated to a kinetic study of the dissolution reaction in acidic media. The relation between structural changes and dissolution mechanism is studied during the dissolution evolution on different chemistry (NMC, LFP, …). The surface composition analyses of residual particles are performed. The electrochemical experiments allows to precise redox reactions taking place at the interface. The limitations of NMC dissolution in acidic media are also identified. Finally, the results allow defining new ways of treatment to reduce the recycling impact.

Improved Sustainability in High-Purity Chemical Production for Lithium Ion Batteries from Raw and Recycled Feedstocks

Thomas Bibienne, Rob Fraser, Mark Machado, Lily So, and Henry von Schroeter

Climate change is driving electrification of transportation and the need for lithium ion batteries (LIBs). The demand for LIBs in the next decade is expected to grow tenfold, which will translate to similar demand growth for battery chemicals, particularly battery grade nickel and cobalt sulphates. Due to the small consumption of LIBs to date, the supply of high-purity battery chemicals is primarily based on modifications of existing metallurgical plants. Additionally, most existing and proposed facilities produce soluble by-product salts like sodium sulfate that pose marketability and/or environmental challenges. This paper introduces a battery chemical refining process suitable for raw and recycled feedstocks with expected improvements to capital and operating costs, that avoids the production of problematic by-products.

Hydrometallurgy

Effective Treatment of Domestic US Cobalt Ores and Concentrates

Andy Tomaka and Corby G. Anderson

The global battery revolution currently relies upon battery technology requiring increasing demand for cobalt. As such cobalt is an essential Critical Material. Currently, there is very little US production of cobalt and most of this is as a by product copper and nickel production. The primary barrier to production is that primary cobalt deposits are mostly contained in the arsenic bearing minerals cobaltite, CoAsS, and scudderdite, CoAs. This current project will demonstrate the effective selective separation of cobalt from the arsenic in these minerals. A clean cobalt concentrate is produced. Arsenic can be stabilized as a ferrihydrite, scorodite of a patented cerium based technology. Potential flowsheets and pertinent economic estimates will be provided.

Separating and Recovering Cobalt and Iron from Co, Fe-bearing Metallurgical Slag via Acid Leaching Process

Yuanbo Zhang, Yikang Tu, Zijian Sua, and Tao Jiang

Co-containing metallurgical slag has been regarded as one kind of potentially recyclable Co resources. Pyrite cinder is a typical metallurgical slag produced in sulfuric acid industry, which contains a significant grade of valuable metals, such as iron, cobalt, etc. However, it is very difficult to utilize the pyrite cinder due to its low surface activity. In this study, an acid leaching process was used to separate iron and cobalt from the pyrite cinder. The results indicated that the leaching efficiency of cobalt achieved more than 98% under optimal conditions. Under the phosphate system, most of iron formed insoluble iron phosphate precipitate, which was easily separated from cobalt by filtration. The cobalt was recovered by organic extraction and the battery-grade precursor of iron phosphate was finally obtained through purification.

High-value Nickel, Copper and Cobalt Products from Bulk Sulfide Concentrates Using Albion Process™ Atmospheric Leaching

Daniel Mallah, Paul Voigt, Mike Hourn, and Glenn Stieper

In the ore beneficiation value chain, a bulk concentrate product is often selected as a low technical risk option at the expense of marketability. It can also be a high value-addition product in the right circumstances. In integrated processing, a successful example is Glencore's Raglan Mine (Nunavik, Canada). Since 1997, the site has been producing a bulk Ni-Cu concentrate, which is further processed to Ni matte at Glencore's Sudbury smelter. More typically, a hydrometallurgical flowsheet is well suited for nickel ores with poor flotation selectivity, resulting in bulk Ni-Cu-Co concentrates. Currently buoyed by demand from the battery industry, Glencore Technology has successfully tested numerous bulk sulfide concentrates for treatment in the Albion Process™. A new process has been developed for the recovery of Ni, Cu and Co as saleable products from solution. In this paper, several case studies of bulk concentrate treatment enabled by Albion Process™ are presented.

PHREEQC Modeling of Fluoride Removal from Nickel Sulfate Solutions Using Calcium Compounds and Experimental Validation

Elbert Muller Nigri, Stefan Lakemond, James Vaughan, and Sônia Denise Ferreira Rocha

This work aims to remove the residual fluoride present in the concentrated nickel sulfate solution obtained after removing manganese, cobalt, and magnesium from the mixed nickel-cobalt hydroxide precipitate (MHP). Fluoride removal will occur by precipitation using calcium salts. However, excess calcium can generate extra solids and scale due to the formation of sulfate salts. The PHREEQC geochemical program was used to simulate precipitation reactions based on the saturation index of fluorite, anhydrite, and gypsum. The simulations indicate that the use of calcium in the stoichiometric dosage has saturation indices of 2.94–3.00, 0.19 and 0.05 for fluorite, gypsum, and anhydrite, respectively, with a residual fluoride of 22–19 mg/L. And, with the dosage reduced to 52% of the stoichiometric mass, these saturation indices became 2.80, −0.01 and −0.15, with a residual fluoride of 24 mg/L. The results indicate that this dosage can be used in the process without the formation of anhydrite and gypsum.

An Innovative Approach for Nickel Recovery from a Copper Tankhouse Electrolyte

Tracy Morris

Nickel is a common impurity in copper anodes that can build up in the electrolyte of copper electrometallurgical plants that needs to be controlled. In the past evaporation was a frequently used method to produce a nickel sulfate product by first de-copperizing the electrolyte before evaporator units were used to precipitate the nickel as nickel sulfate. With the advent of a process referred to as APU (Acid purification Unit) it is now possible to separate nickel, copper and other base metals away from the sulfuric acid from which the nickel can be recovered as a nickel carbonate product. This paper will delve into how the process was incorporated within the ASARCO- Amarillo Copper Refinery using exiting tanks and piping without additional capital and also review how the middling and other products were handled so that no metal loss or environmental issues arose.

Starved Acid Leaching Technology for Nickel and Cobalt Recovery from Lean Resources

David Dreisinger

The Starved Acid Leaching Technology (SALT) uses reduced amounts of sulfuric acid to leach a portion of the nickel and cobalt from lean ore materials. The use of acid must be economic with respect to the nickel and cobalt extracted. Lean resources of saprolite nickel laterite ores may become economic through application of SALT. The SALT process can be tailored to recover various intermediate products including mixed hydroxide precipitate (MHP) or mixed sulphide precipitate (MS). The current development of the SALT process and the prospects for refining MHP or MS to high value nickel and cobalt products will be presented. The SALT technology applied to lean resources may increase the supply of nickel and cobalt for battery material manufacture to meet the demands of electrification in the transport sector.

Markets

A Combined Material Flow and Value Chain Analysis as a Tool to Improve the Understanding of Cobalt Criticality in the European Union

Raphael Danino-Perraud

We will investigate the criticality of cobalt in the European Union with a value chain analysis with a material flow analysis from 2008 and 2017. This research aims at a better understanding of the economical concerns about this metal, identifying the vulnerability of its supply and filling the data gap. The analysis of the structure of EU cobalt imports and exports allows a better understanding of the vulnerability of the cobalt value chain in Europe. In that regard, the geopolitics of stocks and flows should be a determining element of our analysis. Identifying them within the EU also helps to determine the potential of the urban mine and contribute to the diminution of the EU vulnerability regarding cobalt. In addition, the analysis of the value chain highlights the race to move-up the value chain because of the higher added value of the transformed products and the acquisition of new technologies.

Global Electrification of Electric Vehicles and Intertwined Material Supply Chains of Cobalt, Copper and Nickel

Ruby Nguyen, Roderick Eggert, Corby G. Anderson, and Mike Severson

Electric vehicles (EVs) are deemed a solution to a low-carbon economy. Future EV adoption faces high uncertainty resulting from battery material supply risk, diverse battery technologies, and an incomplete understanding of potential impacts on power and other systems. Various EV projections have been made, but research into supply with market dynamics to support decision-making is inadequate. Using a dynamic market model, we estimate global cobalt, copper and nickel production given different EV growth scenarios from 2020 to 2040. Results suggest that EV targets might not be met if the market and commercial battery technology deployment stay the same for the next two decades. Potential projects, including the Idaho Cobalt Project and Sconi, could add 2% to EV sales during the simulation. While shortening battery lifetime and improving collection and recycling rates increase secondary supply, the overall effects on EV growth are mixed.

Materials

Effect of Fluorine on the High Temperature Oxidation Behavior of Nickel-based Alloys

Alexander G. Donchev and Mathias Galetz

The Al content of conventional Ni-based alloys usually is not enough to form a protective outer alumina layer. Instead a chromia scale develops during high temperature exposure in oxidizing environments with local internal alumina formation. This chromia layer exhibits some disadvantages, e.g. rather high growth kinetics, permeability to nitrogen or Cr oxide evaporation above 900 °C. On the other hand, alumina does not show such disadvantages. In a new attempt fluorine is applied on the surface of several conventional Ni-alloys to promote the formation of a protective alpha-Al2O3 layer via the so-called fluorine effect. The almost exclusive formation of gaseous aluminum fluorides in the surface zone of the components and their subsequent transformation towards aluminum oxide stimulates the formation of an outer alumina layer during high temperature service. Results of high temperature exposure tests of untreated and fluorinated specimens of several technical Ni alloys will be presented and discussed.

Effect of Electrodeposition Parameters on Chemical and Morphological Characteristics of Ni-Co Electrodeposited Alloys

Morteza Salehi, Ali Saidi, Mahdi Ahmadian, and Keyvan Raeissi

The effect of pulse and direct current in a sulfate electrolyte solution was presented in this paper especially on the surface morphology, microstructure, composition, and grain size. Electrodeposition parameters including current density, electrolyte pH and pulse times in an electrolyte bath were changed. The analytical characterization of Ni-Co alloy was investigated by scanning electron spectroscopy (SEM), X-ray diffraction (XRD). The chemical composition of the deposited alloys was determined using Energy Dispersive Spectroscopy (EDS) technique. The Result showed that the structure of the alloys depends on Co content and changed from single-phase structure (fcc) to dual-phase structure (fcc + hcp). The deposited alloy obtained from direct and pulse current process exhibited a clear acicular and granular morphology respectively. No significant change in chemical composition was observed in length and thickness of deposited alloy. The grain size of deposited alloys obtained between 24–58 nm and 15–21 nm by applying direct and pulse current respectively.

Study on Recycling of Used CM247 LC Ni-base Superalloy

Krushna Kumbhar and Dipak Das

New generation superalloys contain significant quantities of rare earth elements such as Re, Ru and Ta. Because of their low abundances in earth's crust, long term supply of these elements for superalloy manufacturing business may be uncertain. In view of this, recycling of superalloys can provide sustainability of their use and reduce the dependence of fresh elements for superalloy manufacturing. Present study examines feasibility of recycling of Ni-base superalloy (CM-247LC). Directionally Solidified (DS) rods were produced using both fresh (virgin) and one-time recycled alloy. Rods were given heat treatments. Subsequently, creep specimens were fabricated and tested. No appreciable difference in composition and microstructure after heat treatment for two cases. Creep properties were found to be very similar in both the alloys. Based on this study, it can be concluded that these recycled alloy can be considered for reuse for gas turbine engine components.

Nickel/Cobalt Alloy Production from Secondary Materials

Ian Ewart, William Gower, and Shijie Wang

Nickel/cobalt are not as easy to recycle as copper, lead or aluminium do to the fact that they are often used as alloys and it is difficult to prudence the metal in a pure and useable form. While there are several recycle routes being developed for used lithium ion batteries and other nickel/cobalt streams that generate pure chemicals for battery purposes it is complicated and requires certain feedstocks. We will go over a slightly different and more flexible approach to produce a relatively clean nickel/cobalt alloy material that can either be directly used for some metallurgical applications or sent for further processing. The advantage of this approach in some cases is the flexibility of feed materials. A case study producing nickel/cobalt alloy form recycling of aerospace alloys is presented.

The Increasing Use of Nickel in the 21st Century

Gary E. Coates

The use of nickel between 2010 and 2019 increased at a rate of 5.7% per annum. Nickel-containing alloy steels, stainless steels and nickel alloys are finding new uses in pollution control, energy efficiency, long lasting, sustainable and resilient infrastructure, safe food and water, and even the new generation of vehicles that will take us to the moon and beyond. The use of nickel in batteries, a key and rapidly expanding market, will be mentioned, but details will be presented in the Batteries session. The presentation will concentrate on new and innovative applications involving nickel that will contribute to the United Nations Sustainability Development goals.

Mineral Processing

Selective Froth Flotation of Pentlandite from Pyrrhotite By Pulsed-power Technology Using Repetitive High-voltage Nanosecond Pulses

Irina Anatolievna Khabarova, Igor Zh, Bunin

The experiment results on the effect of high-power nanosecond electromagnetic pulses (HPEMP) on the chemical surface composition (SEM–EDX and AFM), electrohemical and flotation properties of pyrrhotite and pentlandite are presented. HPEMP caused a different change in the electrode potential of minerals: an increase in the negative value of the electrode potential of pyrrhotite and an increase in the positive value of the pentlandite potential. The shift of the pentlandite electrode potential to the region of more positive values increased the anionic collector adsorption and the hydrophobization of the mineral surface. The transition of the pyrrhotite potential to the region of negative values prevented the xanthate adsorption and decreased the mineral floatability. For monomineral flotation of pyrrhotite and pentlandite, we established the optimal mode of preliminary electromagnetic pulsed processing of minerals ($t = 10$ s, $N = 10(3)$), at which the contrast of their flotation properties increases in the mean on 20%.

Improved Beneficiation of Nickel from a Saprolite Laterite by Interaction of CaSO4 and Ores Matching with Transition Layer Nickel Laterite

Hongyu Tian, Jian Pan, Deqing Zhu, Zhengqi Guo, and Congcong Yang

Complex nickel bearing mineral composition, high magnesium content, ultrafine dissemination and maldistribution of nickel in the saprolite laterite will inevitably go against the reduction of nickel bearing minerals and growth of ferronickel alloy particles during the reduction process, resulting in low grade and recovery of nickel.

To improving the beneficiation of nickel from a saprolite laterite, additive as CaSO4 and ores matching with transition layer nickel laterite were adopted as strengthening measures in a selective reduction-wet magnetic separation process. Results showed that interaction of additive and ores matching can accelerate destruction of the nickel-bearing minerals structure and generation of liquid phase, which were beneficial to the release of nickel and the mass-transfer efficiency, simultaneously enhance the reduction, aggregation and growth of metallic particles to improve the beneficiation of nickel. Under optimal test conditions, the Ni grade of the ferronickel concentrate reached 6.83 wt.% with corresponding recovery rate of 92.19%.

An Innovative Beneficiation Process Developed for Jinchuan Nickel Ore Resources

Shijie Wang

In order to comprehensively utilize Jinchuan nickel ore resource, an innovative process of flotation—high-pressure oxidation leaching was developed. The process first produced high-quality concentrates (high-nickel and low-magnesium) for nickel flash smelting furnace; then it treated the flotation intermediate product (nickel-containing pyrrhotite concentrates) with oxidized leaching at high temperatures to solubilize the remaining non-ferrous metals in the solution. Magnesium thiosulfate was used to precipitate copper and nickel separately from the leaching solution. A high quality magnesium (by-product) was ultimately obtained from solution purification and crystallization. This paper will report the study of froth flotation that include mineralogy study and chemical reagent experimental tests. The lab and pilot flotation tests and the results will also be presented.

Pyrometallurgy

Study on CFD and Oxygen Lance Injection Technology of High Nickel Ternary Cathode Material Roasting Process in Roller Hearth Furnace

Zhong Ling (Rocky) Wei, Gang Zhang, Xu Qian, and Heng Zhu

The roller hearth furnace (RHF) is applied in high nickel cathode roasting process. Base on CFD analysis and operating data, the temperature and atmosphere inside furnace is not even in heat-up zone and holding zone. In order to improve homogenous and efficiency, oxygen lace and atmosphere control system was developed. The roasting process is examined and optimized by tests in RHF with different lance positions and oxygen injections. It can be concluded that the oxygen injections significantly enhance the reactions on the high nickel material. It is possible to draw a reaction picture inside the furnace, and one realizes that the reaction is the strongest at the beginning, and in the course of which the internal reactions increase. These subsequent reactions can also be influenced and optimized via the lances. The results prove the roasting process have space for further upgrade and reduce cost by utilize energy for heat up.

Development of Thermodynamic Database of Ni-Co-based Alloy System for the Applications of Pyrometallurgical Extraction Process

Min-Kyu Paek, Junmo Jeon, Daniel Lindberg and Pekka Taskinen

Nickel and cobalt alloys are widely utilized for Lithium-ion batteries (LIBs) and Nickel Metal Hydride (Ni/MH) batteries, superalloys, and high-entropy alloys. In order to overcome the current issue of the supply challenges of the critical resources, there have been several attempts for recycling of Ni and Co from industrial battery scrap in the Ni slag cleaning stage of the Ni flash smelting furnace. In the pyrometallurgical extraction process, the carbon and sulfur can be dissolved in the liquid Ni-Co-based alloys from the Ni slag, battery scrap, and reducing agents such as cokes and coal. Therefore, in order to find the optimum condition of the high-yield of alloying elements with a low S distribution ratio, as a core system, the Ni-Co-C-S system was thermodynamically assessed using the Modified Quasichemical Model (MQM) to describe the strong attraction between alloying elements and impurities in the liquid alloy.

Real-time Fe End-point Determination at Sudbury INO Smelter Finishing Converter Using Thermodynamic Process Simulation

Tanai L. Marin-Alvarado, Brett MacKinnon, Arina Moraes, Kurt Westhaver, Phil Nelson, Nicolas Lazare, Vince McIver, and Sari Muinonen

Glencore's Sudbury Integrated Nickel Operations (Sudbury INO) Smelter produces Ni/Cu/Co Bessemer matte, which must meet strict Fe target limits. To provide a real-time Fe estimation, XPS developed a thermodynamic process simulator using a custom thermodynamic database. A proof-of-concept model implemented in MS-Excel and OSIsoft PI-Datalink reads measured values of matte mass, composition, totalized air, oxygen, flux, reverts and custom feed transferred to the finishing converter. The model identifies all the inputs to the vessel by correlating crane events, mass inputs and assays then predicts required amounts of oxygen and flux for each blow to achieve the target Fe, displaying the estimated Fe composition, slag quality and SO2 produced every minute. After initial successful plant-trial, the model, thermodynamic-library and database were ported to Emerson's KNet analytics software for full integration into the plant control System. This article describes in detail the thermodynamic and convert model, plant trial results and KNet implementation.

Sulfation Roasting of a Nickel Laterite Ore

Eduardo Brocchi, Mariana Tavares, Victor Oliveira, and Rodrigo Souza

Chemical alternative to remove nickel from laterite ore, leaving the iron behind, is a matter of interest. The present work investigates a pre-leaching sulfation roasting method aiming to obtain water soluble sulfates and iron oxide. The roasting was carried out by mixing the ore with iron sulfate III and based on its decomposition into iron and gaseous sulfur oxides, this being the agent for the formation of the Ni, Co, Mg and Ca sulfates. The method was evaluated by changing the mass ratio (ore/sulfate) and temperature through different reaction times in a muffle furnace. It was observed that a 700 °C roasting, with 30% excess of Fe2(SO4)3 for 30 min, followed by a 1 h water solubilization at 80 °C, can provided a liquor with nickel, cobalt and magnesium, fully separated from the insoluble iron oxide. The recovery being 34%, 68% and 58%, respectively, which under other operational conditions can be improved.

About the Editors

Corby Anderson is a licensed professional engineer with over 40 years of global experience in industrial operations, corporate-level management, engineering, design, consulting, teaching, research, and professional service. He is a native of Butte, Montana, USA. His career includes positions with Thiokol Chemical Corporation, Key Tronic Corporation, Sunshine Mining and Refining Company, H. A. Simons Ltd., and CAMP-Montana Tech. He holds a B.Sc. in Chemical Engineering from Montana State University, an M.Sc. from Montana Tech in Metallurgical Engineering, and a Ph.D. from the University of Idaho in Mining Engineering-Metallurgy. He is a Fellow of both the Institution of Chemical Engineers and of the Institute of Materials, Minerals, and Mining. He shares 14 international patents and 4 new patent applications covering several innovative technologies, 2 of which were successfully reduced to industrial practice. He currently serves as the Harrison Western Professor in the Kroll Institute for Extractive Metallurgy as part of the Mining Engineering Department and the George S. Ansell Department of Metallurgical and Materials Engineering at the Colorado School of Mines. In 2009 he was honored by the Society for Mining Metallurgy and Exploration with the Milton E. Wadsworth Extractive Metallurgy Award for his contributions in hydrometallurgical research. In 2015 he was awarded the International Precious Metals Institute's Tanaka Distinguished Achievement Award. In 2016 he received the Distinguished Member Award from the Society for Mining, Metallurgy, and Exploration, the Outstanding Faculty Award from the George S. Ansell Department of Metallurgical and Materials Engineering at Colorado School of Mines, and also became a Distinguished Member of the University of Idaho Academy of Engineering. In 2017 he received the EPD Distinguished Lecturer Award from The Minerals, Metals & Materials Society (TMS). In 2019 he was named as a Henry Krumb Distinguished SME Lecturer. In 2019 he was also appointed as a Visiting Faculty within the Minerals Engineering Department of Central South University in China, the largest program of Mineral Processing and Extractive Metallurgy in the world.

Graeme Goodall is the Furnace Integrity & Refractory Superintendent at the Glencore operated Koniambo Nickel SAS smelter located in New Caledonia. Dr. Goodall is a Professional Metallurgical engineer (PEO) with master's and doctoral degrees from McGill University in Montreal, Canada. He has been active in the metallurgical industry for nearly 20 years with experience in process design, furnace integrity, thermodynamics, refractory systems, piloting, and welding.

Sumedh Gostu is a senior research scientist with a special focus on Hydrometallurgy for Air Liquide. Dr. Gostu works on devising innovations in the field of extractive metallurgy utilizing Air Liquide's core expertise of industrial gases and services. Dr. Gostu is involved in commercial projects, pilot, and bench scale projects related to gas injection to liquids for gold, copper, nickel, lithium, and cobalt extraction. Dr. Gostu has worked extensively in oxidative leaching, gas to liquid mass transfer, pressure oxidation, flotation, and chemical precipitation.

Prior to Air Liquide, Dr. Gostu worked with the Center of Resource Recovery (CR3) and recycling as a Ph.D. student. He received his Ph.D. in Materials Science from Worcester Polytechnic Institute and master's from the Kroll Institute of Extractive Metallurgy at Colorado School of Mines. Dr. Gostu's work in CR3 focused on developing a metallurgical flowsheet to recover valuable products from Red-mud or bauxite residue. The work is now being commercialized by an organization focused on valorizing bauxite residue.

Dr. Gostu is actively involved in The Minerals, Metals & Materials Society (TMS), Society of Mining, Metallurgy and Exploration (SME), and The American Society of Materials (ASM).

Dean Gregurek is a senior mineralogist in the RHI Magnesita Technology Center Leoben, Austria since 2001. Dr. Gregurek received his M.Sc. degree at the University of Graz in 1995 and his doctorate degree in Applied Mineralogy from the University of Leoben in 1999. Prior to RHI Magnesita, he worked for two years for Luzenac Europe in talc business. His current research interests and technical expertise are focused on chemical and mineralogical studies related to interactions between refractories, molten metals, and slags from pyrometallurgical furnaces. Dr. Gregurek has been a TMS member since 2012, *JOM* advisor (2014–2017), chair of the Pyrometallurgy Committee (2018–2020) and a co-organizer for the 7th–11th International Symposium on High-Temperature Metallurgical Processing (TMS Annual Meetings 2016–2020).

Mari Lundström is strongly focused on the circular economy of metals and holds a chair of Hydrometallurgy and corrosion in Aalto University. She is currently acting as the head of 22 M€ Finland-based battery metals ecosystem (BATCircle) and acts as the vice chair/Sherpa in Batteries Europe WG2 Raw Materials and Recycling. Her expertise lies in hydrometallurgical recycling, battery metals, gold, sustainable primary production, as well electrochemistry and works closely in the interphase between academy and industry. After 7 years in industry, she was appointed as assistant professor at Aalto University (2015), and she has built in a short time a productive research group of around 25 members, published more than 130 peer-reviewed scientific papers, and supervised 9 Ph.D.s to completion. She has raised more than 9 M€ funding and earned several patents. In 2019 she received the Tapani Järvinen technology award.

Christina Meskers is currently senior research fellow and leader of the Circular Economy Systems group at the Helmholtz Institute for Resource Technology in Freiberg, Germany, as well as Chair of the Extraction and Processing Division and Member of the Board of Directors of The Minerals, Metals & Materials Society (TMS), USA.

She graduated from Delft University of Technology with an M.Sc. (2004) in resource engineering and a Ph.D. (2008) in materials science and engineering, including stays at McGill University, the Norwegian University of Science and Technology, and the University of Melbourne. At UMICORE she was a Senior Manager Open Innovation and a Senior Manager Market Intelligence & Business Research. Through innovation and research, her aim is to ensure the product and materials value chain becomes more sustainable. The United Nations' International Resource Panel report, *Metal Recycling-Opportunities, Limits, Infrastructure* (2013) is a key publication that she co-authored.

Dr. Meskers is a 2020 Brimacombe Medalist, corecipient of the 2014 Ondernemers voor Ondernemers Award, and the 2013 Belgian Business Award for the Environment and received in 2008 the TMS Young Leaders Professional Development Award. She has a strong passion for innovation, strategy, and partnerships coupled with a focus on connecting people and ideas across disciplines, industries, organizations, and value chains. Dr. Meskers served on the industrial advisory boards of international master programs and innovation networks. She has accumulated over 15 years of experience in the (raw) materials sector.

Stuart Nicol graduated from the University of Queensland with a BE (Chemical and Metallurgical Engineering) in 2011 and, after working in industry for a number of years, he received a Ph.D. in Extractive Metallurgy in 2019. He has worked on fundamental research, plant design, and plant operations and has experience in Australia and around the world. Dr. Nicol has experience in a wide range of commodities. Currently, he is working for Gopher Resource, a secondary lead smelter and is assisting in the optimization and upgrading of existing facilities and the testing and piloting of new processes. Dr. Nicol has been a member of the AusIMM since 2009 and TMS since 2019.

Esa Peuraniemi is a development manager at Boliden Harjavalta in Harjavalta, Finland

Fiseha Tesfaye is a senior researcher with the title of Docent in metallurgical thermodynamics in the Johan Gadolin Process Chemistry Centre of Åbo Akademi University, Finland. He received his M.Sc. degree in materials processing technology and Ph.D. degree in metallurgy from Aalto University, Finland. His research activities are focused mainly on the thermodynamic investigation of inorganic materials as well as rigorous theoretical and experimental investigations for promoting improved valuable metals and renewable energy production. In 2018, Dr. Tesfaye was also appointed as a visiting research scientist in Seoul National University, South Korea.

Dr. Tesfaye is a regular contributor and active member of The Minerals, Metals & Materials Society (TMS) and is a winner of the 2018 TMS Young Leader Award. He serves as a subject editor for JOM, the member journal of TMS, and has edited several scientific research books. His personal research achievements include remarkable improvement of experimental research applying the solid-state EMF technique for thermodynamic investigations of inorganic materials, as well as noticeable contribution for promoting the move toward the circular economy. In his research areas, Dr. Tesfaye has published over 55 peer-reviewed publications.

Prabhat K. Tripathy is a research scientist with more than 20 years of research experience in the fields of materials chemistry, process metallurgy of reactive and refractory metals, high temperature electrochemistry, synthesis and characterization of advanced materials, energy-efficient manufacturing process, chemical processing of waste (including nuclear) materials, and lean ore bodies. He earned his M.Sc. (Chemistry), M.Tech. (Materials Science & Engineering), and Ph.D. (Chemistry) degrees from Utkal University (Bhubaneswar, India), Indian Institute of Technology (Kharagpur, India), and University of Mumbai

(Mumbai, India), respectively. He started his professional career at Bhabha Atomic Research Centre (Mumbai, India) and subsequently spent stints at the University of Cambridge (UK) and Massachusetts Institute of Technology (USA) before joining Idaho National Laboratory. Currently, he is carrying out R&D activities in the fields of inert anode technology, molten salt science and technology, materials recycling, and energy-efficient manufacturing processes. He also holds an Adj. Professorship in the Materials Science and Engineering Department at the University of Utah (USA). The scientific underpinning of his research activities has been to understand metals and materials through the application of novel techniques. He has about 92 publications and two US patents to his credit. In addition, he played a key role in licensing five of his technologies to a US-based company recently.

Shijie Wang is a principal advisor at Rio Tinto Copper & Diamond. Dr. Wang has been active in extractive metallurgy and has over 30 years of experience and expertise in metallurgical process development, existing operation optimization, and the metals strategy. His work interests include leaching-solution purification, smelting and refining, recycling, waste treatment, metal recovery, process safety, operational efficiency, and profitability. Dr. Wang holds four US patents, has published more than 50 journal papers including nonferrous metals', precious metals', rare metals', and rare earth metals' resourcefulness and recoveries. Dr. Wang received a B.S. degree in mineral process from China and an M.S. degree and a Ph.D. degree in metallurgical engineering from the University of Nevada, Reno, USA. Dr. Wang has been very active in TMS since 1991 and is the former chair of the Hydrometallurgy and Electrometallurgy Committee of TMS from 2011 to 2013 and the co-chair of Extraction 2018 Conference with TMS, SME, and MetSoc in 2018. He has been the lead and co-organizer of seven symposia at the international conferences through TMS. Dr. Wang received the TMS Extraction & Process Division (EPD) Distinguished Service Award in 2017.

Yuanbo Zhang is a full professor in mineral engineering and ferrous metallurgy and also serves as a vice-dean of the Department of Ferrous Metallurgy in the School of Minerals Processing and Bioengineering, Central South University, China. He received his Ph.D. degree from Central South University in 2006. He acted as a visiting professor in Colorado School of Mines in 2017–2018.

His research interests include high-value utilization of tailings/smelting slags, low-carbon clean metallurgy and CO_2 emission reduction, and preparation of functional materials from natural metallic minerals. He has completed more than 40 projects from government and industry, including 4 National Natural Science Foundation of China Funds. He has been awarded the second prize of National Science and Technology Progress. He has published more than 100 peer-reviewed articles in journals such as *Journal of Hazardous Materials*, *ACS Sustainable Chemistry & Engineering*, *Hydrometallurgy*, *Minerals Engineering*, *Mineral Processing and Extractive Metallurgy Review*, *Construction and Building Materials*, *Materials and Design*, *Applied Surface Science*, *and Ceramics International*, *Powder Technology*. He holds more than 50 national invention patents. He has served as a key reviewer for a number of international journals.

Dr. Zhang has been an active member of the Minerals, Metals & Materials Society (TMS) since 2011, acted as a member of the Materials Characterization Committee and Pyrometallurgy Committee of TMS since 2012, and co-chaired symposium sessions of TMS Annual Meeting (2014–2019).

Part I
Plenary

"Around the Nickel World in Eighty Days": A Virtual Tour of World Nickel Sulphide and Laterite Operations and Technologies

A. Vahed, P. J. Mackey, and A. E. M. Warner

Abstract The world production of nickel at 2.36 million tonnes in 2019 ranks fifth in the output of non-ferrous metals after aluminum, copper, zinc and lead. Nickel is produced from both sulphide ores—representing ~30% of world nickel production—and oxide ores—representing ~70% of world nickel production, with nickel now a vital component in batteries which will power much of the new electric transportation age. The top ten nickel smelter producing countries (2019) include in order: China, Indonesia, Russia, Canada, Australia, New Caledonia, Brazil, Finland, Philippines and Japan. The present paper takes the reader on a virtual tour of most of the nickel sulphide and nickel laterite plants throughout the world with a focus on smelter facilities. Operations and technology employed at each plant visited on the virtual tour are discussed including a brief historical sketch. Future trends identified during the world tour are also discussed with respect to technology and products. The present authors considered that enough time was required for a thorough study tour and settled on eighty days—the same as that for the celebrated fictional story of world circumnavigation in the 1870s—a time when nickel was in its infancy and considered a rare metal.

Keywords Nickel laterite · Nickel sulphide · World nickel smelters · Nickel smelter tonnages · Europe · Canada · Russia · Africa · Asia · Latin america · Oceania · Caspian

---

A. Vahed
AV Met Consulting, 2335 North Ridge Trail, Oakville, ON L6H 0A8, Canada

P. J. Mackey (✉)
P. J. Mackey Technology Inc, 295 Kirkland Blvd, Kirkland, QC H9J 1P7, Canada
e-mail: pjmackey@hotmail.com

A. E. M. Warner
AEMW Consulting, 304 Oakwood Court, Burlington, ON L7N 1W9, Canada

C. Anderson et al. (eds.), *Ni-Co 2021: The 5th International Symposium on Nickel and Cobalt*, The Minerals, Metals & Materials Series,
https://doi.org/10.1007/978-3-030-65647-8_1

Introduction

The celebrated French writer Jules Verne published his book *Around the world in 80 days* in English in 1873 and it is the story of Phileas Fogg and his French valet Passepartout circumnavigating the world in 80 days in the fall of 1872 on a £20,000 wager (equivalent to over 2 million pounds today) from the exclusive Reform Club in London. Phileas with his trusted servant Passepartout started from London to Suez (Cairo) to Mumbai to Calcutta to Hong Kong to Yokohama (Tokyo) to San Francisco to New York and back to London, arriving with minutes to spare to eventually claim the wager. This is an interesting period for nickel as during the decade of the 1870s, the source of most of the world's nickel switched from Europe (mainly Finland and Germany) to New Caledonia. At that time, world nickel tonnage was about 500 tonnes/year with the major use in nickel plating and the so-called German or nickel-silver, an alloy of 65% Cu, 18% Ni and 17% Zn [1].

Note: A cupronickel alloy also containing zinc known as *paktong* or "white copper" had long been used in China—perhaps as early as the fourth-century AD—and evidently made by smelting certain nickel ferrous copper ores and later adding zinc to the smelt [2]. Samples brought to Europe by early travellers were then analysed by German and British scientists. Subsequently, alloys corresponding to *paktong* were produced starting in the 1800s and became known as "Nickel silver" or "German silver" throughout much of the eighteenth century. It is also noted that there are reports of the use of iron-nickel metals from meteorites by prehistoric humans [3].

From the 1880s to the early 1900s, nickel produced from New Caledonia laterite ores—as opposed to the nickel sulphide ores of Europe—dominated nickel production, representing up to 70% of nickel supply over that period [4]. Interestingly then, nickel production from nickel sulphide ores largely in Europe originally dominated early nickel production, then for several decades to about 1910, nickel produced from the laterite ores of New Caledonia dominated, followed by over 50 years with nickel largely produced from sulphide ores, largely with nickel produced from the vast sulphide deposits in Sudbury, Ontario, Canada.

For about the last decade, nickel produced from laterite ores from around the world has increasingly dominated nickel output (Figs. 1 and 3). Figure 2 presents nickel production data from 1870 to 1970 using a semi-log scale to illustrate early production data in an expanded view. Since the 1870s–1880s of the time of Jules Verne, nickel production has increased over 2,000 times, from about 1,000 tonnes in the 1880s to nearly 2.4 million tonnes in 2019, Fig. 1.

The period from about 1880 to 1910 shows nickel produced from the laterite ores of New Caledonia dominating nickel production—by about 1910, world nickel production was about 10,000 ton/a, Figs. 1 and 2—when nickel from Sudbury sulphide ore started to dominate. After about 1950, nickel from laterite slowly increased. (Sources: 1870–1920, developed by authors, 1920–1984, developed by authors based on Ni production by the United States Geological Survey (USGS) by country and inferred ore type, [5]; 1984–2019, courtesy of Wood Mackenzie Ltd (2020). See Fig. 3.

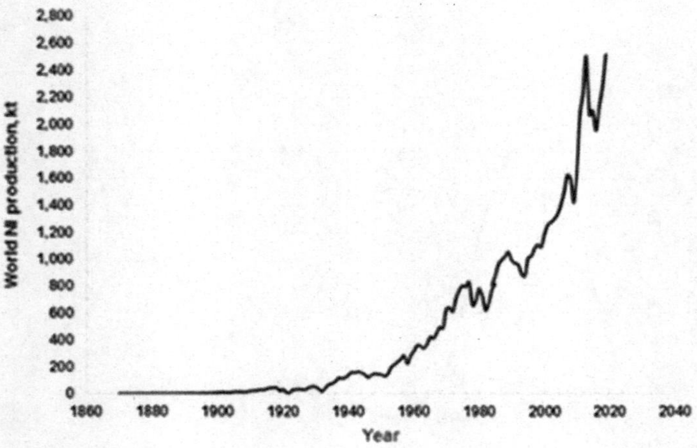

Fig. 1 World output of smelted nickel, 1870–2019. (*Sources* 1870–1900, present authors; 1900–1984, [5] 1984–2019, [6]). (Color figure online)

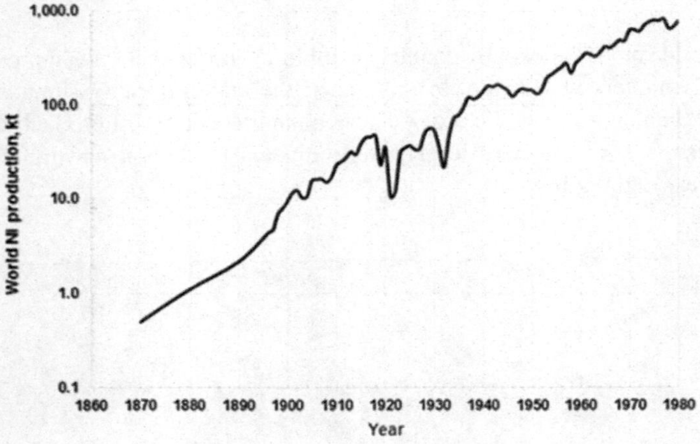

Fig. 2 World output of smelted nickel, 1870–1970 (on semi-log scale for an expanded view). (*Sources* refer Fig. 1). (Color figure online)

Unlike the other non-ferrous metals such as lead, zinc or copper which have been known and smelted for many thousands of years [2], nickel production and use is relatively recent; nickel was only discovered and isolated in the late 1700s, with notable production only starting about the 1880s. In view of the importance of the world nickel industry, this paper reviews the status of nickel smelting today and comments on 35 significant nickel smelters that might be visited on a country to country virtual tour similar to Jules Verne's story starting and ending in London on a route as shown in Fig. 4. The 35 plants visited on the virtual tour produced ~72% of world smelted nickel which amounted to 2.36 million tonnes in 2019 (also

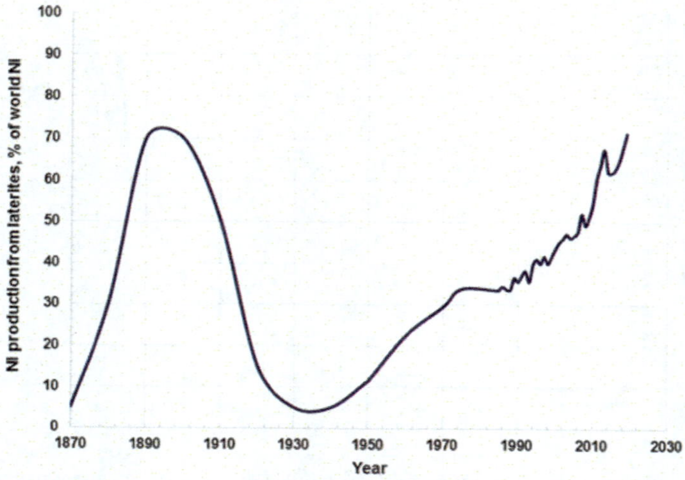

Fig. 3 The proportion of Ni production from laterite since 1870–the present day. (Color figure online)

refer to world smelted nickel by country in Table 1). Today there are approximately 100 nickel smelters worldwide. In this paper, the authors briefly summarize key findings at the plants "visited", provide current and recent nickel production levels, technologies employed, general level of environmental protection, and the challenges for the industry going forward.

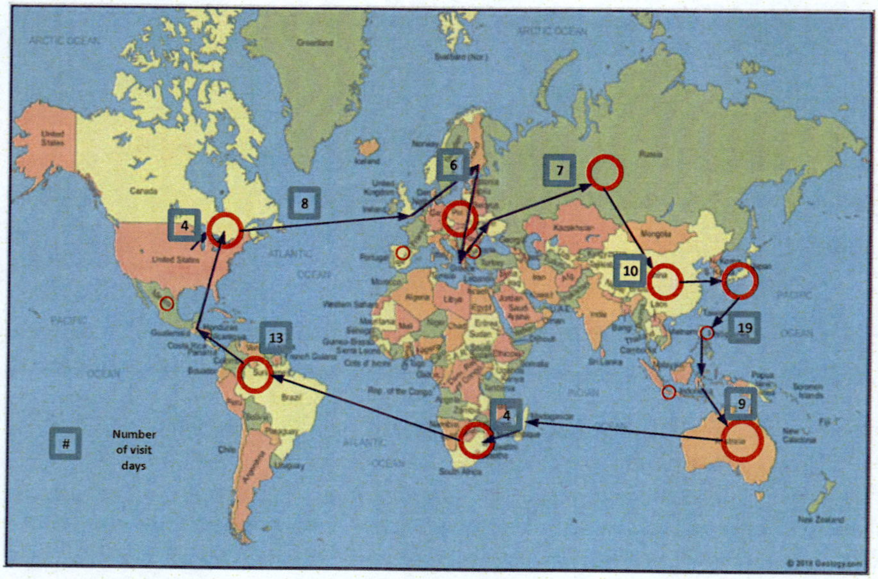

Fig. 4 Tour of world's nickel smelters starting and ending in London. (Color figure online)

Table 1 2019 global primary nickel smelter production by country (kt/a Ni)

Country	2019	Total % of world	Country	2019	Total % of world
Madagascar	34		Austria	3	
South Africa	38		Finland	54	
Zimbabwe	5		Germany	1	
Total Africa	*77*	*3.3*	Greece	12	
			Kosovo	6	
China	711		North Macedonia	15	
Indonesia	470		Turkey	1	
Japan	50		Ukraine	14	
Myanmar	20		*Total Europe*	*106*	*4.5*
Philippines	52		Australia	127	
South Korea	45		New Caledonia	95	
Total Asia	*1348*	*57.1*	Papua New Guinea	35	
Brazil	54		*Total Oceania*	*257*	*10.9*
Columbia	41		Russian Federation	214	
Cuba	48				
Dominican Republic	24		*Total Russia and the caspian*	*214*	*9.1*
Guatemala	15				
Total Latin America	*182*	*7.7*	*Total world*	*2361*	
Canada	177				
Total North America	*177*	*7.5*			

It is acknowledged that much of the statistical data in the above table and included elsewhere in this paper was provided by Wood Mackenzie Ltd.

Europe

As the virtual tour begins in Europe and then moves to many parts of the world where nickel production is relatively recent—say the last 120 years (including parts of Europe, North and South America, Asia, Oceania, etc.)—as compared to say copper or lead for which production stretches back thousands of years, it was felt that a brief summary of the history of nickel, in particular the development of early nickel alloys and the discovery of stainless steel (the major use for nickel today) would be instructive. Much of nickel metallurgy as it is known today essentially began in Europe.

By the 1890s, numerous metallurgists principally in Europe and the USA had determined that steel containing up to about 5% Ni and low in carbon content showed improved strength properties. As reported by Riley in 1889 [6], a rolled and annealed steel containing 5% Ni and 0.2% C was about 40% stronger than a steel without nickel. This property would soon attract the attention of the world navy establishments for its use in armour plate for ships and for ordnance applications; these would become new applications for nickel. Further improvements in alloy composition were made, with chromium additions as well as nickel—however stainless steel was still to be discovered.

In 1912, Harry Brearley, a skilled metallurgist from Sheffield, home to large silver-plated cutlery and related manufacturers, was head of the research laboratory set up by two steelmakers, John Brown and Co. and Thomas Firth and Son. By 1913, Brearley was investigating erosion of gun barrels and experimented with different steel alloys with chromium for improved properties. A knowledgeable metallographer, he noticed that his low-carbon steels containing about 12% Cr did not etch using the usual etching agents such as nital. In a flash of invention, he then proposed the new steel, which would not rust, be used for cutlery—which then became hugely successful. Many other developments by metallurgists principally in the UK, France, Germany and USA occurred around this time, improving alloy compositions—such as adding both nickel and chromium to steel; however it is well recognised that Harry Brearley was the first to see the superior corrosion properties of the alloy and its application to "rustless" cutlery- soon termed stainless steel—and now accounting for about 75% of the world nickel usage. Given this important feature of nickel's development, it seemed appropriate therefore that prior to the voyage, a virtual pre-visit was made to the Sheffield Museum to view displays of the early Sheffield metal work [7]. Sheffield was the world's largest producer of metal cutlery, including using stainless steel in the early part of the twentieth century.

Following the museum visit, the virtual tour began with the first stop after leaving London being Harjavalta, Finland. Later, Terrafame in eastern Finland was visited, and while in Helsinki, a visit to Aalto University where research in nickel metallurgy is undertaken. In total, there are 8 Ni smelters in Europe (several are quite small) producing 106 kt of Ni in 2019 of which we selected 4 smelters for this virtual tour based on technologies and size (Terrafame was included due to its unique bioleaching and Ni refining technologies). These 4 smelters are listed in Table 2, together with 2019 production data and a summary of the main observations. These 4 smelters produced 80 kt Ni in 2019 (75% of the total European nickel smelter production, but only 4.5% of the world production). European mine production was ~75% of the European smelter production. Note for all the content tables: Type—S = sulphide plant, L = Laterite plant. Terrafame operates a large bio heap leaching facility at Sotkamo, Finland. The background of the plant is given in Table 2. Plans are underway to produce nickel sulphate for the battery market with a capacity of 170,000 t of sulphate/a (~36,000 t of Ni/a). While in Helsinki, a virtual visit was made to Aalto University in Espoo, some 20 km west of the city. Aalto University was formed in 2010 by merging three leading Finnish universities. The projects

Table 2 European smelters examined

Country	Smelter, company	Ore type	Production kt/a Ni 2019	Capacity kt/a Ni	Summary of observations
Finland	Harjavalta; Boliden	S	26	45	Modern efficient smelter, first Outokumpu Ni flash when commenced in 1959, now incorporates the DON process. Has high S capture, low fugitives, good energy efficiency
	Terrafame	S	27	35	Terrafame operates a large bio-heap leaching facility and metal refinery at Sotkamo in eastern Finland, 600 km north of Helsinki. The plant began in 2008 as Talvivaara, a state-owned company following extensive piloting of the low-grade ore (0.26% Ni, 0.54% Zn, 0.14% Cu and 0.02% Co). A number of mis-steps and operating difficulties led to Terrafame, a Finnish Government entity, successfully operating the plant since 2015. Presently a mixed Ni-Co sulphide is produced, however plans are in progress to produce nickel sulphate for the battery market, with a capacity of 170,000 t/a of sulphate (~36,000 t/a Ni)
Greece	Larco	L	12	23	Located at Larymna and 55% owned by the Greek state, a highly challenging, high Fe-SiO_2 laterite ore, costly to run, faced with a sale or closure. Commissioned in 1966, the plant employs the RKEF treating paleo-laterite ores high in Fe, SiO_2 and low in MgO producing a low-grade FeNi and a fayalitic type slag

(continued)

Table 2 (continued)

Country	Smelter, company	Ore type	Production kt/a Ni 2019	Capacity kt/a Ni	Summary of observations
Ukraine	Pobuzhsky	L	15	20	RKEF plant located at Pobuzhie in Southern Ukraine, about 200 km north of the Black Sea port of Odesa. Built in the 1970 s by the Soviets, the smelter underwent several changes and was purchased in 2002 by the Solway Group who made improvements. The plant now operates on imported ore with a new dryer, the 4 original kilns and an upgraded six-in-line furnace with Hatch copper coolers operating at 42 MW
North Macedonia	FENI; Cunico Resources	L	15	20	Located 15 km east of Kavadarci, local and imported ores are pelletized and treated in a Lepol grate prior to kiln processing in one of two kilns and two electric furnaces
Subtotal (above)			95	143	Nickel smelter production 75% of total European Smelters
Other European Smelters			11	5	
Total European Smelters			106	148	Nickel smelter production 4.5% of total world Smelters
Total European Mine Production			79		Mined production ~75% of smelter production in Europe

related to nickel metallurgy carried out at the School of Chemical Engineering were of great interest.

The sole nickel sulphide smelter in Europe (Harjavalta) uses best-in-class technology with effective environmental protection. Energy efficiency and costs are generally good. The laterite plants in Europe each have several projects for improvement and modernization which are at different stages. These projects will also likely help lower costs down from the present position around the upper half of the world nickel cost curve (Fig. 5), [8].

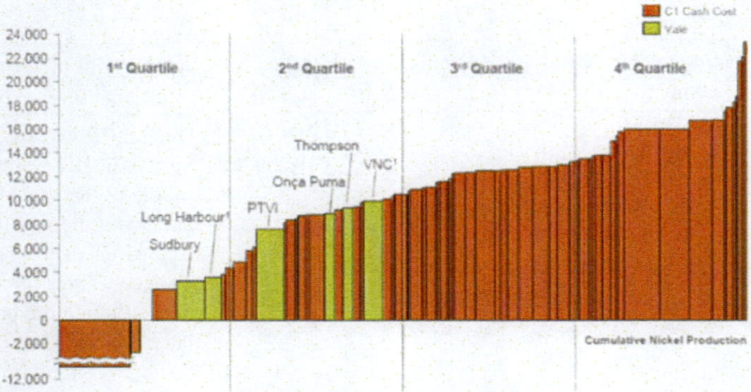

Fig. 5 2015 nickel smelter cost curve [8]. (Color figure online)

Russia

From Ukraine moving on to Moscow to commence a virtual tour of 2 plants in Russia. Russia has a total of 5 nickel smelters out of which the 2 nickel sulphide plants are discussed—both in northern Russia—the large Norilsk plant in Siberia, while the other plant, Pechenga, is located near Murmansk close to the Finnish, Swedish and Norwegian northern borders. It is noted that formerly, there were several laterite nickel plants in Russia, some blast furnaces producing a form of what is now called nickel pig iron. All laterite smelters in Russia have now closed due to the low grade of the ore. The 2 smelters visited produced 213kt Ni in 2019 (99% of the total Russia nickel smelter production). Total nickel production in Russia in 2019 was 214 kt/a nickel, representing some 9% of world smelted nickel in 2019, with Russia ranking third in world nickel smelter output by country after China, and Indonesia, and hence has an important place in the industry. See Table 3.

Asia

From Moscow moving on to six countries comprising Asia to commence a virtual tour of 15 smelters/plants which countries included: China (7 smelters), Japan (3 smelters), South Korea (1 smelter), Philippines (1 hydromet plant) and Indonesia (3 smelters).

Table 3 Russian smelters visited

Country	Smelter, company	Ore type	Production kt/a Ni 2019	Capacity kt/a Ni	Summary of observations
Russian Federation	Norilsk	S	185	210	The first Ni was produced at Norilsk in 1942, at the middle of the WWII, under harsh conditions. Significant expansions over more than six decades to exploit the huge sulphide ore bodies make the now listed company the largest single Ni producer in the world (also one of the world's largest Pd/Pt producer). Currently, the electric furnace-PS converter plant operations have been curtailed due to reported SO_2 emissions investigation which remains under review. A new Ni flash furnace at the nearby Nadezda complex treats the nickel concentrate (started 2016)
	Pechenga	S	28	60	The Pechenga Ni smelter is located about 150 km west of Murmansk, Russia near the Russia-Finnish border in an area known as Petsamo ceded to Russia after World War 2. Built in the 1940 s-1950 s, the plant has two travelling grate roasters, two 11.5 m by 26.5 m six-in-line electric furnaces for smelting and 5, 4 m by 12 m PS converters to produce a low-iron matte which is slow cooled. Due to severe SO_2 pollution conditions, the plant is slated to close at the end of 2020 with Ni concentrate evidently to be treated in new facilities at Monchegorsk south of Murmansk

(continued)

Table 3 (continued)

Country	Smelter, company	Ore type	Production kt/a Ni 2019	Capacity kt/a Ni	Summary of observations
Subtotal (above)			213	290	Nickel smelter production 99% of total Russian Smelters
Other Russian & Caspian Smelters			1	27	
Total Russian & Caspian Smelters			214	317	Nickel smelter production 9% of total world Smelters
Total Russian & Caspian Mine Production			214		Mined production ~100% of smelter production in Russia

Asia (China)

From Ukraine, the virtual tour arrived in China to visit a total of seven smelters from one end of the country to another, listed in Table 4.

The production in 2019 from the seven smelters visited on the virtual tour was 409 kt nickel, or about 57% of China's total annual smelter nickel production at 711 kt.

In total, there are thirty-seven (37) nickel smelters in China, spread throughout the country largely producing nickel pig iron (NPI). Of these, seventeen smelters operate at an annual capacity $\leq$ 15 kt contained nickel, fifteen smelters at $> 15 \leq$ 100 kt and five at > 100 kt, each group amounting to an annual nickel production of 149.9, 433 and 790 kt, respectively. Those in the last category (> 100 kt) are amongst the largest production capacities in the world. In 2019 nickel mine production in China was 13% of smelter production. This infers that 87% of the nickel comes from imported feed.

Many Chinese nickel plants are government owned and, except for Jinchuan and several other smelters, the rest operate a Rotary Kiln Electric Furnace (RKEF) plant producing Nickel Pig Iron (NPI), with or without a captive steelmaking complex, have started only during the last two decades. The advent of the NPI technology using perhaps 200–300 small blast furnaces and its growth to a more sustainable RKEF technology has been presented in an earlier paper in 2015 [9]. As predicted in the paper, the introduction of the NPI technology in China became a game changer in the traditional global supply–demand relationship and it allowed the country to produce most of its steel using NPI with imported laterite feed. Selection of the plants to visit was therefore based on the technology used, size and the focus of their products.

Table 4 Chinese smelters visited

Country	Smelter, Company	Ore Type	Production kt/a Ni 2019	Capacity kt/a Ni	Summary of Observations
China	Delong	L	68	100	A privately-owned enterprise established in 2010. A large, modern steel manufacturing operation, integrating RKEF FeNi production and steel smelting, using laterite ores from the Philippines and Indonesia. Planning for 30 lines of RKEF FeNi -steelmaking production complex for mid-2020, to reach 5000 kt/a 300 series stainless steel
	Fujian Desheng	L	18	60	Fujian Desheng Nickel is a RKEF ferronickel plant near Fuzhou in Fujian province in China. Built with technology imported form the Ukraine, the plant produces ferronickel of about 15% Ni
	Fujian Haihe	L	0	20	The Fujian Haihe smelter is a nickel pig iron plant in Fuan city, Fujian province. The plant has a capacity of about 20,000 t Ni/a as low grade (5–10% Ni) nickel pig iron
	Jilin	S	7	22	Established in 2000, located near Jilin Changbai mountains. By 2015, a large nonferrous, high-tech, international enterprise. Ausmelt TSL installed at the Jilin smelter in 2006. There is also a carbonyl plant at the smelter. Key supplier of several nickel salt products, electrolytic nickel, nickel and other nonferrous metal concentrates. Recent products include green battery materials. Six major subsidiaries, with one Lithium mine in Quebec, Canada

(continued)

Table 4 (continued)

Country	Smelter, Company	Ore Type	Production kt/a Ni 2019	Capacity kt/a Ni	Summary of Observations
	Jinchuan	S	119	160	A state-owned enterprise established after the discovery of the world's 3rd largest Cu/Ni ore deposits nearly five decades ago, in North West China. Third largest nickel producer and fourth largest cobalt producer in the world. A modern nickel smelter, including an Outokumpu Ni flash smelting furnace + EF appendage. The first in the world to use an O_2 enriched top blowing Ni-smelter. By 2029 China is expected to have 40–50% of the global EV and the green industry battery market, and Jinchuan is to supply a major portion of it's nickel
	Shandong Xinhai	L	146	160	Located in the city of Linyi in the Shandong province and with abundant electrical power supply. An important producer of nickel alloys using RKEF technology from imported laterite ores. Investing heavily in Indonesia to be close to laterite feed. Early in 2020, Signed an agreement with the Indonesian Silkroad Nickel to build and operate a 400 kt/a RKEF smelter + a coal power plant on the island of Sulawesi

(continued)

Table 4 (continued)

Country	Smelter, Company	Ore Type	Production kt/a Ni 2019	Capacity kt/a Ni	Summary of Observations
	Tsingshan FuAn	L	51	46	The Tsingshan nickel pig iron plant is located in Fuan city in Fujian province. The plant has six RKEF lines. Tsingshan has pioneered RKEF technology with modules producing about 8,200 t Ni/a. Each module or line includes a 28 MW, 18 m dia. electric furnace producing ~10% Ni nickel pig iron from ~480,000 t/a of 1.8% Ni ore. All Tsingshan's nickel pig iron is processed to stainless steel in the company's mills
Subtotal (above)			409	568	Nickel smelter production 57.5% of total Chinese Smelters
Other Chinese Smelters			302	805	
Total Chinese Smelters			711	1373	Nickel smelter production 30% of total world Smelters
Total Chinese Mine Production			95		Mined production ~13% of smelter production in China

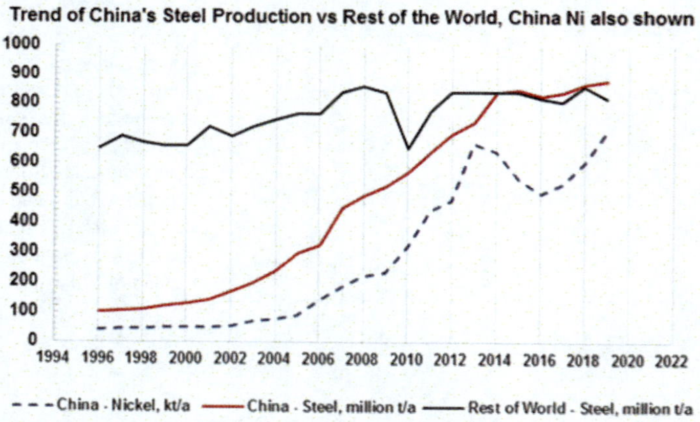

Fig. 6 Trend of China's steel production 1994 to 2020 versus the rest of the world [11]. (Color figure online)

In Fig. 6 China's smelter nickel production is shown from around 1995 to 2018. Production of NPI started in 2005 from about 100 kt/a nickel reaching about 500 kt/a in 2014, just before Indonesia banned its ore exports. Since then, China has been securing enough laterite feed from external sources under special investment agreements with Indonesia and the Philippines to run its domestic RKEF plants [10], and concomitantly started transferring its NPI RKEF technology to Indonesia with significant capital expenditure, which allowed China to eventually outpace the global steel production and lead ever since. These included full ownership and/or partnership installations of modern, integrated RKEF plants feeding steelmaking complexes. The plants in China and Indonesia during this period widely used a higher grade of laterites to produce 6–20% Ni NPI, suitable for producing the higher grade of stainless steels (e.g. 300 series), which have increasingly been replacing the refined nickel as the main nickel contributor. In 2018, NPI contributed 51% of the global nickel supply, and with it for the first time it pushed China's global market share of steel production beyond 50%, as shown in Fig. 6 [11].

China used most of the steel for internal consumption to feed its construction and value-added products and components of machinery, bridges, pipelines, washing machines, etc. [10, 11]. However, starting on January 1, 2020, the fate of the RKEF plants that were dependent on the laterite feed from Indonesia is not so clear. China will likely draw down its laterite stockpiles for now, until possibly such time that they can start up the new RKEF Projects in Indonesia to supply the steelmaking plants with new NPI feed. In the meantime, the steel world's suppliers and producers are watching carefully to better estimate the risk of COVID-19 outbreak in China and including the Island of Sulawesi, in those areas where most of the NPI projects are located. Since 2014, the top story of China has become the story of NPI.

Only Class I nickel (generally electrolytic Ni ~99.8% Ni) is suitable for the electric vehicle (EV) and the green technology power storage (Lithium Ion Batteries). This precludes the NPI as the feed for the battery material. During the last few years, however, China has started taking a keen interest on the EV market. Two of the plants visited, Jinchuan and Jilin, have the technology for production of Class I nickel and intermediate products destined for the battery manufacturing, as shown in Table 4.

When the delegates visited the Jinchuan smelter, they were informed that by 2029 China is expected to have 40–50% of the global EV and the green industry battery market, and Jinchuan is eyeing to supply a major portion of its nickel. Indeed, in a news report in 2019, Reuter announced that China's Jinchuan hope to fill the supply gap left by Indonesia's ban [12]. Since 2018, the top story of China has become the story of NPI and EV batteries.

Asia Excluding China

From China, as listed in Table 5, the virtual tour went to visit the Gwangyang smelter in South Korea, followed by a visit to three smelters in Japan. After leaving Japan, next stop was Coral Bay, an HPAL hydromet plant in the Philippines and then finally to Indonesia to visit three significant nickel smelters. The plants visited on this part of the tour produced 71.4% of the nickel in this region and 27% of the world nickel. Mine production was much higher at 188% of smelter/plant production with a lot of ore being exported to China.

Table 5 Asian smelters visited excluding China

Country	Smelter, Company	Ore Type	Production kt/a Ni 2019	Capacity kt/a Ni	Summary of Observations
Indonesia	PT Aneka—Pomalaa	L	26	26	An Indonesian state-owned company at Pomalaa commissioned in 1975 using the RKEF process making 18 to 21%Ni in FeNi. Have 4 dryers; 4 reduction kilns; 3 EFs and 2 refining lines
	PT Vale—Sorowako	L	71	92	Previously PT INCO. Commissioned in 1976 with 3 RKEF lines making a matte product. Additional dryers and kilns and converters added in 1997. Liquid sulphur added to the calcine at the discharge of the kilns so furnaces make a low sulphur matte that is processed through 3 PS converters
	Tsingshan Indonesia	L	240	300	Located in Sulawesi in the Indonesia Morowali Industrial Park. Uses the RKEF process to make NPI. It was commissioned in 2013 and planning to expand.
Japan	Nippon Yakin—Oheyama	L	5	15	Directly reduces garnierite ore to produce high and low carbon FeNi using two rotary kilns. The FeNi is then used directly in stainless steel production
	Hachinohe; Pamco	L	32	45	The RKEF FeNi smelter is located at Hachinohe near Aomori, Japan. It has 1 dryer, 3 kilns and 3 electric furnaces (60 MW each). It produces FeNi in the range 17 to 23% Ni as ingots and shot

(continued)

Table 5 (continued)

Country	Smelter, Company	Ore Type	Production kt/a Ni 2019	Capacity kt/a Ni	Summary of Observations
	Hyuga; Sumitomo	L	13	22	Hyuga was established in 1956 and receives approximately equal quantities of laterite ore from New Caledonia and from Indonesia. The flowsheet is RKEF to make several grades of ~20%Ni in FeNi as shot. There are two kilns and two electric furnaces
Philippines	Coral Bay; Rio Tuba	L	23	24	In 2001, SMM began construction of a new nickel refining facility on the island of Palawan in the Philippines. This plant utilized the High Pressure Acid Leach (HPAL) method, The plant was commissioned in 2005. A successful ramp-up of the operation to attain 100 percent designed capacity was achieved in less than one year
South Korea	Gwangyang; Posco, SMSP	L	45	54	A joint venture that supplies the metallurgical Gwangyang plant in South Korea, with ore sourced from its mining centres at Ouaco, Poya, Nakety and Kouaoua in New Caledonia. The plant, with a production capacity of 30,000 tonnes of nickel metal contained in ferronickel, conducted an expansion in 2011 to achieve its current capacity
Subtotal (above)			455	578	Nickel smelter production 71.4% of total Asian Smelters (excluding China)
Other Asian Smelters (excluding China)			182	366	
Total Asian Smelters (excluding China)			637	944	Nickel smelter production 27% of total world Smelters
Total Asian Mine Production (excluding China)			1198		Mined production ~188% of smelter production in Asia excluding China

The Société du Nickel de Nouvelle Calédonie et Corée (SNNC) started up its new ferronickel smelter in 2008 reaching its full design production level in 2009 [14]. The No.1 plant is based on the RKEF process, with two rotary kilns and a single large furnace to achieve economies of scale. The furnace is 22.2 m in diameter at 94 Mega Watts (MW)/120 Megavar (MVA). In 2014 No. 2 plant was added [15]. It includes the largest (hearth size 40.8 m by 14.4 m) rectangular Fe-Ni electric smelting furnace in the world (135MVA) with an extensive monitoring system. The plant is located adjacent to POSCO's steelworks in Gwangyang, Korea. POSCO, and SMSP, Société Minière du Sud Pacifique, co-own the facility.

The three smelters the virtual tour visited in Japan are Pamco's Hachinohe smelter near Aomori Japan, the Hyuga smelter owned by Sumitomo in Miyazaki Japan which are both RKEF smelters and Nippon Yakin where ore richer in garnierite is reduced without full melting using two rotary kilns to produce high and low carbon FeNi. The FeNi is then used directly in stainless steel production.

Moving on our virtual travels to the Philippines, we stopped at Rio Tuba's Coral Bay plant. This plant utilizes the High-Pressure Acid Leach (HPAL) method. The plant was commissioned in 2005 and was successfully ramped-up to attain 100 percent design capacity in less than one year [16].

After leaving the Philippines, we then flew to Jakarta and from there to Ujung Pandang in Sulawesi to visit three Nickel smelters on the island. The PT Vale Indonesia plant was originally owned by PT Inco The first RKEF line was commissioned in 1976 followed by two more lines and a hydroelectric plant at Larona. The plant was expanded in 1997 with the addition of another dryer kiln and electric furnace. The capacity was significantly increased over the years with the addition of additional hydroelectric generating capacity and the upgrading of the original furnaces from 45 MW to ~90 MW. The smelter makes a nickel matte by addition of liquid sulphur to the kiln calcine. The furnace sulphur-deficient matte is converted to a converter matte of ~78% Ni, 0.5% Fe and ~20% S. This matte is granulated, and the product is shipped to Japan for further refining.

PT Antam Tbk ("ANTAM") is an Indonesian state-owned company under the Indonesian Department of Mines and Energy and operates a nickel mining and processing plant at Pomalaa in South East Sulawesi, Indonesia. It is one of the largest mines operated by Antam. The Pomalaa plant processes nickel ore into 18 to 21% Ni in ferro-nickel ingots and shots for export. The plant has 4 dryers; 4 reduction kilns; 3 electric furnaces (EF) and 2 refining lines. Power generated at site by diesel (8) generators and coal (2) generators.

Tsingshan Smelter is in Sulawesi in the Indonesia Morowali Industrial Park. It uses the RKEF process to make NPI [17]. It was commissioned in 2013 and has expanded several times (already with over 24 RKEF lines). Tsingshan is one of the world's biggest stainless steelmaker and has made substantial investments in Indonesian nickel. Tsingshan has introduced significant amounts of NPI on-stream including in Indonesia to meet demand, likely dulling nickel prices at the time. Tsingshan evidently now plans to construct an Indonesian HPAL plant to produce nickel-cobalt salts for the battery market from nickel laterite ores. Using previously uneconomic technology Tsingshan says it will transform Class 2 laterite deposits into Class 1 metal for the battery market.

Oceania

From Indonesia, the virtual tour took the delegates to Australia and from there to New Caledonia to visit a total of four smelters, two in each country; Kalgoorlie and Murrin Murrin in Australia and Koniambo and Doniambo in New Caledonia, as listed in Table 6. The total capacity of the four plants is 290 kt annual nickel production, or about 64% of the total Oceania smelters at 454 kt/a.

The Kalgoorlie Nickel Smelter is located close to south Kalgoorlie and is fully owned by BHP Billiton -Nickel West. The smelter started in 1972 under the ownership of Western Mining Corporation (WMC). It includes an Outokumpu flash furnace (FF) with an electric furnace extension (A.K.A. Appendage) for "in situ" slag cleaning, three moderate size converters and an acid plant. BHP Billiton acquired the plant in 2005.

Nickel concentrates are delivered to the smelter from several concentrators. The product matte contains 68% Ni and 2–3% Cu and 1% Co. 70% of the matte product is sent to the Kwinana refinery south of Perth and based on Sherritt technology, where several LME grade marketable products are made and 30% is exported as nickel matte. Total smelter nickel production is at 115 kt/a.

In the 2019 fiscal year, BHP (Nickel West) seized the opportunity to participate in the production of battery grade intermediate materials for the electric vehicles (EV) and other green technology. Accordingly, a nickel sulphate plant is under construction at the Kwinana nickel refinery for Stage 1 production of 100 kt/a nickel sulphate [13].

Murrin-Murrin is a hydrometallurgical plant, one of the first of the more recent laterite projects using sulphuric acid leaching in high-temperature, high pressure, autoclave vessel technology (High Pressure Acid Leaching or HPAL) for the recovery of nickel and cobalt. It is located close to Leonora in Western Australia. First opened in 1999 as a joint venture between the subsidiaries of Anaconda Nickel (later Minara Resources) and Glencore International with a nameplate capacity of 45 kt/a nickel [18]. (Murrin Murrin was one of three laterite plants brought on line in Western Australia, the others were Cawse and Bulong).

In the 1990s, when the laterite ores were dominating the nickel resources, the three plants were touted as having the future low-cost, environmentally sustainable technology. "The feeling was that the sulphide mining faced a difficult and uncertain future" [19].

The three plants faced start-up issues related in part due to design and construction features; Murrin Murrin had a slow start-up [18]; the two other plants closed. In 2011, Glencore acquired ownership of the company, and following some capital improvements the plant reached close to its nameplate capacity, stated to be between 40 and ~45 kt/a.

The Doniambo smelter was first established in 1880 in Noumea, New Caledonia, as the Société Le Nickel (SLN), the plant is the first laterite operation in the world. It became a wholly owned subsidiary of Eramet in 1985 and has expanded and modernized since then, the latest in 2001 with an increase in nickel production to 75 kt/a.

A typical open pit laterite mining operation, the mines are usually located high up on the mountains, between 500 and 1000 meters, and are sometimes visible when driving in the outskirts of Noumea. Doniambo uses conventional RKEF technology to produce FeNi at 25% Ni and shaking ladle technology for refining Until 2016, about 80% of the FeNi was shipped to the market and the rest was sulphidized in

Table 6 Oceania smelters visited

Country	Smelter, Company	Ore Type	Production kt/a Ni 2019	Capacity kt/a Ni	Summary of Observations
Australia	Kalgoorlie; BHP	S	84	115	The smelter started in 1972 under the Western Mining Corporation (WMC), BHP Billiton acquired the plant in 2005. The process includes a flash furnace + EF appendage for slag cleaning, three converters and an acid plant. Delivered nickel concentrate is smelted in the flash furnace. The smelter product matte contains 68%Ni, 2–3% Cu, 1% Co. 70% of the matte is sent to the Kwinana refinery in Perth and 30% is exported. A battery grade nickel sulphate plant for 100 kt/a production in 2019
	Murrin Murrin; Glencore	L	43	45	A hydrometallurgical plant, the first recent plant to use HPAL technology for Ni and Co recovery. Opened in 1999 as a joint venture between the Anaconda Nickel (later Minara) and Glencore International. Reaching nameplate capacity was gradual. In 2011, Glencore acquired full ownership of the company. Eventually reached close to its capacity after significant cost and improvements

(continued)

Table 6 (continued)

Country	Smelter, Company	Ore Type	Production kt/a Ni 2019	Capacity kt/a Ni	Summary of Observations
New Caledonia	Doniambo; Eramet	L	47	75	Built in 1880 in Noumea, New Caledonia, as SLN, the plant is the first laterite operation in the world. In 1985, it was taken over by Eramet, now as Eramet-SLN. Expanded and modernized in 2001 and increased nickel production to 75 kt/a. Operates conventional RKEF + (Shaking) ladle refining until 2016, 80% of the FeNi was shipped to the market, balance was sulphidized with final matte at 75% Ni, 1% Fe and shipped to the Le Havre-Sanduville refinery in France. Final products from the refinery are Fe and Co chlorides, high purity nickel chloride (battery grade) salts and 99.97% Ni purity nickel cathode
	Koniambo; Glencore	L	24	60	In 2013- 2014, Glencore (Xstrata) commissioned and ramped up Koniambo FeNi smelter, located in north west of the New Caledonia island. One of the largest and highest grade of saprolite resources. Two process lines, with a production capacity 60 kt/a Ni in ferronickel. The highly energy efficient plant uses a cyclone flash dryer/calcination process, in situ reduction with coal injection into a fluid bed, feeding directly into a DCAF
Subtotal (above)			198	295	Nickel smelter production 77% of total Oceania Smelters

(continued)

Table 6 (continued)

Country	Smelter, Company	Ore Type	Production kt/a Ni 2019	Capacity kt/a Ni	Summary of Observations
Other Oceania Smelters			59	159	
Total Oceania Smelters			257	454	Nickel smelter production 10.9% of total world Smelters
Total Oceania Mine Production			420		Mined production ~163% of smelter production in Oceania

a special converter to an iron-nickel matte, which then was converted to a matte at 75% Ni and ~1% Fe in a Peirce-Smith (PS)converter before it is shipped to the Le Havre (now Le Havre Sanduville) hydrometallurgical refinery in France [20]. At the refinery, where matte from Europe is now treated, marketable intermediate products of iron, cobalt chlorides, pure nickel chloride (liquid or solid), which can be sold as nickel salt forms (carbonate or nickel hexahydrate) or obtain high quality nickel cathode [21].

"Koniambo is one of the largest and highest-grade undeveloped nickel resources in the world". In 1998 Falconbridge Limited was selected by the Société Meniere du Sud Pacifique (SMSP) as its industrial partner [22]. This was quite fitting for Falconbridge at the time, since the company had been looking for another opportunity to invest in a laterite project. The poor experience of the HPAL plants in Western Australia likely tipped the company's view towards a pyrometallurgical technology and initially developing the saprolite portion of the ore body. Given the high impact of energy costs and emission containment issues encountered in many existing RKEF operations, Falconbridge embarked on developing their innovative Nickel Smelting Technology (NTS) with Direct Electric Arc Furnace (DCAF) by 2005 [19]. The process is considered an upgrade to the Falcondo process.

In 2005, Falconbridge was taken over by Xstrata, but the partnership with SMSP remained unchanged (SMSP 51%, Xstrata 49%), and the company moved with engineering studies. In 2013, Xstrata was absorbed by Glencore, the same year the plant became fully operational. Koniambo was officially inaugurated in 2014 with a design capacity of 60 kt Ni/a (3 million t/a saprolite ore) as FeNi at 35% Ni in two lines, each with a DCAF capacity of 80 MW. The reported mine life is for at least 25 years [22].

Africa

From New Caledonia the virtual tour proceeded to Madagascar and then to South Africa to visit two plants; Ambatovy—a nickel-cobalt hydrometallurgical plant and the Anglo-Platinum Waterval nickel-copper-PGM pyrometallurgy smelter, as listed in Table 7.

The total nickel production in 2019 from the two plants was 59 kt/a, or about 41% of the total African smelters. Mine production was 71% of the smelter production in 2019.

The mine for Ambatovy is located 80 km east of Antananarivo (the Capital) and is amongst the largest lateritic nickel mines in the world, with a production capacity of 60 kt/a in nickel and 5.6 kt/a in cobalt. It is a joint venture and Sherritt International serves as the operator of the facility. Sherritt initially started as the majority shareholder with 40%, but since, this has been reduced. Other significant partners are Sumitomo and Korea Resources Corporation [23]. The reported life of mine is 29 years.

The Ambatovy nickel plant was completed in 2011, based on Sherritt's hydrometallurgical process, and started its first production in 2012, achieving commercial production in 2014. From the mine, the slurried laterite ore is delivered via pipeline to the processing plant. Ambatovy's products are nickel (99.8% pure), cobalt (99.8% pure) and ammonium sulphate—a by-product of the plant that is used in agriculture.

Ambatovy actively pursued goals for good corporate sustainability, environmental preservation and conservation, winning several awards for environmental sustainability and green technology. The cost of the Project increased, like many other large hydrometallurgy plants, and exceeded $6 billion [24]. Furthermore, since the start up, the fluctuations in the nickel market have adversely affected the company. In 2019, the nickel and cobalt output was at 33.7 kt and 2.9 kt, respectively, somewhat below the planned production rates. Production curtailments in 2020 are likely expected due to the COVID-19 pandemic, as the plant was forced to shut during the lock-down. Under these circumstances Sherritt's participation in the JV beyond 2019 is uncertain. "The company is to exit the Ambatovy Joint Venture ...... but its decision will not affect nickel or cobalt production" [25].

The first large electric furnace for Ni-Cu concentrate commenced at Waterval in 1969 and the plant has expanded several times [26]. The plant and is situated in Rustenburg, in South Africa's North West Province. The largest smelter within the three plants in the Anglo Platinum group. The smelter receives wet concentrate to produce crushed, slow-cooled, sulphur-deficient nickel-copper matte rich in platinum group metals (PGMs), gold, and base metals for further processing at the Magnetic Concentration Plant (MCP) at the Base Metals Refinery [26], with a nickel production capacity of 22 kt/a. The smelter initially started with two Elkem electric furnaces, each at 18 MW, four 6.1 m (later extended 7.1 m) Peirce-Smith converters and in 1976 commissioned a single contact acid plant for reduction of SO_2 emissions. In the 1990s, Waterval took on additional modernization initiatives replacing the rotary dryers and pellet feed with flash dryers, pneumatic feed system and a concentrate silo, as well as upgrading both furnaces to 34 MW. To support the added plant capacity, in 2002, the plant developed a customized Top Smelting Lance (TSL) technology, in conjunction with Ausmelt of Australia (now Outotec) to replace the Peirce-Smith converters, called the Anglo Platinum Converting Process (ACP). Copper coolers were incorporated in the 4 m high crucible and high pressure membrane coolers

Table 7 African smelters visited

Country	Smelter, Company	Ore Type	Production kt/a Ni 2019	Capacity kt/a Ni	Summary of Observations
Madagascar	Ambatovy; Sumitomo, Sherritt	L	34	60	The plant was completed in 2011 as a JV. Based on the Sherritt process, with a design capacity of 60 kt/a in nickel and 5.6 kt/a in cobalt. Products from the refnery are high purity nickel (99.8%), cobalt (99.8%) and ammonium sulphate. Smelter production close to but below design capacity in part due to the market conditions. In 2019, the nickel and cobalt productions were at about 34 kt and 2.9 kt, respectively. Sherritt's participation in the JV beyond in 2019 is uncertain
South Africa	Waterval; Anglo American	S	25	22	Part of the Anglo-American companies, a pyrometallurgy plant commissioned in 1969–1972, processing wet Ni-Cu-PGM concentrate to produce a S-defcient Ni-Cu-matte for the recovery of the PGM and Ni and Cu metals. Initially, designed for 22 kt/a with two 18 MW EF and four 6.1 m PS converters. Upgrading in later years included flash dryers, two 34 MW EF and two customized TSL converters, called ACP and a slag cleaning EF. The ACP converter can process about 72,000 t/a of converter matte, corresponding to about 33,000 t/a in nickel. Annual nickel productions for the 2017 and 2018 fscal years are reported at 23 and 26 t/a, respectively
Subtotal (above)			59	82	Nickel smelter production 41.2% of total African Smelters

(continued)

Table 7 (continued)

Country	Smelter, Company	Ore Type	Production kt/a Ni 2019	Capacity kt/a Ni	Summary of Observations
Other African Smelters			84	60	
Total African Smelters			143	142	Nickel smelter production 6.1% of total world Smelters
Total African Mine Production			102		Mined production ~71% of smelter production in Africa

in the upper part of the furnace [26, 27]. Soon afterward a slag cleaning furnace equipped with water-cooled slag-line was installed to treat the ACP slags. With the installation of another ACP unit, as a back-up in 2004, it enabled Waterval to feed all the furnace matte from all three Anglo Platinum smelters to the ACP. A new acid plant was installed to handle the additional SO_2 gases [28]. The ACP technology is the first application of the Ausmelt converting to the Precious Group Metal's (PGM) industry.

In the basic process flowsheet, the wet concentrate is dried in four flash dryers, one on standby, which is stored in a 4,500-t silo, from which the dry concentrate is delivered to the primary EF smelting furnaces. The EF furnace matte is tapped and then granulated, and the furnace slag is milled and floated to generate a concentrate, which is sometimes recycled or stockpiled directly, if low in pay metals. The furnace matte from all Anglo-Platinum smelters is converted in the Waterval ACP converter, where the matte is upgraded through removal of the iron and part of the sulfur to form the Waterval Converter Matte (WCM). The matte is cast in molds, slow cooled over several days, then crushed and forms the final product from the smelter, which is further treated at the Matte Crushing Plant to recover the metallic fraction containing all the PGMs. The base metal recovered from the WCM is processed in the Base Metal Refinery, where the copper and nickel are separated. The converter slag is granulated and recycled to a Slag Cleaning Furnace (SCF).

The ACP converter has the capacity to treat about 72,000 t/a of converter matte. This corresponds to about 33,000 t/a in nickel and 4 million ounces of platinum (as the maximum capacity) [27]. The Anglo-American annual nickel productions for the 2017 and 2018 fiscal years are reported at 23 and 26 t/a, respectively.

South America and the Caribbean

The next stop on the virtual tour was South America and the Caribbean. The tour visited four countries namely Brazil, Columbia, Guatemala and the Dominican Republic and five smelters. These five smelters produced 111kt Ni in 2019 which is ~61% of the Ni produced in this continent. The South American and Caribbean region produced ~7.7% of the world Ni in 2019. Mine production was 123% of smelter production. See Table 8.

The vast laterite nickel deposits of Brazil have been known since the beginning of the last century, or even earlier. The first nickel production in Brazil began in the 1930s by Companhia Niquel do Brazil producing FeNi at the Liberdade deposit [29]. This was followed by the commencement in 1962 of a rotary kiln electric furnace (RKEF) smelter by Morro do Niquel SA located near the town of Pratapolis in the

Table 8 South American and Caribbean smelters visited

Country	Smelter, Company	Ore Type	Production kt/a Ni 2019	Capacity kt/a Ni	Summary of Observations
Brazil	Barro Alto; Anglo American	L	34	40	The Barro Alto smelter, which commenced in 2011 with a design capacity of 36,000 t Ni/a, includes the following main plant units: a crushing plant; two dryers, 4.6 m dia. by 32 m, two rotary kilns, 6.1 m dia. by 185 m; and two, six-in-line electric furnaces nominally rated at 83 MW. There are also two refining stations rated at 9 MW to refine the 30% Ni ferronickel product before shipping. The plant had a partial shutdown in 2012–2013 to rebuild the furnaces with improved copper cooling. Pilot plant test work on upgrading the ROM ore is on-going
	Onca-Puma, Vale	L	12	26	The On9a Puma smelter, which commenced in 2011 with a design capacity of 58,000 t Ni/a, now includes the following main plant units: a crushing plant; two dryers, 6 m dia. by 45 m; two rotary kilns, 6 m dia. by 135 m; and one, up-graded six-in-line electric furnace nominally rated at 100 MW (with capability to 120 MW). One of the two original 83 MW furnaces was upgraded with new copper coolers in 2012–2013 (the 2nd furnace was closed). There are also two refining stations rated at 9 MW to refine the 25% Ni ferronickel product before shipping

(continued)

Table 8 (continued)

Country	Smelter, Company	Ore Type	Production kt/a Ni 2019	Capacity kt/a Ni	Summary of Observations
Columbia	Cerro Matoso; South 32	L	41	55	Located in northwest Colombia in the municipality of Montalbano, was discovered in 1940. Plant was commissioned in 1982 with a design capacity of 22.7 kt/a nickel as 42% FeNi, using a RKEF technology. In 2001–2002, the production capacity increased, by adding a second RKEF line, to 55 kt/a. Key equipment operating include, two rotary kilns at 185 m and 135 m in length, and two EFs of 70–75 MW. The ore head grades have been steadily declining, affecting the production, totaling 41 kt/a in 2019
Dominican Republic	Falcondo; American Nickel	L	24	35	Located in the interior of the Dominican Republic, near the town of Bonao, the smelter was built in 1970 and commissioned in 1971 with a design capacity of 35 kt/a nickel in FeNi. Falcondo is the first laterite operation designed and built by Falconbridge Nickel Mines (now Glencore). In 2015, Glencore sold the company to the American Nickel Ltd, a private Greek company. The process includes a rotary drier, vertical reduction shaft furnaces, 6-inline rectangular 55.5 MVA, high intensity EF, and ladle refining facilities. The plant operated its own 200 MW power pant with imported bunker C oil on-site. The process is sensitive to energy cost. The American Nickel has made some changes to the process. The production in 2019 is reported at 24 kt/a

(continued)

Table 8 (continued)

Country	Smelter, Company	Ore Type	Production kt/a Ni 2019	Capacity kt/a Ni	Summary of Observations
Guatemala	Fenix; Solway	L	20.3	25	The RKEF plant known as Exmibal was built by Inco to produce matte, however it only operated from 1977 to 1980 when it was closed and mothballed citing high oil costs. The Solway Investment Group purchased the property from Hudbay and Skye in 2011, re-opening the plant as Fenix in 2014 producing ferronickel rather than matte. The power plant and the electric furnace have since been upgraded. The facility mined the equivalent of 41,000 t Ni in ore in 2019 (about 2.7 million dry tonnes ofore), with about 26,000 t of Ni in ore exported, mostly to the Pobuzhsky smelter in the Ukraine, and the balance treated at Fenix
Subtotal (above)			131.3	181	Nickel smelter production 61% of total South American and Caribbean Smelters
Other Asian Smelters (excluding China)			51	127	
Total Asian Smelters (excluding China)			182	308	Nickel smelter production 7.7% of total world Smelters
Total Asian Mine Production (excluding China)			224		Mined production ~123% of smelter production in South America and the Caribbean

state of Minas Gerais [30]; a second furnace was built in 1969, bringing the capacity to 2,500 t Ni/a. The Tocantins limonite deposit in the state of Goiás has been known since the early 1900s, and exploitation began in the 1930s by German interests; in the early 1940s, the American Smelting and Refining Co. of the USA carried out test work on the deposit, however a commercial plant was not built [31]. Commercial production at Tocantins began in 1981 with the commencement of the Caron plant with a rated capacity of 11,000 t Ni/a once the two lines were operational [32]. Brazil's third nickel plant commenced in 1982 with the start-up of the Codemin RKEF smelter located near Tocantins treating saprolite ore. With a design concept based in part on that at Morro do Niquel (including the furnace located at a lower elevation relative to the feed and dryer), Codemin had an initial capacity of 6,000 t Ni/a.

In Brazil, the virtual tour included visits to Barro Alto and Onça Puma. Anglo American commenced mining at Barro Alto, located about 15 km west of the town of Barro Alto in the state of Goiás in 2004, with ore shipped to Codemin some 150 km south-west for smelting and owned by Anglo American; construction of the Barro Alto smelter began a few years later.

The Barro Alto smelter includes the following main plant units: a crushing plant; two dryers (4.6 m dia. by 32 m), two rotary kilns (6.1 m dia. by 185 m); and two, six-in-line electric furnaces nominally rated at 83 MW. There are also two refining stations rated at 9 MW to refine the 30% Ni ferronickel product before shipping. The plant commenced operations in March 2011, with a design capacity of 36,000 t Ni/a.

With production below design (the plant produced 21,600 t Ni in 2012—about 60% of design), the furnaces were modified over the period 2012–2013 [33]. The plant has operated well since, producing 34,000 t of Ni in 2019 from ore averaging 1.69% Ni. Anglo-American reported in 2019 that testing of an ore upgrading process was under way at Barro Alto which if successful, would bring the grade of the ore smelted to 1.94% Ni and allowing for an increase in nickel output of some 5,000 t Ni/a [34].

The Onça Puma nickel laterite smelter located some 400 km south-west of Marabá, and about 10 km from Tucumã in Pará state, commenced operations in March 2011. The Onça Puma smelter includes the following main plant units: a crushing plant; two dryers, 6 m dia. by 45 m; two rotary kilns, 6 m dia. by 135 m; and two, six-in-line electric furnaces nominally rated at 83 MW. There are also two refining stations rated at 9 MW to refine the 25% Ni ferronickel product before shipping. The original design capacity of the plant was 58,000 t Ni/a [35]. A furnace re-build program began in mid-2012—one furnace was upgraded to a nominal rating of 100 MW (with capability to 120 MW), and the plant re-commenced operations in 2014 with two dryers, two kilns and one electric furnace [36]. The plant has generally ran well since that time, however production in 2019 was only 12,000 t Ni.

Moving on to visit Cerro Matoso whose deposit in northwest Colombia is in the municipality of Montelibano, was discovered in 1940. The Cerro Matoso nickel mine and smelter was built by a partnership with Hanna Mining, Billiton and others using RKEF technology. Following about a decade of engineering and pilot plant studies, the construction began in 1979 and the plant was commissioned in 1982 with a design capacity of 22.7 kt/a nickel as 42% FeNi [37]. Billiton, subsequently BHP-Billiton, increased its share of the project over a period of time; since 2015, South 32 has operated the plant. Production capacity was increased in 2001 by adding a second RKEF line to 55 kt/a. The smelter currently operates two rotary kilns at 185 m and 135 m in length and two large EF of 70–75 MW. In 2017, the smelter production was at 41 kt/a. Current estimate for the mine life is reported to be 10 years, as of January 2019. A key reason is that Cerro Matoso's ore head nickel grade has been steadily declining from a high of around 3.2%Ni during the startup [36] to a low of 1.66% Ni by 2019 [38], which likely will continue to decline.

The Falcondo smelter is in the interior of the Dominican Republic, near the town of Bonao. Built-in 1970 and commissioned in 1971 with a design capacity of 35 kt/a nickel in FeNi. Falcondo is the first laterite operation designed and built by Falconbridge Nickel Mines (now Glencore). The development of the process began with studies in 1955 at the Falconbridge Research Laboratories in the northern outskirts of Toronto and continued with several years of piloting the reduction and smelting steps in the Dominican Republic near the current plant site [39]. The seminal high voltage shielded arc furnace was developed during these campaigns and patented by Frederick Archibald and Gerry Hatch [40], both inductees to the Canadian Mining Hall of Fame.

The process flowsheet has been described in detail by the present authors in a previous paper [9]. In summary, partially dried ore briquettes are fed to 12 specially designed shaft furnaces for gaseous reduction and transferred to three six-in-line 55.5 MVA rectangular, shielded arc, electric furnace by means of transfer cars. Most of the reduction takes place in the shaft furnace, but traces of residual carbon left in the calcine complete the reduction in the electric furnace, resulting in a 38% FeNi grade. The plant operates its own 200 MW power plant on-site with imported bunker C oil. Downstream processes for handling the EF slag and FeNi refining are the same as the traditional RKEF process. The application of the high intensity shielded arc technology revolutionized the electric furnace smelting of laterites in the last 30–40 years and the ever-increasing furnace sizes and power ratings of today's and what is envisaged for the future [41] is a testimony to its success.

In 2015, the company was acquired by the American Nickel Ltd, a privately owned company. On account of dependency on imported oil, the Falcondo plant is quite sensitive to oil cost. The new company has since made several changes to the plant; however, oil remains the main fuel. The production in 2019 is reported at 24 kt/a.

There is one operating nickel laterite smelter in Guatemala, now referred to as Fenix is now operated by the Solway Group. Originally under the company Exploraciones y Explotaciones Minerais Izabel SA, or Exmibal, majority owned by Inco of Canada, the Exmibal RKEF smelter, with a design capacity of 11,300 t Ni/a as a 76% Ni nickel matte, commenced operations in early 1977, with the first matte converter charge blown in August 1977. The plant included the following units: a crushing plant; one 3.6 m by 45 m oil-fired dryer, one 5.5 by 100 m oil-fired kiln with oil injected through lances for reduction plus liquid sulphur injection by lance for sulphurization (at about 1% of kiln feed); one 18 m dia. Elkem electric furnace rated at 45 kVA; and two, 4 m (13 ft.) by 7.6 m (25 ft.), Peirce-Smith converters to blow the iron level to ~0.6% Fe in finish matte (~77% Ni). The site was also equipped with a 61.3 MW Bunker C-fired power plant plus up to 10 MW additional emergency power supplied by diesel generators. The plant only produced 7,000 t Ni in matte in 1979 (~62% of design). In November 1980, the smelter was closed in part due to rising costs of oil (oil represented ~60% of opex); the plant was placed under care and maintenance in 1981 and remained so for many decades.

The property was purchased by Skye Resources in 2004, with ownership subsequently transferred to Hudbay of Canada in 2008 with plans to re-open the smelter.

The Solway Investment Group then obtained the property in 2011 now referred to as Fenix; it re-opened in 2014 producing ferronickel rather than matte as in the original Inco design. The power plant and the electric furnace have since been upgraded, with Fenix producing 15,000 t Ni in 2019. The facility mined the equivalent of 41,000 t Ni in ore in 2019 (about 2.7 million dry tonnes of ore), with about 26,000 t of Ni in ore exported, mostly to the Pobuzhsky smelter in the Ukraine, and the balance treated at Fenix.

North America

The final leg of the virtual tour took us to North America to visit two nickel giants in Canada in the Sudbury Basin namely The Sudbury Integrated Nickel Operation in the town of Falconbridge and Vale's Copper Cliff operations and smelter. These two smelters produced 141 kt Ni in 2019 ~80% of the North American nickel production and 7.5% of the world nickel production. Mine production was 115% of smelter production in 2019. See Table 9.

The original smelter located in the town of Falconbridge, Ontario was owned by Falconbridge Ltd until the 2000s. By 2002, Noranda had increased its ownership of Falconbridge and in 2006 the company was acquired by Xstrata. In 2013, Glencore bought Xstrata and the operation at Falconbridge was renamed the "Sudbury Integrated Nickel Operations".

The smelter started operations in 1930 [42] and for the first three years of operation, the smelter solely processed hand-sorted ore (Fig. 4), until the Falconbridge Mill began operation. In 1932 a sinter plant was constructed to make an agglomerated feed for the blast furnace. By 1960 there were 5 sinter plants and 3 blast furnaces. The Bessemer matte from the smelter has and is being sent to the Nikkelverk refinery in Norway for refining. In 1978 the smelter flowsheet was changed to include two fluid bed roasters and two rectangular electric furnaces with the Peirce Smith converters being retained. Initially the degree of roasting was ~50% but has been increased to ~80% to increase gas capture. Today there are four Peirce Smith converters two of which have been modified as "Hybrid" vessels [43] and two are used as finishing vessels. The hybrid operation allows for both furnace matte conversion but also slag cleaning which is important for increased metal recovery.

The Copper Cliff Smelter commenced operations in 1930 as the International Nickel Company of Canada. In 1976, the name was changed to Inco Ltd which was then bought by Vale in 2006. By 1961, the original smelter had two blast furnaces, 42 multi-hearth roasters, seven reverberatory furnaces for nickel concentrate, two oxygen flash furnaces for copper concentrate, 24 converters (three for copper) and two electric melting furnaces for matte separation products [42]. By the early 1970s, operating equipment diminished from 42 to 24 multi-hearth roasters and from 6 conventional air-coal fired reverbs to 2 oxy-fuel fired reverbs. Copper concentrate was processed through a fluid bed dryer and Inco Flash furnace.

In the 1980s, Inco undertook a major R&D program to investigate alternative smelting routes. This program focused on two alternative smelting routes, roast-reduction smelting of nickel concentrate (RRS) and oxygen flash smelting of a

Table 9 North American smelters visited

Country	Smelter, Company	Ore Type	Production kt/a Ni 2019	Capacity kt/a Ni	Summary of Observations
Canada	Sudbury Integrated Nickel Operations; Glencore	S	71	75	The smelter began in 1930, and until 1978 operated with sinter plants, blast furnaces and Peirce Smith converters. The smelter flowsheet was then changed to improve environmental performance with the installation of two fluid bed roasters with 50% S removal, two rectangular electric furnaces and acid plant. The plant now operates at 80% S removal, one furnace with two hybrid special converter vessels, and two finish converters
	Vale Sudbury; Vale	S	70	80	The smelter was commissioned in 1930 and operated until 1994 as a copper nickel smelter with multi hearth roaster, reverberatory furnaces, a copper INCO fash furnace and Pierce Smith converters. In 1994 the flowsheet was changed to reduce operating costs and SO_2 emissions with 2 lines of fluid bed dryers, INCO flash furnaces and PS converters along with a new acid plant. Copper was processed as a chalcocite concentrate (MK) in a reactor and multi purpose vessels and anode furnaces to make blister copper. Since 2018 the smelter has been simplified again to a single flash furnace line. Copper processing was also discontinued

(continued)

Table 9 (continued)

Country	Smelter, Company	Ore Type	Production kt/a Ni 2019	Capacity kt/a Ni	Summary of Observations
Subtotal (above)			141	160	Nickel smelter production 79.7% of total North American Smelters
Other NA Smelters			36	107	
Total NA Smelters			177	267	Nickel smelter production 7.5% of total world Smelters
Total NA Mine Production			204		Mined production −115% of smelter production in North America

bulk copper-nickel concentrate [44]. The bulk smelting route was chosen and by November 1993, the Copper Cliff smelter flowsheet was altered with changes to milling to produce a Cu-Ni bulk concentrate, shutdown of the reverberatory furnaces and construction of two Inco flash furnaces for processing the bulk concentrate, construction of an oxygen top-blowing nitrogen bottom-stirring vessel for the processing of the copper sulphide-rich material produced during matte cooling and subsequent milling and referred to locally at the plant as "MK" (the vessel treating this material was referred to as the "MK reactor") and erection of a new 2,900 t/day sulphuric acid plant for treating most of the new flash furnace and MK Reactor off-gases.

In order to reduce emissions and increase productivity further, the smelter flowsheet has been simplified such that since 2018, the smelter is now operating with a single Inco flash furnace, two Peirce Smith converters with converter gases going to the acid plant with gases from Matte Processing fluid bed roasters (FBRs) and flash furnace. The MK reactor has been repurposed as a converter slag cleaning vessel to increase cobalt recovery.

Back to London and Home

Following the return trip by ship in Jules Verne's classic story, the present virtual "Around the Nickel World" trip was able to complete the journey from New York to Southampton UK on Cunard's legendary Queen Mary 2. The seven-day virtual Atlantic crossing afforded time to reflect on the visits of the past 80 days and prepare the report. The virtual tour ended on time at the Reform Club, London.

Conclusions

The virtual tour over 80 busy days to 35 nickel smelters and plants around the world has given the authors interesting insights to the world primary nickel industry. They witnessed were some well-designed and operated plants in parts of Europe, Japan and elsewhere to more aging plants still undergoing modernization including environmental controls in parts of Europe, Russia and elsewhere, and a range of facilities in between. China's dominance in nickel metal production (30% of world Ni metal)—with much produced as NPI from imported laterite ores as well as being the world's largest consumer of nickel (about 50% of world Ni) was clear. It was also seen that with the greater emergence of renewable energies requiring batteries and the projected growth in electric vehicles, nickel remains an important metal today and, in the future. Hence it is considered that demand for nickel will remain strong. Based on the tour, the stainless-steel sector, the largest industrial consumer of nickel, will remain robust. There will, however, be a shift to more nickel produced in the form of nickel sulphate and Class 1 Ni (essentially electrolytic nickel ~99.8% Ni content) and likely proportionally less Ni as NPI.

The range of technologies employed and the manner of operation, maintenance practices as well as local factors such as labour and energy shows up both in the considerable range of operating costs and general plant conditions observed on the virtual tour. The nickel industry is undergoing great change with respect to the new demands for nickel as a battery material and at the same time it's on the edge of a technology revolution with a greater use of automation, robotics and artificial intelligence at metallurgical plants and other innovative technologies under development around the world. Nickel production has always been an energy intensive metal to produce, especially that from nickel laterite. With more and more nickel now produced from laterites as observed on the virtual tour, energy usage will be important (at present, about 70% of world nickel is produced from laterites). It is estimated that a typical ferronickel plant treating limonitic ore of average grade requires fossil fuel (generally mostly as coal) totalling 307 GJ/tonne of nickel in the form of ferronickel, and an electrical power requirement of 33,000 kWh/tonne of nickel in the form of ferronickel.

On the other hand, a typical HPAL plant needs a power requirement of 10,000 kWh/tonne of finished nickel metal, plus 12 t of sulphur/t of finished nickel metal (to produce sulphuric acid for leaching and other requirements when acid is not available), 28 t of limestone/t of finished nickel metal, and considerable quantities of water [45]. Increasing energy and labour costs worldwide will drive plants to higher efficiency and greater use of automation to help keep costs down.

The virtual world tour led the group to wonder if enough research and development work is being undertaken to sustain new production levels for the plants of the future; it is noted however that interesting pockets of technology work were underway. It is hoped that the good plants witnessed today on the virtual tour will in fact continue to upgrade and remain in the "competitive category" over the next decade. As noted, the authors saw a wide range in the level of support for research and development in the different countries and plants themselves, some excellent work in Europe, Japan and elsewhere as examples—while in some countries it was thought too low.

As noted, China now consumes about 50% of world nickel production, hence events there will impact the nickel industry far beyond its borders. It is considered that three key factors may potentially have a profound impact on China's supply/demand dynamics and hence world nickel:

i. The worldwide challenges brought about by COVID-19 that introduced new risks to the nickel market;
ii. If the Indonesian ban on exporting laterite ore can significantly starve the Chinese RKEF plants, this will significantly curtail its own NPI production;
iii. The slow pace of developing battery grade intermediates due to the fear of capital investment needed f or Class I nickel.

Still, in general, the world nickel industry is found to be a strong and vibrant one today and it is the authors' view that with due attention to all the aspects discussed here, the future looks bright indeed. Time will tell!

Acknowledgements The authors wish to gratefully acknowledge Wood Mackenzie Ltd., London for kindly providing and allowing the use of much of the statistical data used in this paper.

References

1. Grant A (2016) A British history of "German silver", Part II. J Antique Metalware Soci 24:48–77
2. Tylecote RF (1976) A history of metallurgy, The Metals Society, UK
3. Rosenberg SJ (1968) NBS Monograph 106, Nickel and its alloys, National Bureau of Standards, Washington, DC, pp 1–6
4. Holloway GT (1917) Report of the Royal Ontario Nickel Commission: with appendix. Ontario, Canada, Toronto
5. USGS (2020) Nickel 1932-present, https://minerals.usgs.gov/minerals/pubs/commodity/nickel/
6. Riley J (1899) Alloys of iron and nickel. J Iron Steel Institute 35:45
7. Sheffield City Council (2020) Sources for the study of Harry Brearley and stainless steel, Sheffield Libraries Archives and Information 2011 (v.1.0). www.sheffield.gov.uk/archives
8. Vahed A et al (2017) Meeting today's challenges in nickel production: reducing operating costs at nickel sulphide and laterite plants. Paper presented at the 56th annual conference of Metallurgists—2017 international nickel-cobalt symposium, 27–30 August 2017, Vancouver, Canada. ISBN 978-1-926872-36-0
9. Vahed A et al (2015) A review of technology developments in nickel pyrometallurgy and nickel production trends over the last three decades. Paper presented at the 54th annual conference of metallurgists, 23–26 August 2015, Toronto, Canada. ISBN 978-1-926872-32-2
10. International Nickel Study Group Insight, April 2012-No 16
11. Richter W (2020) Wolf Street crude steel production: China blows the doors off rest of the world, wolfstreet.com (6 June 2020)
12. Reuters staff (2019) China's Jinchuan confident it can full Indonesia nickel ore void. https://www.reuters.com/article/us-china-nickel-jinchuan/chinas-jinchuan-confident-it-can-fill-indonesia-nickel-ore-void-idUSKBN1XF17J (5 November 2019)

13. MiningLink Newsletter (2019) bhp Billiton: Kalgoorlie nickel smelter. www.bhpbilliton.com (F8 2019)
14. Rodd L et al (2010) SNNC: A new ferronickel smelter in Korea. Ferronickel smelting, Paper presented at the Infacon XII: International ferro-alloys congress, 6–9 June, 2010, pp 697–708, Helsinki, Finland
15. Jeoung HH et al (2018) The features of the SNNC's largest FeNi smelting furnace. Paper presented at the Infacon XV: International ferro-alloys congress, 25–28 February 2018, Cape Town, South Arica
16. Coral Bay Nickel Corporation website
17. Tsingshan Indonesia websites information
18. Mining Technology (2018) Future of mining: Anaconda Nickel Ltd nickel and cobalt mine. www.mining-technology.com
19. King, M.G. (2005) Nickel laterite technology—finally a new dawn? JOM 57:35–39
20. Eramet newsletter (2020) SLN: A long-standing player in nickel. www.eramet.com/en/SLN. Accessed 4 July 2020
21. Warner AEM et al (2006) World nonferrous smelter survey, Part III: Nickel: laterites. J Metals, 11–20
22. Xstrata Nickel Newsletter (2007) Koniambo Project 17 October 2007
23. Sherritt International—Operations—Metals—Ambatovy Joint Venture (2019). https://www.sherritt.com/English/operations/metals/Ambatovy-JointVenture
24. Conic Metals Corporation (2020) Investor presentation April 2020. www.concmetals.com
25. Metal Bulletin Fastmarkets (2020) Ambatovy "exit" will not affect nickel cobalt output in Madagascar. http://www.metalbulletin.com, 26 August 2020
26. Jacob M (2006) Process description and abbreviated history of Anglo Platinum's Waterval Smelter. Paper presented at the South African pyrometallurgy 2006, Johannesburg, South Africa 5–8 March 2006
27. Jones R (2005) An overview of the South African PGM smelting. Paper presented at the 44th annual conference of metallurgists, Calgary, Canada, 21–24 August 2005
28. Hundermark RJ (2011) The smelting operations of Anglo American's platinum business: an update. Paper presented at the South African pyrometallurgy 2011, Johannesburg, South Africa 6–9 March 2011
29. Boldt JR, Queneau P (eds) (1967) The winning of nickel, its geology, mining, and extractive metallurgy. Longmans, Canada Ltd
30. Langer B (1979) Ferronickel production at Morro do Niquel, Minas Gerais, Brazil, Paper presented at the TMS international laterite nickel symposium, Chapter 22, pp 397–411, AIME-SME
31. Pecora WT (1944) Nickel-Silicate and associated nickel-cobalt-manganese-oxide deposits near Sao Jose do Tocantins, Goiaz, Brazil, US Geological Survey, Bulletin 935-E, Washington, 1944
32. The Editor (1981) Brazil's second nickel producer goes on stream at 5,000 mt/yr., Eng. Min. Journal, December, pp 123–124
33. Stober F et al (2015) Optimization of capital cost and smelter revenue-integrated team rebuild of Barro Alto furnaces, Paper presented at the conference of metallurgists, Toronto, August 2015. The metallurgical society of CIM, Montreal, Quebec, Canada
34. Moore P (2019) Anglo updates on Barro Alto bulk sorting project, trial about to commence
35. International Mining, 12 Nov 2019
36. Rodrigues RL et al (2010) The Onça Puma nickel project, Paper presented at the conference of metallurgists, Vancouver, August 2010. The metallurgical society of CIM, Montreal, Quebec, Canada
37. Southall S et al (2012) Rapid redesign and restart of the Vale Onca Puma nickel smelter. Paper presented at the Infacon XIV-the fourteenth international ferroalloys congress, Kiev, Ukraine
38. Kift RJ et al (2004) A historical overview of Cerro Matoso operation 1980 to 2003. Paper presented at the TMS international Laterite Nickel symposium, Charlotte, North Carolina 2004

39. Mining Data Solutions-Major mines and projects: Cerro Matoso. https://miningdataonline.com/property/336/Cerro-Matoso-Mine.aspx
40. Thornhill PG et al (1979) Development of Falconbridge ferronickel process. Paper presented at the TMS international laterite nickel symposium, 1979
41. Archibald FR Hatch GG (1973) Electric arc furnace operation, US Patent 3,715,200. 6 February 1973
42. Walker C et al (2009) Nickel laterite rotary kiln-electric furnace plants of the future. Paper presented at the 48th annual conference of metallurgists, international symposium-pyrometallurgy of nickel and cobalt, pp 33–50, Sudbury, Canada, August 2009
43. Kerfoot DG et al (2011) The Nickel Industry in Canada 1961 to 2011. The Canadian Metallurgical and Materials Landscape 1960 to 2011. MetSoc
44. Muinonen S et al (2017) The Sudbury integrated nickel operations converting practice—An update, paper presented at COM2017, The conference of metallurgists, Vancouver, BC, Canada, The metallurgical society of CIM, and Montreal. Quebec, Canada
45. Diaz C et al (2011) The Nickel Industry in Canada 1961 to 2011. The Canadian Metallurgical and Materials Landscape 1960 to 2011. MetSoc
46. Sherritt International Corporation (2014) NI 43-101 technical report on the Ambatovy nickel project in Madagascar, September 21. Fort Saskatchewan, Canada

A Review of Nickel Pyrometallurgy Over the Past 50 Years with Special Reference to the Former Inco Ltd and Falconbridge Ltd

A. Vahed, P. J. Mackey, and A. E. M. Warner

Abstract In the 1970s, Canada was the world's leading nickel producing country and Canadian firms had big expansion plans. The International Nickel Company of Canada later renamed Inco Ltd had major in-house research programs underway developing technologies for laterite deposits overseas and for the nickel and copper refineries in Sudbury, Ontario. Two new laterite smelters were commissioned in the 1970s in Indonesia and Guatemala simultaneously based on the Rotary Kiln Electric Furnace Process. Falconbridge undertook similar programs and commissioned the Falcondo laterite smelter in the Dominican Republic based on the company's proprietary technology in 1971. In the 1980s–1994 the emphasis on the Canadian nickel smelters was mainly driven by environmental pressures to reduce SO_2 emissions. From the 1990s up to the present time, significant environmental operational improvements have been achieved in Canada, and the development of technologies for the laterite deposits for Koniambo and Goro projects in New Caledonia were shaped primarily by energy efficiency and environmental considerations. This paper will review these important developments that enabled the technology advances that were achieved by these companies. Other developments in nickel sulphide and laterite pyrometallurgy will also be discussed.

Keywords Nickel laterites · Nickel sulphides · Nickel smelting · Nickel converting · Key developments

---

A. Vahed
AV Met Consulting, 2335 North Ridge Trail, Oakville, ON L6H 0A8, Canada

P. J. Mackey
P. J. Mackey Technology Inc, 295 Kirkland Blvd, Kirkland, QC H9J 1P7, Canada

A. E. M. Warner (✉)
AEMW Consulting, 304 Oakwood Court, Burlington, ON L7N 1W9, Canada
e-mail: aemwarner@cogeco.ca

C. Anderson et al. (eds.), *Ni-Co 2021: The 5th International Symposium on Nickel and Cobalt*, The Minerals, Metals & Materials Series,
https://doi.org/10.1007/978-3-030-65647-8_2

Overview

Nickel is an essential metal to modern society, primarily through its use as an important ingredient in stainless steel. Thus, stainless steel finds vital applications in food preparation facilities, as well as important uses in transportation and industrial applications. Nowadays, electric vehicles are becoming a key nickel demand driver for nickel products that meet this application.

In 1970, Canada was leading the world in Nickel mining and production with ~40% of the world mine production of ~675 kt nickel [1]. Of this the International Nickel Company produced ~85% of the Canadian Nickel and Falconbridge ~14%. At this time sweeping technical innovations were being developed which shaped the industry for many years that followed.

About 70% of the estimated world land-based nickel reserves are nickel laterites of which about 40% are suitable for smelting [2]. However, the prominent production of nickel in 1970 was from nickel sulphide ores with much development work being performed to bring laterite deposits into operation. By 1988 ~30% of world nickel was produced from laterites. This has steadily changed in favor of laterites as world nickel production has increased ~40% by 1994, and 51% by 2010. In 2020 ~70% of the world's nickel production is from laterite ores. Figure 1 shows the steady increase in Nickel production over the last 50 years [3]. The effects of SO_2 from nickel sulphide ores, has stressed eco-systems resulting in governments issuing successive control

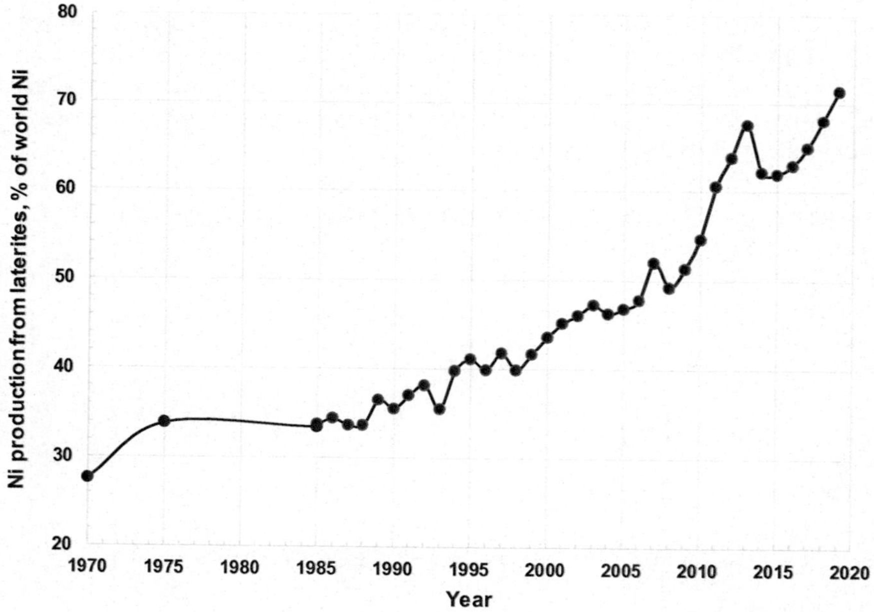

Fig. 1 Ni production from laterite ores 1970 to 2020. (by authors 1970–1984, [3] 1985 on). (Color figure online)

orders to reduce these emissions. Emphasis also shifted to include other atmospheric emissions such as dusts and toxic metals and some other gases. This has contributed also to the shift for increased nickel production coming from laterites.

Canada has always played an important role in nickel production and nickel metallurgy. In the 1970s Canada supplied ~40% of the world's nickel. However, this proportion has declined considerably to around 7% of the world nickel supply today. This is shown in Fig. 2. Still, as shown in Fig. 3, Canada remains an important nickel supplier and contributor to nickel extractive technology. In fact, Canadian nickel production has remained relatively flat over the last 50-year period with the rest of the world picking up the increased production.

Due to the importance of nickel to society, nickel demand is continuing to rise. Nickel companies have continually been under pressure to contain production costs amid cyclic languishing nickel prices. With ore grades dropping, while labour and energy costs, which are key cost contributors are rising, actions to lower production costs have resulted in some innovative approaches and technology developments by nickel companies to increase productivity and lower production costs.

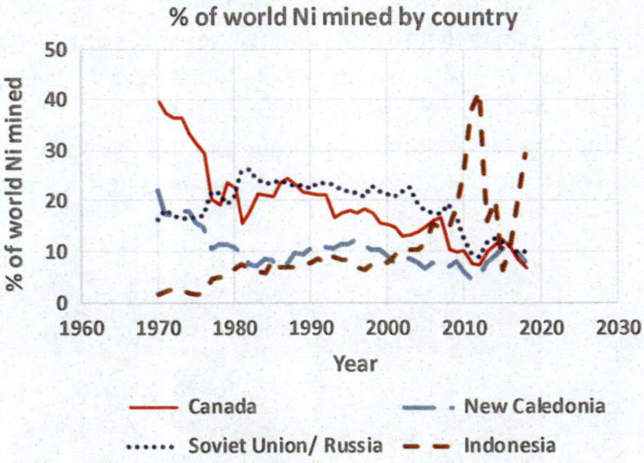

Fig. 2 % of world Ni mined by Canada, New Caledonia, Soviet Union/Russia and Indonesia since 1970. (Color figure online)

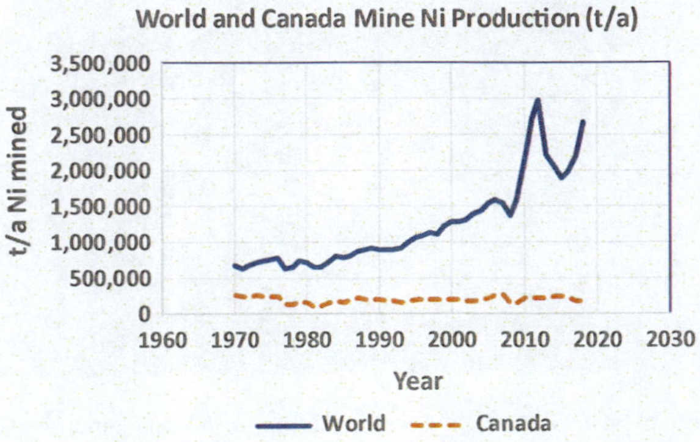

Fig. 3 World and Canada mine nickel production since 1970 [1]. (Color figure online)

The three authors have between them 150 years of experience working in smelters, refineries and research with nickel, copper and cobalt having joined their respective companies around 1970. Hence this paper reviews nickel pyrometallurgy and developments since 1970 to the present which the authors have witnessed and had made some contributions.

Canadian Nickel-Cobalt Operations 1970 to 2020 (Sulphides)

Inco Now Vale Operations

Port Colborne Research

Scenically located on the shores of Lake Erie, the three stations of Inco's Port Colborne research complex appear in this photograph in the order of their construction (Fig. 4).

No. 1 station, in the foreground was built in 1959, No. 2 (back left) in 1960, and No. 3 in 1967. To this unique scientific establishment came bulk samples of nickel ores from many parts of the world and were used in the development of new extraction processes on a pilot plant scale [4].

Shaping the destiny of the International Nickel Company of Canada's metallurgical operations both at home and in many countries abroad, the Inco research station

Fig. 4 Inco's Pilot Plant Research Complex in Port Colborne. (Color figure online)

complex at Port Colborne was the proving ground for sweeping innovations in the nickel industry [4].

A new nickel refinery installed in Copper Cliff (1972) was developed as a result of innovations in nickel smelting and carbonyl technology developed at the research stations. Hydrometallurgical and pyrometallurgical piloting of laterite nickel oxide ores from Guatemala, New Caledonia, Australia and Indonesia, leading to two new smelters completed by the mid-1975 in Indonesia and Guatemala.

In 1976, The International Nickel Company of Canada officially changed its name to Inco Ltd and was then bought by Vale in 2006. For many decades, the two main nickel producers for Inco in Canada were the Sudbury and Thompson operations. The Voisey's Bay Mine and mill started in 2005 and the Long Harbour refinery in 2014.

Sudbury Operations

In Sudbury in the 1970s, Inco continued to expand with the commissioning of the new Clarabelle Mill, capable of treating 35,000 tons of ore a day, in 1971 and the opening of the new mine and mill complex at Shebandowan, Ontario in 1973 [5]. At the smelter the 1250 ft superstack was built with some new gas cleaning equipment to improve the local environmental quality. In between the early 1970s and the mid-1980s, the Copper Cliff Smelter nickel circuit operating equipment diminished from 42 to 24 multi-hearth roasters and from 6 conventional air-coal reverbs to 2 oxy-fuel fired reverbs. Copper concentrate was processed through a fluid bed dryer and the Inco Flash furnace. At one time there were 24 Peirce Smith converters in the converter aisle with three to four being used for copper. A significant energy conservation related technology change at the Copper Cliff Smelter was the substitution of roof mounted oxy-oil burners for end wall air-oil burners in reverberatory furnaces in 1979 [6]. No. 5 reverb furnace went on line, equipped with 12 oxy-oil burners. The oxy-oil fired furnace had 45% higher smelting capacity, and 50% lower energy consumption per tonne of dry solid charge. In 1984 as a result of the lower cost of natural gas versus oil and government incentives, natural gas rather than oil was used for converting the No. 2 reverb furnace to full oxy-fuel roof firing in 1984 [7].

Increased nickel losses had demonstrated the limitations of additional pyrrhotite rejection in the mills as a means of further decreasing emissions. Inco undertook a major Research and Development (R&D) program to investigate alternative smelting routes. This program focused on two alternative smelting routes, roast-reduction smelting of nickel concentrate (RRS) and oxygen flash smelting of a bulk copper-nickel concentrate [8]. The RRS process implied retention of copper concentrate flash smelting, probably coupled with continuous conversion. Inco progressed the RRS concept from bench scale experiments, through piloting, to commercial testing at the Thompson Smelter in two years [9, 10]. Two commercial-scale campaigns were conducted in 1981–1982. Subsequently the technical viability of the bulk copper-nickel concentrate flash smelting route was established in two commercial scale campaigns that were conducted in the Copper Cliff copper flash furnace in 1985 and 1987.

With the bulk smelting process, the sole feed for production of blister copper is finely comminuted, nickel containing Cu_2S concentrate (MK in Inco's vernacular) that is produced in Matte Processing from Bessemer matte separation. New technology was required for processing this MK. In the mid-1980s, at the Copper Cliff Smelter, a Peirce-Smith converter was adapted for flash converting wet MK [11] and piloting of flash conversion of dry MK was conducted at Inco's Port Colborne Research Stations (PCRS) [12]. Semi-blister copper, i.e. sulphur-saturated copper, was produced; however, both the commercial and pilot plant campaigns showed that flash converting of MK would result in unacceptable high levels of dusting. Following the successful pilot scale demonstration of Inco's top oxygen blowing-nitrogen bottom stirring process for the continuous conversion of bulk Cu-Ni matte [13], the PCRS equipment was modified to investigate the applicability of this technology for converting MK to semi-blister. This was successful, and this technology was finally patented and adopted for processing MK in the modernized Copper Cliff Smelter [14]. The development and use of porous plugs in this and other vessels was a key enabling technology also patented by Inco for non-ferrous applications.

As a result, by November 1993, the Copper Cliff smelter flowsheet was changed with changes to milling to produce a Cu-Ni bulk concentrate, shutdown of the reverberatory furnaces and construction of two Inco flash furnaces for processing the bulk concentrate, construction of an oxygen top blowing nitrogen bottom stirring vessel for the processing of MK (MK Reactor); and erection of a new 2,900 tonnes/day sulphuric acid plant for treating most of the new flash furnace and MK Reactor off-gases. With the completion of the nickel sulphide fluid bed roaster gas treatment project in 2007, these gases have also been cleaned in the acid plant.

In order to reduce emissions and increase productivity further, the smelter flowsheet has been simplified such that since 2018, the smelter is now operating with a single Inco flash furnace, two Peirce Smith converters with converter gases going to the acid plant with gases from matte processing fluid bed roasters (FBRs) and flash furnace. The MK reactor has been repurposed as a converter slag cleaning vessel to increase cobalt recovery.

Nickel is refined in Sudbury at the nickel refinery and in Wales at the Clydach refinery with feed from the Sudbury operations. Voisey's Bay also produces nickel at the Long Harbour refinery from concentrates produced at Voisey's Bay. Cobalt is refined at the Port Colborne refinery from Sudbury refinery streams and at the Long Harbour refinery.

Thompson Operations

The integrated nickel operations with mine, mill, smelter and refinery at Thompson, Manitoba, was commissioned in 1961 [15]. The original Thompson smelter contained three fluid bed roasters feeding directly to three 18 MVA electric furnaces (EFs) and four Peirce Smith (PS) converters. By 1969, the smelter had two additional roasters connected to two additional 30MVA electric furnace with three more converters added for a total of seven Peirce Smith converters. Early 1970 was the only time that the smelter saw all five roaster-furnaces operating at the same time processing a nickel concentrate with ~7.5% Ni.

As milling practices changed to reject pyrrhotite and gangue, nickel concentrate grades increased, the number of operating lines in the Thompson smelter decreased such that the focus was using the two larger furnaces (#1 and #2). In 1981/1982 the #4 and #5 roasters were modified to slurry feeding with #5 EF modified and used for the plant scale Metal Recovery test (MRT) campaigns. #1 and #2 furnace lines remained the work horses of the smelter until it was shut down permanently in 2018.

Falconbridge then Xstrata Now Glencore Operations

Sudbury Integrated Nickel Operations

In 1970, the former Falconbridge smelter, now referred to as the Sudbury Smelter-Sudbury Integrated Nickel Operations of Glencore produced 38,000 t of nickel in finish matte shipped to the company's nickel refinery at Kristiansand, Norway for refining. While the smelter was considerably larger in 1970 than when the plant first started operations in February 1930 (a blast furnace for smelting and Peirce-Smith converters), the technology in 1970 was essentially similar. Over the period 1930–1932, annual nickel output averaged some 2,300 t of Ni [16, 17]. Falconbridge's new President, Thayer Lindsley had secured the Kristiansand refinery in Norway in 1928 to handle Falconbridge matte since all other known nickel refining technologies at the time were controlled by the former Inco. The Falconbridge matte was therefore cast, crushed and shipped for refining, this arrangement remains in place to this day. A brief review of the Kristiansand refinery is included below.

Recognizing the need for technical development and support, the company opened the Falconbridge Metallurgical Laboratory at Thornhill north of Toronto in 1952, merging in 1986 with the technical Laboratory located at the smelter site. Subsequently this building was replaced by a new adjacent facility in 1997 [20]; since 2006 this facility has been known as XPS (Fig. 5).

Ground-breaking work carried out at these laboratories led to the commercialization of a number of processes for the Company including: The Falconbridge Iron Ore Process operated at the smelter from 1956 to 1972 [21–23]; the Falconbridge Nickel Iron Refinery operated unevenly at the smelter from 1969 to 1973 [23, 24]; the Falcondo Nickel Laterite Process began in 1971 [25], Falconbridge's first venture in treating nickel laterite ores dating back to 1955, remains in operation by Americano Nickel; test work for the Falconbridge Matte Leach Process, plant commenced at Kristiansand in 1975 [26]; development of the Falconbridge Nickel Smelting Technology (Koniambo) plant started in 2013 and remains in operation [27]. These laboratories also contributed to the success and growth of the company in numerous ways.

Fig. 5 Aerial view of XPS building in Falconbridge, Ontario (formerly the Falconbridge Technology Centre). (Color figure online)

Of interest, the technical approach adopted at the Falconbridge smelter was generally somewhat different to that used at the six-times larger (in Ni output) Copper Cliff plant of the former Inco now Vale located just some 30 km south in Copper Cliff. Falconbridge, the smaller of the two companies was developing an innovative culture which would remain with the company. Examples of technology differences are the continued use of blast furnaces for smelting for nearly 50 years at Falconbridge, whereas Inco soon phased out blast furnaces at Copper Cliff, installing a large number of multi-hearth roasters and fuel-fired reverberatory furnaces in 1930 [28].

The early smelters in Canada employed tall stacks to disperse sulphurous gases. The Falconbridge smelter was first built with a 53 m stack when sulphur dioxide emissions are estimated by the present authors at about 28,000 t SO_2/a based on data in [16]. This had risen to over 300,000 t SO_2/a by the early 1970s [23, 29] when increasing concern about the so-called "acid rain" and related environmental issues led to technology changes with Falconbridge developing its own sulphur-capture technology.

Drawing in part on roasting and smelting technologies at benchmark smelter plants at Thompson (Inco, Manitoba-roasting/smelting) and Mt. Isa (Mount Isa Mines, Australia—roasting), and on its own experience with fluid bed roasting [21], Falconbridge re-vamped the smelter based on fluid bed roasting with SO_2 capture as acid and electric furnace smelting.

Following extensive test work during the early 1970s, including operating a 0.42 m dia. pilot roaster and a 1 MW pilot electric furnace, a new commercial plant commenced in 1978 [16]. This included two slurry-fed 5.6 m dia. fluid bed roasters operating at 620 °C, with calcine treated in two, 27 MW six-in-line electric furnaces sized 9 m by 30 m. Roaster gases representing 50% sulphur elimination were directed to a new 1,180 ton/d acid plant, lowering sulphur dioxide emission to 112,000 t SO_2/a, meeting Ontario Government Control Order guidelines [22,

23, 29–32]. The new plant ran well, and by the mid-1980s, nickel tonnage (which had dropped between the mid-1970s and 1982) was 30,000 t Ni/a. Within a few years the company embarked on a plan which involved higher and higher degrees of roasting to meet new Ontario Government guidelines [22, 23, 29, 30]. Subsequently, sulphur dioxide emissions dropped to 50,000 t SO_2/a by 1996. By 2018 this had been reduced to 27,000 tonnes with technology implemented for improved control of the atmosphere in the electric furnace to limit gaseous emissions [23, 30, 33, 34].

In 1986, the Falconbridge Converter Slag Cleaning vessel was commissioned [34]. This process was developed to treat converter slag instead of using the electric furnace. Even though the furnace was operated under reducing conditions by coal addition, metal losses required further attention. Falconbridge has always strived for low metal losses—the original choice of blast furnace for smelting was in part to keep metal losses low. It was reported in 1934, that Falconbridge had close to the lowest Ni level in slag amongst Sudbury smelters [19, 35]. Further improvement at the converters was achieved with the introduction of the Falconbridge Slag Make Converter in 1999 using Savard/Lee shrouded tuyeres, called the Air Liquide Shrouded Injector or ALSI [34, 36]. This vessel was able to oxidize the bulk of the iron in the increasingly sulphur-deficient matte (typically 50% Ni plus Cu, 32% Fe and 17% S), with the final step carried out in the finish converters [34, 37]. These developments allowed the plant to treat the available feed and meet target levels for sulphur dioxide emissions while operating with high metal recoveries.

The Falconbridge smelter changed to a single furnace operation in 1994 with an upgrade to the No. 2 furnace to operate up to 45 MW and closing the No. 1 furnace [38]. Raglan concentrate was introduced to the smelter in 1998 when the rich Raglan mine located in Northern Quebec, known since the 1930s, commenced operation. Studies by Falconbridge in the early 1970s had considered a new smelter to be located at Douglas Harbour located on the Ungava Peninsula at latitude ~61°N [39], however this was soon shelved; of interest, with the Thompson smelter now closed, the Sudbury smelters at latitude ~46°N remain Canada's most northern nickel smelters. Also in 1998, matte casting/solidification was changed to matte granulation [34, 40]. Further improvements to converter hooding and the feed system for the slag make converter was streamlined, thereby improving environmental conditions, a new larger slag cleaning vessel 16.8 m in length was installed in 2014 [33, 34].

The plant today operates with the roasters achieving 80% sulphur elimination, the single electric furnace and two so-called hybrid converting vessels combining converting and slag cleaning and two finish converters [33]. Falconbridge handles considerable quantities of scrap nickel and cobalt material and a 3.1 m dia. by 14.3 m kiln to dry and calcine scrap material commenced in 2007 [33, 41]. The Falconbridge smelter now has the capacity to produce over 70,000 t Ni/a and 2,500 t/a Co in a vastly different smelter compared to the plant that began over 90 years ago. The smelter, now referred to as the Sudbury Integrated Nickel Operations produced 71,000 t Ni in 2019.

Kristiansand Nickel Refinery

The Nikkelverk Kristiansand refinery has been in operation since 1910 when it was established with guidance from H.V. Hybinette who had developed a new process for nickel-copper separation by a roast-leach-electrolysis technique. Declining ore reserves there and the interest of Falconbridge to secure nickel refining technology led to the purchase of Kristiansand in 1928 [17, 18, 42]. The plant involved matte roasting, leaching the calcine in sulphuric acid to dissolve the copper and processing the residue for nickel recovery by electrowinning [42]. In 1975, some 65 years after Hybinette's process had commenced and following a decade of laboratory and pilot testing, the new Falconbridge chlorine matte leach process was introduced [26]. While expanded since 1975 along with several other improvements, this process remains in place today.

Other Sulphide Operations 1970 to 2020

In 1970, ~75% of nickel production came from sulphide ores with Canada producing ~60% [43] as shown in Fig. 6 with Inco and Falconbridge smelters producing ~95% of this production. The other producer in Canada being Sherritt Gordon Ltd which delivered 9.8 kt Ni in 1970. Outside of Canada, the USSR was the second major producer, mainly from Norilsk at 121 kt Ni in 1970 (~24% of the world Ni from sulphides). Other but smaller nickel production came from Australia (Western Mining), some intermediate sulphide operations in Cuba, the USA and South Africa.

Figure 7 shows how the situation has changed over the last 50 years. The nickel production from sulphides in 2019 [3] was 732 kt (156% increase since 1970) which now is ~31% of the world nickel produced. The three countries with the highest production in 2019 were the Russian Federation (29%; 214 kt Ni vs. 110 kt Ni in 1970), China (27%; ~200 kt vs. 0 kt Ni in 1970) and Canada (24%; 177 kt Ni vs. 277 kt Ni in 1970). In the Russian Federation Norilsk is now the largest single nickel

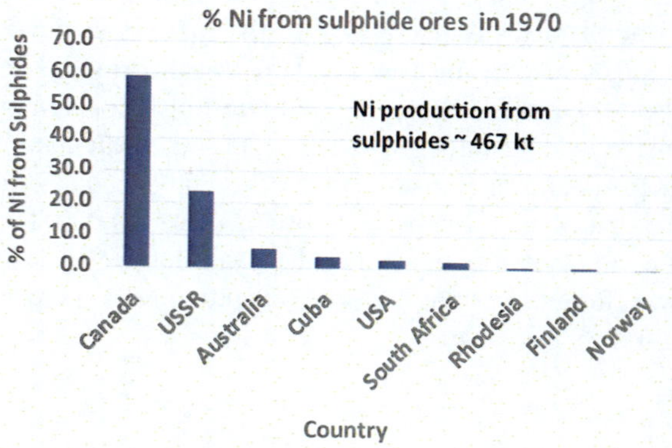

Fig. 6 % nickel production by country of total nickel production from sulphide ores in 1970 [3]. (Color figure online)

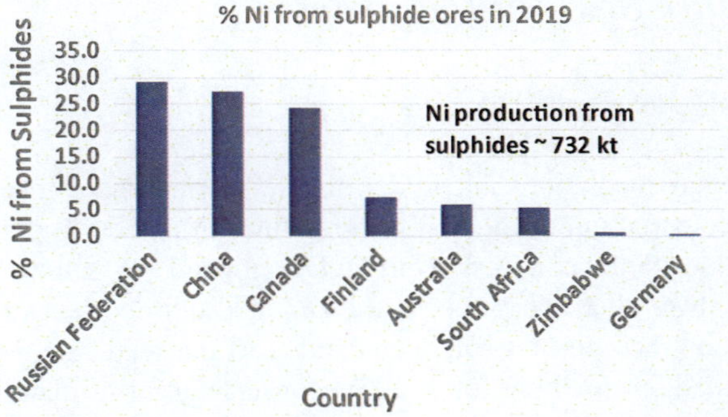

Fig. 7 % nickel production by country of total nickel production from sulphide ores in 2019 [3]. (Color figure online)

producer in the world. China is now the world's largest producer of nickel metal, primarily from laterites as discussed later but also from sulphide smelters such as Jinchuan. Other countries producing nickel from sulphide ores but to a lesser extent are shown in Fig. 7.

Laterite Operations 1970 to 2020 (Canadian Developments)

The first processing treatment for recovering nickel from laterites was developed in 1879 in New Caledonia, based on iron blast furnace technology with added gypsum to produce a matte. Production of nickel from laterites has grown slowly since that time.

The standard laterite pyrometallurgical flowsheet typically referred to as RKEF process (Rotary Kiln Electric Furnace) consists of the following stages, each with a specific objective [45]:

- Drying: elimination of most or a substantial portion of the free moisture content of the ore.
- Calcining-Reduction: elimination of the remainder of the free ore moisture and of its crystalline water, preheating of the ore, and reduction of a substantial portion of the nickel and a controlled portion of the iron.
- Electric furnace smelting: completion of reduction of the nickel and separation of the product ferronickel from the gangue that reports to a ferromagnesium silicate slag.
- Refining: elimination of undesirable minor elements from the ferronickel to meet market specifications.

Inco Now Vale Operations

PT Inco Now PTVI

Inco, the major shareholder of P.T. International Nickel Indonesia (PT Inco) started construction of the first of three lines of an RKEF plant and infrastructure in 1975 and started commissioning in December 1976 with furnace power from an oil-fired power plant. The process at PT Inco is an important variation of the standard RKEF flowsheet in that sulphur is added to the feed calcine a produce sulphur-deficient matte in smelting followed by converting of this material to a low-iron nickel matte product which is granulated and shipped primarily to Japan for final refining [46].

A second and third line were approved to be added along with a 110 MW hydro-electric plant nearby on the Larona river with commissioning in 1978. In 1980 PT Inco annual operating capacity was downgraded from the original 50,000 tons of Ni/a to 37,000 tons of Ni/a. This resulted from the corrosion of the refractory lining of the furnaces due to the acidic nature of the higher-grade Ni ores initially being processed. A lower grade, more alkaline ore was blended with this ore and copper cooling devices were added to the sidewalls of the electric furnaces.

Over the years, major improvements were achieved in the process plant [47, 48]. In the dryers, improved lifter design, the higher percentage of East Block ore processed, and direct firing of a portion of the fuel decreased costs by 40%. Other cost reductions were achieved in the reduction kilns by eliminating the side burners and changing the temperature profile to burn more of the combustible gases in the kiln. Large cost reductions were also achieved in the converting area by replacing two of the original rotary Kaldo converters with Peirce-Smith converters, which dramatically shortened the converter heat times and the high mechanical repair costs and refractory consumption of the Kaldo converters were eliminated.

PT Inco expanded the process plant in 1998 in a project called the "Fourth Line Expansion" adding a dryer, kiln, furnace and converter plus some infrastructure. It was reported that by 2000 capacity increased by 23 kt Ni/a. [49]. Further expansions since then have been under consideration. In 2019 from PTVI (new name after Vale bought Inco) the production from the smelter was 71 kt Ni.

To be able to use the additional power on the furnaces, it was necessary both to increase the transformer size and to increase the operating voltage on the furnaces. The original 45 MW transformers had a maximum voltage with the secondary connected in delta of 560 V. The furnaces were operated at the top voltage tap to control the matte temperature, and the electrodes were essentially just at the slag interface, or in the brush-arc mode. Falconbridge pioneered the use of high voltage, generally arc smelting, for ferronickel production in the Dominican Republic on six electrode furnaces. In 1987, the PT Inco three electrode furnace was connected in "wye", which permitted operating voltages up to 970 V. At this voltage the furnace is operating in the smothered arc mode. More power was available, and smelting rates were increased by 15%. As a result, new larger transformers capable of supplying much higher power, at much higher voltages were installed on the existing furnaces. Today the furnaces operate at average power of 55–60 MW [45].

In summary, PT Inco Indonesia, with its high-grade lateritic nickel deposits, its hydro-electric power, and stable-experienced Indonesian workforce, is now one of the lowest cost nickel producers in the world. In 2019 it produced 71 kt nickel.

Onca Puma

The Onça Puma nickel laterite smelter located some 400 km south-west of Marabá, and about 10 km from Tucumã in Pará state, commenced operations in March 2011. The Onça Puma smelter includes the following main plant units: a crushing plant; two dryers, 6 m dia. by 45 m; two rotary kilns, 6 m dia. by 135 m; and two, six-in-line electric furnaces nominally rated at 83 MW. There are also two refining stations rated at 9 MW to refine the 25% Ni ferronickel product before shipping. The original design capacity of the plant was 58,000 t Ni/a [50]. The company carried out a furnace re-build starting mid-2012. One furnace was upgraded to a nominal rating of 100 MW (with capability to 120 MW), and the plant re-commenced operations in 2014 with two dryers, two kilns and one electric furnace [51]. The plant has generally run well since that time, however production in 2019 was only 12 kt Ni.

Falconbridge then Xstrata Now Glencore Operations

Some 20 years after starting the Falconbridge smelter handing sulphide ores, the company began investigating nickel production from laterites in 1952 at the newly established Falconbridge Metallurgical Laboratory located at Thornhill north of Toronto. Less than 15% of the world's nickel was produced from laterites in the early 50s, however venturing into laterite treatment was an opportunity for expansion. In 1955, Falconbridge commenced studies on the nickel laterite deposits near Bonao in the Dominican Republic. Initial scoping laboratory scale tests on the Caron process and on nickel segregation were carried out on limonitic samples of the ore body which was still under exploration [25]. It was found that reduction carried out using a small shaft furnace operated with reducing gas from partial combustion of natural gas gave an interesting calcine product.

At the time, the blast furnace was well established for laterite smelting while kiln reduction was relatively new [18]. The results of the laboratory work plus the continuing geological studies which were proving the substantial reserves of saprolite led to the decision to recover nickel as ferronickel and develop the shaft furnace concept for reduction. To obtain process design data for a commercial plant, Falconbridge wisely decided to conduct pilot-scale reduction tests on fresh ore in briquetted form at the site, with calcine smelting tests carried out in a pilot electric furnace at the Company's Nikkelverk unit in Norway; the work was coordinated by the Toronto laboratory over the period 1958–1962 [17, 26]. Further stages of piloting were carried out over the period 1965–1968 using a Shell oil-fired Gasifier to generate reducing gas with a 2 MW rectangular electric furnace 2.5 m wide by 3.7 m long (inside) installed at site.

Falcondo

Subsequently with favorable results, the decision was made to proceed with a commercial ferronickel plant based on the main features as the pilot plant with an initial design capacity of 29,900 t of Ni/a treating ore averaging about 1.5% Ni. This plant now called Falcondo included: an ore preparation, drying and briquetting facility, 12 shaft furnaces of proprietary design having a hearth area of 7.5 m^2 (a 4 times scale-up) to treat up to 25 dt/h of ore operating with one Shell naphtha-fired gasifier per shaft furnace, calcine transport in 4 t refractory canisters by rail, and three, 7.3 m by 22.9 m (inside), 44 MW six-in-line electric furnaces. A conventional Fe-Ni refining facility completed the process. Power was provided by a new 200 MW Bunker C-fired power plant, along with a 70 km oil pipeline from the port at Santo Domingo to site. The first electric furnace commenced operations in June 1971, with all furnaces operational by early 1972. The well-planned and executed pilot work paid off, with production in 1973, about one year after start-up, reaching 25,400 t of Ni (~88% of design).

However, unknown to Falconbridge when the plant was designed, world events would soon lead to huge increases in the price of oil, seriously affecting nickel economics at Falcondo. When the plant started operations oil was about $22/barrel-roughly where it had been since the mid-1960s when pilot work began. The oil price started to rise due to Middle East conflicts in early 1974, reaching $56/barrel by February 1974, peaking at over $100/barrel by 1980 and by about 1986, and settling to about $40/barrel for a period. Falcondo's operations would thereafter be closely tied to the price of oil.

Falconbridge had worked closely with Canadian engineering companies since 1963. This long association led to pioneering developments in electric furnace technology, including innovative robust copper coolers for protecting the refractory walls from slag attack, and electric arc operation to provide stable and efficient smelting. As a result of this work, furnace life at Falcondo was prolonged and the plant has operated at one of the lowest unit power consumption rates as kWh/t calcine in the nickel industry.

In 2015, Glencore sold the plant to the American Nickel Ltd, a privately-owned company. The Falcondo facility is quite sensitive to oil cost, which at times has taxed the plant. The new owner has made several changes to the plant; however, oil remains the main fuel. The production in 2019 is reported at 24 kt/a.

Koniambo

In 1994, some 20 years after Falcondo had commenced operations, and following changes in corporate ownership, Falconbridge Ltd. carried out a strategic planning exercise covering all areas of the company's operations. This study brought focus to the growth potential seen for nickel and opportunities in laterite processing. As discussed by King et al. in 2005 [25], studies further identified the potential for improved saprolite technology over existing processes, involving fluid bed reduction of dry ore with coal instead of the oil-fired shaft furnace technology used at Falcondo, or a rotary kiln as employed on a number of ferronickel operations built and operated over the two previous decades, along with Direct Current (DC) smelting technology to handle the fine reduced ore. This approach, called the Falconbridge Nickel Smelting Technology was investigated at the laboratory and pilot plant scale [26]. At about the same time, the company examined potential laterite deposits of interest and by about 1998 formed a joint venture with Société Minière du Sud Pacifique (SMSP) in New Caledonia, with the objective of building a new ferronickel smelter treating ore from the Koniambo deposit. Known since the early part of last century, this deposit in the northern part of the island is a large laterite deposit of which the saprolite zone is of order of 300 million tonnes averaging 2.2% Ni.

Following a series of laboratory and small pilot-scale tests, demonstration testing of the new reduction step using powdered coal as the reductant was carried out on a 330 kg wet ore/h pilot plant in Germany, with calcine shipped to South Africa for pilot smelting in a pilot scale DC electric furnace over the period 1998–2001.

With favorable results and completion of a Bankable Feasibility Study, a 60,000 t Ni/a ferronickel plant was completed in 2006. Plant construction began soon after with operations commencing in 2013 [27]. The start-up of this large plant has been steady but gradual, with 2019 production of 24 kt Ni.

Other Laterite Operations

In 1970, ~25% of nickel production came from laterite ores with SLN in New Caledonia producing ~70% [43] as shown in Fig. 8. SLN was formed in 1880 following the discovery of large nickel deposits in New Caledonia. Since 1958, SLN made FeNi and matte through electric furnace smelting of kiln- calcined laterites. Other countries producing Ni from laterite ores in 1970 were Cuba, Indonesia, Greece and Brazil. Total nickel production from laterites that year was only 154kt.

During the last 50 years, nickel production from laterites has significantly increased as shown in Fig. 9 The nickel production from laterites in 2019 [3] was ~1629 kt (1060% increase since 1970) which now is ~69% of the world nickel produced. The two countries with the highest production from laterites in 2019 were China (31%; ~511 kt Ni vs. 0 kt Ni in 1970) and Indonesia (29%; ~470 kt vs. 18.5 kt Ni in 1970).

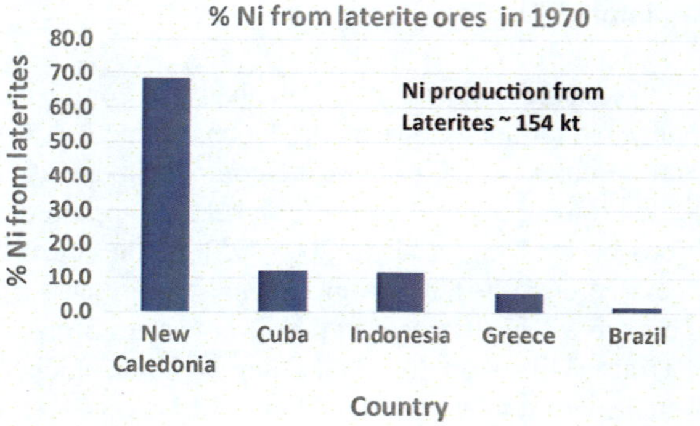

Fig. 8 % nickel production by country of total nickel production from laterite ores in 1970 [3]. (Color figure online)

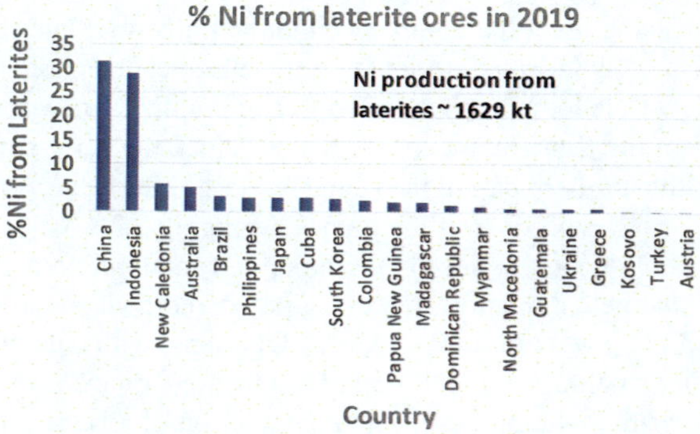

Fig. 9 % nickel production by country of total nickel production from laterite ores in 2019 [3]. (Color figure online)

There are now many other countries producing nickel from laterites but at much lower production levels (100 kt/a) as shown also in Fig. 9.

Other Nickel Production Developments 1970 to 2020

Nickel Pig Iron (NPI)

Beginning around the year 2000, the global consumption of nickel started to gradually increase, driven by strong demand for stainless steel in China and the rest of the world. According to the media reports at the time, the trend became so steep that by 2007 the global warehouse nickel stocks reduced to less than a fraction of a day's supply! Concomitantly, this strong demand for steel pushed the nickel prices higher than \$50,000/t by 2007, before it dropped back to \$15,000–25,000/t levels because of the financial crisis of 2008–2009 [52]. What effectively controlled and stabilized the nickel prices for the following years to come was the development of NPI technology in China starting in 2005 using limonite/saprolite ores exported mostly from Indonesia and the Philippines.

Many Chinese nickel plants are government owned and, except for Jinchuan and few other smelters, the rest operate a RKEF NPI smelter, with or without a captive steelmaking complex, have started only during the last two decades. The advent of the NPI technology using 200–300 small blast furnaces and its growth to a more sustainable RKEF technology has been presented in an earlier paper in 2015 [53]. As predicted in the paper, the introduction of the NPI technology in China became a game changer in the traditional global supply–demand relationship and it allowed the country to produce most of its stainless steel using NPI with imported laterite feed.

The traditional process for recovering nickel from limonite had been limited to hydrometallurgical processing until 2000, when the NPI process started using a "backyard" pyrometallurgical process based on blast furnace (BF) know-how for pig iron production. By 2011–2012, NPI had become widely used by the Chinese stainless steel industry to produce low-grade stainless steel (200 and 300 series). As the result, the NPI technology became a game changer in the traditional global supply–demand relationship and it allowed the country to produce most of its stainless steel to supply its highly active construction industry and consumer products [53]. NPI production initially started with fifteen small, old blast furnaces that were destined to close under the Chinese government's energy efficiency policy. The production amounted to 4–6 kt/a to 30 kt/a contained nickel, totaling 178 kt/a nickel with grades ranging from as low as 1–2% Ni and up to 10–15% Ni. By the end of 2007, there were a staggering 300 BF NPI producers. However, the production of NPI in a BF suffered from excessive slag volume, high slag viscosity due to Cr_2O_3, and severe refractory corrosion due to the addition of slag modifiers, such as fluorites, which made the process unsustainable. The blast furnaces were eventually discontinued and replaced, by the order of the Chinese government, with RKEF technology for all NPI production. Initial electric furnaces installations were small (5 MW to 20 MW), but today, modern RKEF units are operating throughout China using submerged electric arc furnace (EAF) producing from a total of 20 kt/a to 80kt/a contained nickel with typical recoveries exceeding 90% [54]. The NPI produced in the EAF during this period typically contained 7–8%Ni, and high levels of Si and C [55]. The NPI production is low cost and has a "payability" similar to (30% Ni) FeNi [62].

By 2014, just before Indonesia banned its ore exports, China had secured enough laterite feed from external sources and under special investment agreements with Indonesia and the Philippines to run its domestic RKEF plants and increase its nickel production rate up to about 500 kt/a [56]. Concomitantly, also started transferring its NPI RKEF technology to Indonesia with significant capital expenditure, which allowed China to eventually outpace the global steel production [56, 57]. The plants in China and Indonesia during this period widely used a higher grade of laterites (limonite/saprolite mix, nickel intermediate products) and scrap to produce 6–25% Ni NPI, suitable for producing higher grade steels (e.g. 300 series), which have increasingly been replacing the refined nickel as the main contributor. In 2018, NPI contributed 51% of the global nickel supply, and with it for the first time it pushed China's global market share of steel production beyond 50% [53, 56].

Green Battery Power Storage

Only Class I nickel is suitable for the electric vehicle (EV) and the green technology power storage (Lithium Ion and other Green Batteries). This precludes the NPI and any other Class II nickel as a suitable feed for battery material. During the last few years, however, the key global players, particularly China, have started taking a keen interest on the EV market. The global EV sales are expected to grow from about 2% in 2020 to 30% by 2030. The growth considers the environmental issues and decarbonization initiatives, accelerated by the geopolitical efforts to meet the Paris Accord. Such an increase in the EV alone leads to enormous demand for nickel and cobalt [59] with the most promising lithium ion (LI) battery designs of today. This analysis does not, however, consider other materials potentially replacing nickel and/or cobalt in the construction of future batteries.

In their April 2020 report, the Conic Metal Corporation, quoting from a Glencore strategic report, indicates that to meet the nickel demand for the expected EV production by 2030, the industry will have to have the capacity to produce about 1300 kt/a nickel and 340 kt/a cobalt, 140% and 270%, respectively, over and above the 2020 production rates of 920 kt/a nickel and 115 kt/a cobalt [60]. These estimates do not include the green technology and the battery charging infrastructures. Considering the scale and investment costs of the existing plants, Conic Metal estimates that the industry would need a capital investment in the order of \$50-\$100 billion to ensure the supply of the required nickel (Sulphides + Laterites, Pyromet or Hydromet) [59]. This will pose a serious challenge, not only to China, in the long term, but also for the global nickel industry. Given the growth market that is a given for the Class II products (FeNi, NPI) and a potential, albeit large but not immediate, opportunity for the battery grade Class I products, the decision to supply the Class I or Class II nickel for the long term would be a milestone to watch for the direction that the industry will take.

Here is the caveat. It is argued that the key determinant of the nickel industry's future will be the extent and speed that the industry will adopt to meet the demand of the EV and green energy technologies for this decade. On the other hand, there are factors that could have profound impacts on the supply/demand dynamics for Class II or Class I nickel for the longer-term investment in green battery materials. Some of these factors include, a) if the uncertainties in the market brought about by COVID-19 lasts for long, b) the fate of the low cost RKEF NPI producers in China with continued ban of Indonesian laterite, c) potential for the replacement of Ni with other materials or simply a lost opportunity for investment, if the pace of developing battery grade nickel/intermediates were too slow due to the fear of capital investment in Class I Nickel.

New Projects Under Construction or Consideration

There are several new projects that are already on the way or being considered. These include new NPI projects coming on-line, mostly in Indonesia, by Tsingshan group, one of the largest Chinese investors in steelmaking (up to ~210 kt by the end of 2019). Additional NPI production (117 kt/a) is also expected for commissioning in 2020 in Weda Bay, a joint venture between Tsingshan and Shenshi Eastern Special Steel Co, and further production in a similar installation as in Weda Bay co-owned by Eramet and Tsingshan [58].

Nickel sulphate, a key material for two dominant batteries Li-Ion-Ni-Co-Mn (NCM) and Ni-Co-Al (NCA) is largely produced and consumed in China. In 2018, the global production of nickel sulphate amounted to 700 kt. About 450–500 kt was produced in China. Other battery grade nickel installations are planned in Indonesia by PT Aneka Tambang in cooperation with Shandong Xinhai and Huaou Cobalt to produce for EV batteries, expecting to invest between $12- $60 billion [58].

The estimate for the Class I battery grade nickel production around the world by HPAL is about 220 kt/a, 70% of which is deployed in Indonesia. Also, PTVI has planned to invest $5 billion in the coming years, including $2.5 billion in battery grade nickel in partnership with Sumitomo Metal [58]. Others include BHP-Nickel West reporting significant progress in the construction of a 100 kt/a nickel sulphate plant in 2019 and expects to continue to grow the battery material business [61].

Conclusions

Following initial nickel mining and production in Sudbury, Ontario, nickel production expanded west to Manitoba, Saskatchewan and British Columbia, and east to northern Quebec (Ungava peninsula) and Newfoundland—all contributing to Canadian nickel output. It is considered that further discoveries of nickel in the Ungava region—where rich ores are now mined (~2–3% Ni and ~0.8% Cu)—are likely. The Ring of Fire area in Northern Ontario is still untapped, while with up to 100 million tonnes of 0.8% Ni pyrrhotite stored in Sudbury, recent authors [24] wondered if such tailings can be considered a future resource—likely they can be. Canada therefore has a good chance to continue to supply a significant proportion of the world's nickel moving forwards. As discussed in this paper, Canadian companies have also contributed significantly to the technology and production of nickel from laterites, which such deposits today supply over 70% of the world's nickel. One hopes that the level of innovation in nickel mining and metallurgy at companies and universities in Canada will rise from where it is today to better capture these opportunities.

Given trends to more nickel usage forecast in the greater use of electric storage batteries and the like, as well as continued need for stainless steel (the present largest use for nickel), it is considered the long-term prospects for nickel are generally good. As discussed in a second paper at this Symposium by the present authors [63], China now consumes about 50% of world nickel, hence it was felt that events in China will impact the nickel industry far beyond its borders. Paraphrasing the comments in that paper, it is therefore considered that three factors can likely have an important impact on China's supply/demand dynamics and hence the world and Canadian nickel:

i. The worldwide challenges brought about by COVID-19 introduced new risks to the nickel market;
ii. If the Indonesian ban on exporting laterite ore can significantly starve the Chinese RKEF plants, this will significantly curtail its own NPI production;
iii. The slow pace of developing battery grade intermediates due to the fear of capital investment needed for Class I nickel.

Still, in general, the Canadian and world nickel industry is robust, and the long-term future looks bright. It is thought that the next two years will be critical as to whether this prediction comes about.

Acknowledgements The authors wish to gratefully acknowledge Wood Mackenzie Ltd., London for kindly providing and allowing the use of much of the statistical data used in this paper.

References

1. Top nickel producing countries. 1970 to 2019. You Tube video
2. Mackey PJ et al (2012) Nickel production technology 2012—update on new and emerging technical developments. November 2012; America Nickel Conference, Toronto
3. Wood Mackenzie Limited (2019) Nickel production data 1984 to 2019. Wood Mackenzie Limited, Guildford, Surrey, United Kingdom
4. Inco Triangle. February 1970. (Available on the internet)
5. Kerfoot DG et al (2011) The nickel industry in Canada 1961 to 2011. The Canadian Metallurgical & Materials Landscape 1960 to 2011
6. Blanco J et al (1980) Oxy-fuel smelting in reverberatory furnaces at Inco's Copper Cliff smelter: 50th congress of the Chilean institute of mining engineers, Santiago, November 1980
7. Antonioni T et al (1985) Energy conservation at Inco's Copper Cliff smelter: AIME annual meeting, New York, USA, February 1985
8. Inco Staff (1994) Environmental programs at Sudbury. In: Champigny N, Dillon P (eds) Proceedings of the 96th AGM of CIM and the 1994 mineral outlook conference: CIM, Montreal, QC, Canada, pp 62–77
9. Díaz C et al (1994) Inco roast-reduction smelting of nickel concentrate: CIM Bulletin, June 1994, pp 62–71
10. Díaz C et al (1994) Deep roasting of nickel concentrate: CIM Bulletin, June 1994, pp 72–78
11. Blanco J et al (1986) Productivity improvements at Inco's Copper Cliff smelter: AIME annual conference 1986, New Orleans, USA
12. Davies H et al (1993) Flash converting of chalcocite concentrate at Inco's pilot plant. In: Reddy RG et al, Queneau PE (eds) International symposium, extractive metallurgy of copper, nickel and cobalt 1993, v. I, Fundamental aspects: the minerals, metals and materials society, Warrendale, PA, USA, pp 623–637
13. Warner A et al (2000) Continuous conversion of copper-nickel mattes. In: Kaiura G et al, Toguri JM (eds) Symposium on fundamentals of metallurgical processing: the metallurgical society of CIM, Montreal, QC, Canada, 2000, pp 513–526

14. Díaz C et al (1998) Low-dusting pyrometallurgical ore processing using top-blown, bottom-stirred converting vessel having porous plugs at the base, for sulphide ores, concentrates and mattes: US Patent No. 5,853,657, 29 December 1998
15. Diaz C et al (2011) Innovation in nonferrous metals pyrometallurgy—1961 to 2011, The Canadian Metallurgical & Materials Landscape. MetSoc
16. Gill JB et al Nickel-copper smelter at Falconbridge, Trans CIM, vol 35, pp 315–322
17. Crawford GA (1989) Falconbridge Ltd.—60 years in a century. CIM Bull 82(925):93–96
18. Boldt JR, Queneau P (1967) The winning of nickel, its geology. Mining and Extractive Metallurgy, Longmans Canada Limited, Toronto
19. Reed GB et al (1964) Converting practice at Falconbridge Nickel mines smelter, paper presented at the CIM annual meeting, Montreal, April, 1964
20. Lakshmanan VI et al (2011) Leading-edge technologies—mining, metals and materials research organisations in Canada-1960–2011. In: Kapusta J, Mackey P, Stubina N (eds) The Canadian metallurgical and materials landscape 1960–2011. The metallurgical society of CIM, 2011, pp 429–439
21. Thornhill PG et al (1965) The production of by-product iron ore at Falconbridge. In: Anderson JN, Queneau PE (eds) Pyrometallurgical processes in non-ferrous metallurgy the metallurgical society conferences, vol 39, The Metallurgical Society of AIME, New York, 1965, pp 351–359
22. Marcuson S et al (2007) The changing Canadian nickel smelting landscape late 19th century to early 21st century. Can Metall Quart 46(1):33–46
23. Rezaei S et al (2017) Canadian pyrrhotite treatment: the history, inventory and potential for tailings processing. Can Met Quart 56(5):410–417
24. Peek E et al (2011) Nickeliferous pyrrhotite—waste or resource. Min Eng 24:625–637
25. Thornhill PG et al (1979) Development of the Falconbridge ferronickel process, paper. In: Evans DJI, Shoemaker RS, Veltman H (eds) International laterite symposium, 1979. The society of mining engineers, AIME, pp 127–151
26. Stensholt EO et al (1985) The Falconbridge chlorine leach process, paper in, Extraction metallurgy '85, 9-12 September, 1985. Inst Mining Metall London UK 1985:377–397
27. Eccleston E et al (2017) Fluidized bed reduction of nickel laterite at Koniambo Nickel, paper presented at The conference of metallurgists, Vancouver, BC, Canada, 2017. The metallurgical society of CIM, Montreal, Quebec, Canada
28. The Staff (1937) Operations and plants of international nickel company of Canada limited. Can Min Jnl 58:581–683
29. Ontario Ministry of the Environment (1982) Sudbury environmental study synopsis 1973–1980, prepared for the Ontario ministry of the environment. Ontario, Toronto
30. Lehder Environmental Services Limited (2019) Glencore Sudbury integrated nickel operations-Sudbury smelter, toxics accounting report, 2019 reporting year, June 2019. Ontario, Canada
31. McKague AL et al (1984) Operation of Falconbridge's new smelting process. CIM Bull, 86–92
32. Norman GE et al (1984) Environmental improvements at Falconbridge Limited. CIM Bull, 94–98
33. Muinonen S et al (2017) The Sudbury Integrated Nickel Operations converting practice—an update, paper presented at The conference of metallurgists, Vancouver, BC, Canada, 2017. The metallurgical society of CIM, Montreal, Quebec, Canada
34. Salt B et al (2009) Evolution of the converter aisle at Xstrata's Sudbury smelter, paper. In: Kapusta J, Warner T (eds) International Peirce-Smith converting centennial, TMS, pp 135–149
35. Gronningstater A et al (1934) Principles and practice of metallurgy at Falconbridge. Trans CIM, 219–250
36. Kapusta J et al (2005) High oxygen shrouded injection at Falconbridge-five years of operations. In: Converter and fire refining operations, (A. Ross, T. Warner and K. Scholey Eds.), TMS, pp 47–60
37. Coursol P et al (2009) Optimization of the Xstrata Nickel Sudbury Smelter converter aisle using discrete event simulation, paper presented at The 48th conference of metallurgists, Sudbury, Ontario, Canada, 2009. The metallurgical society of CIM, Montreal, Quebec, Canada

38. Tisdale DG et al (1997) Adapting to one furnace at Falconbridge, paper presented at The 36th conference of Metallurgists, Sudbury, Ontario, Canada, 1997. The metallurgical society of CIM, Montreal, Quebec, Canada
39. Pugsley TF et al (1996) Raglan project optimization and economics. CIM Bull 89(1005):84–86
40. Schonewille R et al (2005) Update on Falconbridge's Sudbury nickel smelter. In: Donald J, Schonewille R (eds) Nickel and cobalt 2005-challenges in extraction and production, 479–498
41. The Editor (2008) Custom feed calciner-arch and structural project, Sudbury, Ontario, Sudbury Mining Solutions Journal, 1 June, 2008, Sudbury, Ontario, Canada
42. Archibald FR et al (1962) The Kristiansand nickel refinery. J Metals, 648–652
43. Reno HT (1970) Nickel, US bureau of mines, minerals year book-metals, minerals and fuels, 1970, pp 783–792, Washington, DC, USA, 1970. http://digicoll.library.wisc.edu/cgi-bin/EcoNatRes/EcoNatRes-idx?type=article&did=EcoNatRes.MinYB1970v1.HReno2&id=EcoNatRes.MinYB1970v1&isize=M
45. Warner A et al (2006) JOM World Nonferrous Smelter Survey, Part 111: Nickel: Laterite, JOM April 2006
46. Musu R et al (1979) PT Inco's Indonesian Nickel Project, Laterite Symposium Chapter 16
47. Guiry JD et al (1985) P.T. Inco's Indonesian Nickel Project: an update, International seminar on laterite
48. Bell MC et al (1990) P.T. Inco (Indonesia) Plant Expansion, CIM Bulletin January 1990
49. Dalvi AD et al (2004) The past and the future of Nickel Laterites, PDAC presentation
50. Rodrigues RL et al (2010) The Onça Puma nickel project, Paper presented at the conference of Metallurgists, Vancouver, August 2010. The metallurgical society of CIM, Montreal, Quebec, Canada
51. Southall S et al (2012) Rapid redesign and restart of the Vale Onca Puma nickel smelter. Paper presented at the Infacon XIV-the fourteenth international ferroalloys congress, Kiev, Ukraine
52. Bloomberg Finance LP (2017) Great Financial Crisis, www.bloomberg.com
53. Vahed A et al (2015) A review of technology developments in nickel pyrometallurgy and nickel production trends over the last three decades. Paper presented at the 54th annual conference of metallurgists, 23–26 August 2015, Toronto, Canada. ISBN 978-1-926872-32-2
54. Guo XJ (2009) Chinese nickel Industry-projects, production and technology. Paper presented at the annual conference of Metallurgists, August 2009, Sudbury, Canada, pp 3–21. ISBN i-894475-88-7
55. Rao M et al (2013) Carbothermic reduction of nickeliferous laterite ores for nickel pig iron production in China: A review. JOM 65(11):26
56. Richter W (2020) Wolf Street crude steel production: China blows the doors off rest of the world, wolfstreet.com, 6 June 2020
57. International Nickel Study Group Insight, April 2012-No 16
58. Fastmarket-MetalsBulletin (2019) 2019 Review, NPI to extend its dominant role in nickel supply, http://www.metalbulletin.com, November 2019
59. Conic Metals Corporation (2020) Investor presentation April 2020. www.conicmetals.com (8 AV)
60. Glencore Strategic Report (2017) Electric vehicles- a disruptive force underpinning our commodities, December 2017, www.glencore.com
61. BHP Billiton (2019) Kalgoorlie nickel smelter-Nickel West. www.bhpbilliton.com, F2019
62. FPX Nickel Corporation (2020) Investor's presentation, September 2020. www.fpxnickel.com
63. Mackey PJ et al (2021) "Around the nickel world in 80 days": A virtual tour of world nickel sulphide and laterite operations and technologies, TMS March 2021

Establishing a Domestic Cobalt Supply Chain: Unlocking Challenging Feedstocks

Frank Santaguida

Abstract Cobalt as a natural resource is common in many ore deposits throughout North America, yet most of the metal is imported for use in manufacturing batteries, steel, and alloys. In part, this is due to complexity of minerals typically hosting cobalt as well as the association in these minerals with toxic metals such as arsenic and antimony. New technologies funded through U.S. and Canadian government initiatives are being developed to improve the feasibility and minimize the environmental impact of processing these ores. As a result, new processing facilities will be capable of meeting the increasing domestic demand for cobalt as use becomes more prevalent.

Keywords Cobalt · Mineralogy · Feedstock · Processing · Supply-chain

Extended Abstract

At present, a domestic integrated cobalt supply chain does not exist despite all of the natural resources and processing capacity potential in North America. Cobalt is a key component within lithium ion batteries that power electric vehicles providing thermal stability to longer range energy cells. As a result, the demand for cobalt has grown significantly over the past 10 years and predicted to at least double in the next 10 years. The global supply chain is dominated by China, which produces 80% of the cobalt material used in battery manufacturing and controls a large proportion of the mines in the Democratic Republic of Congo where approximately 70% of the world's cobalt ore resources occur and is mined.

F. Santaguida (✉)
First Cobalt Corp. Vice President Exploration, Toronto, Canada
e-mail: fsantaguida@firstcobalt.com

C. Anderson et al. (eds.), *Ni-Co 2021: The 5th International Symposium on Nickel and Cobalt*, The Minerals, Metals & Materials Series,
https://doi.org/10.1007/978-3-030-65647-8_3

As ore material, cobalt largely occurs as a byproduct metal to nickel-copper deposits and to specific types of copper deposits. In Canada, several nickel-copper mines produce cobalt along with platinum and palladium, but most is refined in Europe and customized for use in steel and alloy metals rather than for battery manufacturing. Similar resources are known to exist in the U.S. in Minnesota and Michigan yet remain to be fully developed. The largest known resources of cobalt in the U.S. occur within central Idaho where several deposits have been discovered in which cobalt is the primary metal. The Blackbird Mine in Lemhi County, Idaho was in production intermittently between 1902 and 1968. Peak cobalt production was achieved in the 1950s where in 1954 the Blackbird operations contributed over 50% of the U.S. cobalt ore supply. The Idaho cobalt deposits also contain copper as well as gold in places. The deposits tend to be small in size but contain high grades of cobalt that occur in clustered areas. Some larger deposits are also known at lower cobalt grades but also have high copper potential. Many deposits remain to be developed for mining; therefore, the area is slated to become an important domestic supply in the future. Vast resources such as these in Minnesota and Michigan have not been developed, in part, due to complexity of the minerals containing cobalt (Fig. 1); most containing iron or arsenic which are costly to remove during mineral processing.

The refinery capacity for cobalt material for battery manufacture in North America is set to change with the First Cobalt Corp. operation to open in Ontario in the near future. This existing facility is small but effective in size and capable of yielding a significant amount of cobalt material. At peak capacity, the facility could supply up to 5% of the World demand for cobalt. The plant is an example of an end-product

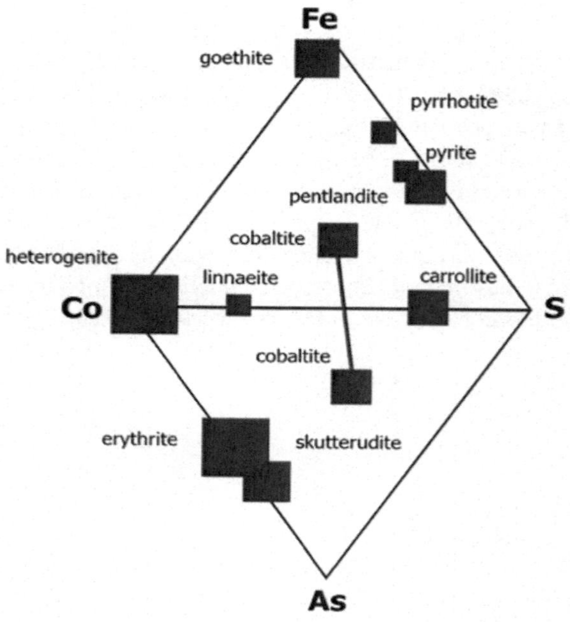

Fig. 1 Cobalt ore mineral compositions. Size of the symbol reflects relative importance to current global cobalt supply. The feasibility of processing Fe-rich or As-rich cobalt ores has been limited traditionally. (Color figure online)

processing operation that can be suitably developed elsewhere should these other domestic resources be brought into production.

Government initiatives play an important role in advancing the schedule towards a domestic cobalt supply chain. The U.S. and Canada have signed agreements to work together to establish a robust domestic market for cobalt and other Critical Minerals. The Department of Energy, through the Critical Materials Institute has supported a wide range of several research and development projects to advance new technology to improve the feasibility of producing Critical Materials, specifically cobalt, in the U.S. The U.S. Geological Survey within the Department of Interior has also supported directed programs to improve mineral exploration and development of key areas; including central Idaho to discover new cobalt resources.

A domestic cobalt supply chain is at hand, but still requires effort by industry and government to achieve this goal. The modern methods of mining and mineral processing have greatly reduced environmental impacts and energy consumption demonstrating a responsible approach to establishing this chain is also achievable.

Sustainable Developments in the Nickel Recovery Process

John Quinn, Dennis Burger, and Shijie Wang

Abstract The conventional practice for controlling impurities and excess copper in the electrolyte at the copper refinery is to bleed a certain volume of electrolyte from the tankhouse and treat this bleed solution in a copper-removal circuit, i.e., deposit out or liberator cells. The decopperized electrolyte is then fed to the nickel sulfate plant, where crystallized nickel sulfate and black acid are produced from the evaporators. Aiming at sustainable developments, a new process to recover the nickel, was innovated and commercialized to eliminate the production of black acid and to cut the elevated energy cost at the FMI–El Paso Operations. This new environmentally friendly process is narratively described, viz., advanced decopperization, efficient recovery of sulfuric acid, effective removal of arsenic, antimony, and bismuth impurities, and economical recovery of nickel carbonate. Significant sustainable developments in the brown field, especially focusing on saving energy consumption and reduction of carbon dioxide (CO_2), are also rationally discussed in this paper.

Keywords Copper bleed electrolyte · Electrowinning · Acid purification unit · APU · Solution purification · Cathode copper · Sulfuric acid · Nickel carbonate

Introduction

To make electrolytic copper of commercially acceptable purity, both the concentration of impurities and of copper in the refinery electrolyte must be controlled. Conventional practice for impurity and copper control is to bleed a certain volume of electrolyte from the tankhouse and treat this bleed solution in a copper-removal circuit, i.e. liberator cells. The decopperized electrolyte is then fed to the nickel sulfate

J. Quinn · D. Burger
Freeport McMoRan – El Paso Operations, 897 Hawkins Blvd, El Paso, TX 79915, USA

S. Wang (✉)
Rio Tinto Kennecott – Smelter & Refinery, 4700 Daybreak Pkwy, South Jordan, UT 84095, USA
e-mail: Shijie.Wang@riotinto.com

C. Anderson et al. (eds.), *Ni-Co 2021: The 5th International Symposium on Nickel and Cobalt*, The Minerals, Metals & Materials Series,
https://doi.org/10.1007/978-3-030-65647-8_4

plant to control the concentration of nickel in the electrolyte. Currently, crystallized nickel sulfate is produced from evaporators and black acid, an environmentally hazardous waste, is generated from the process.

To reduce the waste of black acid and energy consumption in the evaporator operation of the nickel sulfate process, sulfuric acid has to be removed from the decopperized electrolyte and nickel sulfate in the spent electrolyte needs to be converted to nickel carbonate; which initiates a new scheme for treatment of tankhouse bleed electrolyte. After a literature review, Eco-Tec Technology [1], acid purification unit (APU), was innovative developed in this experimental study to remove and recover sulfuric acid.

To produce a marketable nickel carbonate product, impurities in the bleed electrolyte, such as copper, arsenic, antimony, bismuth, iron, magnesium and zinc, have to be removed and the solution purified. Based on the differential solubility of base metals at different pH levels, chemical reagents, such as sodium carbonate (soda ash) and di-ammonium phosphate (DAP), are used in this experimental study to achieve the separation of the impurities and nickel carbonation in two filtration stages.

This paper covers the results of electrowinning pilot tests, APU ion exchange pilot tests, and the nickel carbonate plant commissioning for the tankhouse bleed electrolyte processing. An enterprise performance system, viz., understand customer requirements, innovate and employ the best processes, prioritize and leverage the actions you take to add value, and reduce waste and perform metal recovery friendly and efficiently, is also described for sustainable development in this paper.

Process Description and Pilot Test Results

The process developed for the treatment of tankhouse bleed electrolyte is described under the headings of the important steps with the necessary process conditions and some reasons for their selection being included for each stage. A simplified process flow diagram is shown in Fig. 1. The major operations are:

- Removal of copper,
- Recovery of sulfuric acid,
- Weak acid solution purification,
- Recovery of nickel.

Removal of Copper

Electrowinning (EW) from tankhouse bleed electrolyte is a well-established technique for recovery of the copper. The electrowinning reactions may be described by the following general half reactions:

$$\text{Cathode reaction: } Cu^{++} + 2e^{-} \rightarrow Cu \quad (1)$$

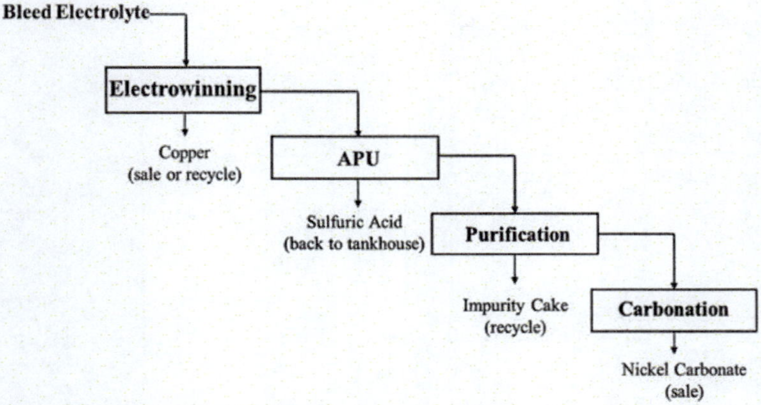

Fig. 1 A simplified process flow block diagram

$$\text{Anode reaction: } H_2O \rightarrow 1/2O_2 + 2H^+ + 2e^- \tag{2}$$

$$\text{Overall liberator cell reaction: } CuSO_4 + H_2O \rightarrow 1/2O_2 + 2H^+ + Cu \tag{3}$$

The harvested copper cathodes can be either fed to rod casting or blended in anode casting in the plant.

For its bleeding, electrolyte is withdrawn from the electrorefining tanks and the copper is removed in the test cell (Fig. 2), using lead inert anodes and copper starting sheet cathodes. As shown in the table, the feed and final copper and sulfuric acid are 41.5 gpl and 4 gpl, and 194 gpl and 267 gpl, respectively. Good solid deposit with rough surface for recycle, while the final solution is sent to the APU pilot test for

Fig. 2 Copper removal pilot plant and EW test cell. (Color figure online)

Table 1 Operating data for EW pilot test

Customer solution: type volume	Tankhouse bleed (Cir. 5) 50 gallons	Cell data number type anode cathode		1 Rectangular 5 4			Pump: setting – 5gpm	
Date	Time	Rectifier output		Solution data				
		Amps	Volts	Temp °F	Sample No EWP1–0	Cu C g	Acid gpl	Notes
3/4	8:48AM							Start power on
	9:00AM	150	2.20	144	EWP1-0	14.5	194	
	2:30PM	150	1.90	144				
3/5	7:00AM	150	1.81	150	EWP1-1	39.8	210	
	4:00PM	150	1.81	150	EWP1-2			
3/6	7:00AM	150	1.90	145	EWP1-3	20.2	267	
3/6	12:15PM	150	1.90	135	EWP1-4	13.0	244	
3/6	4:00PM	150	2.00	130	EWP1-5	8.30	250	
	5:00PM	150	2.00	134	EWP1-6	4.00	267	
Avg./total	48 h	150	2.00	132				
			1.91	151				Shutdown havest

Copper cathode bottom sample chemical analysis:
Element: As Sb Te Bi Se Pb Fe Sn Ni Zn Mn Cd Si
ppm: 3.1 0.36 1.1 6 0.75 6.0 4.01.2 4.4 < 0.2 < 0.2 < 0.2 5.5

sulfuric acid recovery. A typical test analysis of the electrolyte for copper removal in the tankhouse and the copper cathode bottom sample are shown in Table 1.

Recovery of Sulfuric Acid

The low copper but high acid content solution is processed through the APU for recovery of sulfuric acid, which employs weak base ion exchange resins that have the ability to sorb acids while largely excluding metallic sulfates.

Copper cathode bottom sample chemical analysis:

Element As Sb Te Bi Se Pb Fe Sn Ni Zn Mn Cd Si ppm 3.1 0.36 1.1 63 0.75 6.0 4.0 1.2 4.4 < 0.2 < 0.2 < 0.2 5.5

The APU, originally developed at the University of Toronto and called Recoflo™, was introduced by Eco-Tec in 1977 for purification of sulfuric acid anodizing solution, which is characterized by the use of short beds and fine mesh ion exchange resins and operated in a counter-current flow scheme.

The APU is employed in the new process of treatment of tankhouse bleed electrolyte. The sulfuric acid that is recovered from APU can be used either as make-up acid in the stripper electrolyte in the tankhouse or in the autoclave leaching process at the Special Metals Plant.

Weak Acid Solution Purification

The weak acid solution from the APU pilot test (Fig. 3) is then purified by soda ash (sodium carbonate) and di-ammonium phosphate (DAP) for removal of impurities. These impurities must be removed so that a marketable nickel carbonate can be produced.

In both qualitative and quantitative chemical analysis, it is often necessary to take advantage of differing solubility to remove only one or several salts from solution. Calculations using solubility products can tell us when a separation of this type is possible.

Sodium carbonate (or soda ash) was chosen for pH adjustment rather than sodium hydroxide (or caustic) because it is cheaper. It was also chosen because the carbonate precipitate is a more granular material which is easier to filter. The dominant chemical reactions are given as follows, (at 25 °C):

$$NiCO_3 \rightarrow Ni^{2+} + (CO_3)^{2-} K_{sp} = 1.42 \times 10^{-7} \quad (4)$$

$$CuCO_3 \rightarrow Cu^{2+} + (CO_3)^{2-} K_{sp} = 2.49 \times 10^{-10} \quad (5)$$

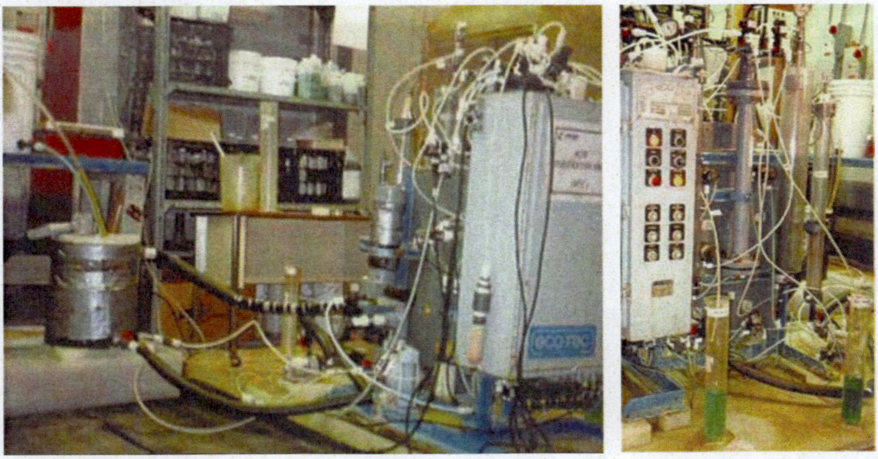

Fig. 3 Recovery of sulfuric acid pilot plant and APU test set up. (Color figure online)

Table 2 Effect of di-ammonium phosphate on impurities removal

	Di-Ammonium Phosphate			
Element	0	4.1	12.3	20.5
Ag (ppm)	<0.15	<0.15	<0.15	12
As (ppm)	0.6	17.8	34.2	53.6
Bi (ppm)	<0.2	<0.8	<0.8	<0.8
Ca (ppm)	114	109	86.1	84.8
Cr (ppm)	<0.15	<0.6	<0.6	<0.6
Cu (ppm)	408	472	148	179
Fe (ppm)	1	6.46	<0.2	0.62
Mg (ppm)	268	184	170	161
Ni (g/L)	15.8	13.7	11	9.5
Sb (ppm)	–	16	13.2	11.1
Se (ppm)	<0.25	<1.0	<1.0	<1.0
Te (ppm)	<0.4	<0.4	<0.4	<0.4
Zn (ppm)	221	200	75	66

[a]lbs. DAP/lbs. dissolved impurity in the solution

$$ZnCO_3 \rightarrow Zn^{2+} + (CO_3)^{2-} K_{sp} = 1.19 \times 10^{-10} \quad (6)$$

$$FeCO_3 \rightarrow Fe^{2+} + (CO_3)^{2-} K_{sp} = 3.07 \times 10^{-11} \quad (7)$$

The process makes use of the differential solubility of the above metals at different pH levels. Copper, iron and zinc can be partly removed from solution by an appropriate pH adjustment using sodium carbonate. The remaining impurities can then be selectively separated from the solution by adding DAP to precipitate the respective phosphates (Table 2).

Impurity salts are removed by filtration resulting in a filter cake that can be recycled to the smelter.

Recovery of Nickel

In the final step, the pH is raised to the optimum level of 9 by use of sodium carbonate (Fig. 4), causing precipitation of the remaining nickel, which is separated and recovered from the solution by filtration. The final solution (the filtrate) is sent to Tank 461 for water treatment in the Metal Recovery Plant.

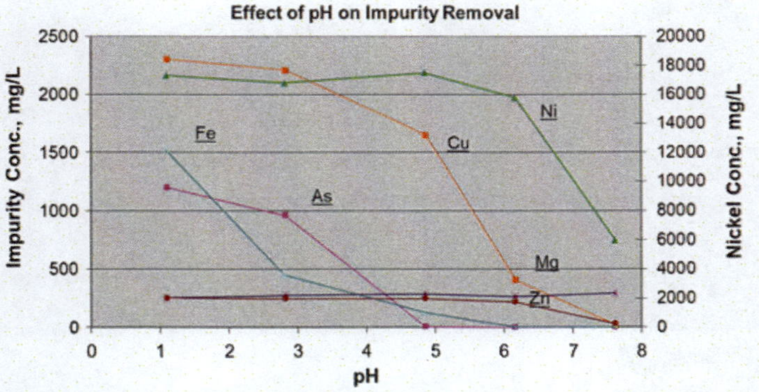

Fig. 4 Effect of pH for impurity removal. (Color figure online)

Innovative and Development Processes for Sustainable Development

New Environmentally Friendly Process

The old nickel sulfate plant removed the copper from tankhouse bleed electrolyte by electrowinning in the liberator cells. The decopperized solution was then evaporated and the nickel sulfate salts fell out of solution. Black acid, an environmentally hazardous waste, was simultaneously generated from the nickel sulfate process.

In order to eliminate the black acid and to stop the energy intensive evaporation, an innovative process was developed, viz., recovery of the sulfuric acid in an acid purification unit (APU), impurity removal by the selective precipitation, and recovery of the nickel through nickel carbonation reactors. The nickel sulfate plant was successfully transformed to the nickel carbonate plant, as shown in Fig. 5.

The process involves four major operations: removal of copper, recovery of sulfuric acid, removal of impurities, and recovery of nickel (Fig. 6). In the first stage, electrolyte is withdrawn from the electrorefining tanks in the tankhouse and the copper is removed in electrolytic cells using insoluble anodes and copper start sheets. The copper cathodes can be sold or recycled, and the copper-free solution is processed through the second stage that employs weak base ion-exchange resins that have the ability to absorb acids while largely excluding metallic sulfates. The sulfuric acid that is recovered is used as make-up acid in the stripper electrolyte in the tankhouse. The sulfuric-acid-free solution is then treated in the purification circuit (the third stage), where a residue containing arsenic, antimony, bismuth, etc., is removed. In the final stage, the purified nickel sulfate solution is processed and the nickel carbonate is separated and recovered from the solution by filtration. The final solution (the filtrate) is sent to the water recovery plant.

Fig. 5 Nickel carbonate plant—filtration-APU-purification-carbonation. (Color figure online)

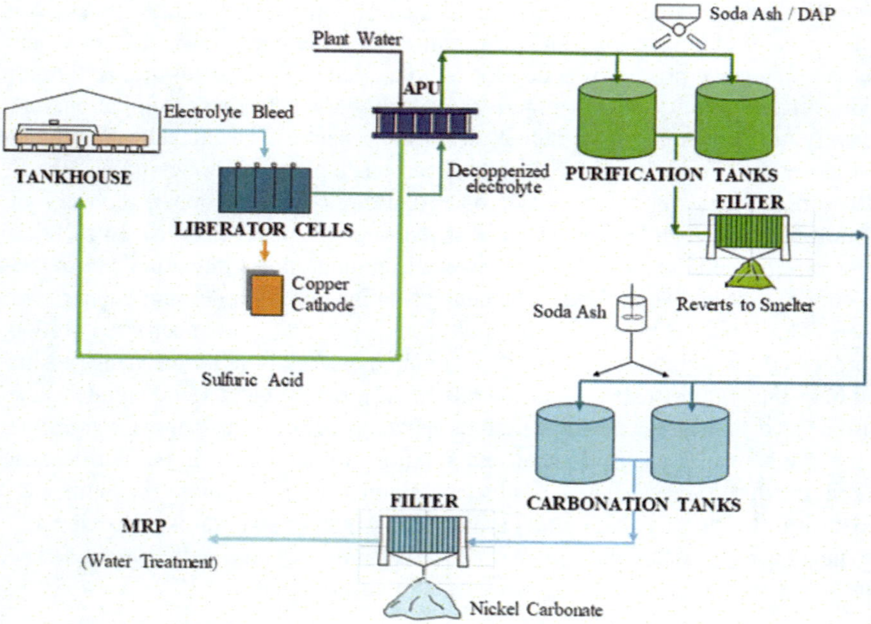

Fig. 6 Treatment of tankhouse bleed electrolyte. (Color figure online)

Nickel Carbonate Plant Engineering and Commission

A research report [2] titled "A new process development for treatment of tankhouse bleed electrolyte" was submitted in August 2002. Based on the successful nature of the results obtained, a preliminary economic evaluation was made on the promise of treating 5000 gal/day of tankhouse bleed electrolyte.

Primarily due to the elimination of black acid, evaporator operation and the recycled sulfuric acid, the financial benefits were found to be attractive. Results of the economic evaluation were summarized, process and financial bases and assumptions for the proposed commercial operation were then presented to the senior management for approval.

Based on the substantial environmental and economic benefits which would be provided by the process, the APU project was approved and the engineering report was later finished and submitted in November 2003. The Nickel Carbonate Plant was designed and the construction and equipment installation were finished by June 2004. The APU and Nickel Carbonation and the NCP commissioning were then carried out in sequence and finished in early August 2004.

With the new process, black acid is eliminated. Since nickel is recovered as a nickel carbonate filter cake at room temperature, evaporator operation for the crystallization of nickel sulfate with its accompanying high energy consumption is eliminated. In addition, 70% of the arsenic is returned with the recovered sulfuric acid back to the tankhouse, ensuring that enough arsenic is present in the electrolyte. Production at the Nickel Carbonation Plant (NCP) commenced in late August 2004.

Nickel Carbonate Plant Optimization

The NCP optimization team was formed in April 2004 with the objective of utilizing the six points of continuous improvement during construction, testing and start-up of the new plant [3, 4]. The team used Quest for Zero tools, identified Key Performance Indicators (KPIs), and established measurements tied to Standard Operating Procedures (SOPs) and Job Safety Analyses (JSAs). There were many incredible things that happened during this period. This team faced challenges, overcame difficulties, and achieved goals. All operators were trained and the MOC was approved during the team activities. There was such tremendous success that the Six Points of Continuous Improvement were achieved, as follows:

Safety: No safety incident; PPE assessments were conducted; the LOTOTO and confined space specific procedures were also reviewed; eyewashes, safety showers, and fire extinguishers were located, checked, and/or installed; 16 SOPs and 16 JSAs were written, reviewed, and logged in the internal website system.

Environmental: No significant spills or releases; eliminated the risk of spilling hazardous waste (black acid); significantly reduced steam use resulting in fewer emissions from Co-Gen; nickel product will have no free acid contained in it making shipping easier; industrial hygiene regulations were reviewed and an MSDS for Nickel Carbonate is being written; an environmental checklist was written and checked.

Quality: KPIs and operational parameters were studied and generated; operational log sheets were generated; APU performance test 100% completed; greater than 90% acid separation by the APU was achieved; dynamic air performance test 100% completed; three operator trainings were conducted; the nickel carbonate filter cake, containing 45% Ni (dry basis) was produced.

Production: Nickel carbonate production—batch sequence was created and used in the operation; we are approaching the designed production rate of treating 10,000 gallons of electrolyte a day; to date, 5 bags of impurity sludge and 22 bags of nickel carbonate, each weighing about 1700 lbs (at about 70% moisture) have been produced.

Cost: Cost reduction and revenue enhancement in the following areas: (1) black acid disposal savings, (2) black acid car rental and shipping savings, (3) sulfuric acid savings, (4) steam savings, and finally (5) difference in nickel metal return of $1.14 MM/year.

Stakeholder Relations: Employees—safer process; community—cleaner technology; stakeholders—good investment; overall—it is the right thing to do for all our stakeholders!

Conclusion

1. El Paso electrolyte components are typical of traditional tankhouse electrolyte. The El Paso electrolyte shows considerable variability in nickel content that needs to be controlled by daily bleeding a certain volume of electrolyte from the tankhouse.
2. As compared to the current nickel sulfate process in the treatment of tankhouse bleed electrolyte, the new developed process recovers copper, sulfuric acid and purified nickel (as nickel carbonate) that results in eliminating black acid generation and evaporator operation.
3. This innovative environmentally friendly process is first globally developed in the copper refinery. It is a significant sustainable development in the brown field that focuses on saving energy consumption, reduction of carbon dioxide (CO_2), and elimination of hazardous material (black acid) handling.

References

1. Sheedy M (1998) Recoflo® ion exchange technology. In: Mishra B (ed) EPD congress 1998, pp 137–149
2. Wang S (2002) A new process development for treatment of tankhouse bleed electrolyte. Phelps Dodge Refining Corp., Technical Services Department Report, EP-1, August 6, 2002
3. Wang S (2004) Final APU performance test. Internal report, July 25, 2004
4. Wang S (2004) El Paso refinery nickel carbonate plant optimization team report. Internal report, September 9, 2004

Part II
Batteries

Recovery of Ni as Tutton's Salts from Simulated Battery Leach Solutions

Antti Porvali, Vivek Agarwal, Helena Angerla, and Mari Lundström

Abstract Recycling of battery materials is a topic with room for improvements in the recycling of the materials present in current Li-ion and NiMH batteries. In this research, a recovery route for Ni as Tutton's salts ($M^{+}2\ M^{3+}(SO_4)_2 \cdot xH_2O$) was investigated, of which Mohr's salt ($(NH_4)_2Fe(SO_4)_2{*}6H_2O$), a commonly utilized titration reagent, is an example. Similarly, both Ni and Co can form barely soluble compounds with NH_4^+ ions in an acidic sulfate media. The behavior of battery metal Ni was investigated as a function of NH_4^+ ion concentration, pH and temperature in single parameter variation experiments. Ni, along with Co, was obtained with ease at a low pH over other metals that could be considered impurities (e.g. Fe). However, the usage of NH_4^+ ions carries its own problems, e.g. in the subsequent downstream processes.

Keywords Precipitation · Tutton's salts · Battery recycling

Introduction

In consumer products and electrical vehicles (EV), the majority of the batteries are currently based on Li-ion battery (LIB) technology. Some consumer products still use other secondary battery technologies, such as nickel metal hydride (NiMH) batteries. However, they have been surpassed by LIBs in nearly every way. Regardless, as long as these batteries are in circulation, it is important to consider how their raw materials could be best re-utilized. Both battery types contain large quantities of Ni, Co, Mn and Li, along with similar impurities or valuable metals in relatively minor quantities, such as Fe, Cu, Al [1, 2]. Presently, some 5% and 50% of global Ni and Co production, respectively, is going into batteries [3, 4]. Depending on the future cell chemistries

A. Porvali · V. Agarwal · H. Angerla · M. Lundström (✉)
Laboratory of Hydrometallurgy and Corrosion, Department Chemical and Metallurgical Engineering, Aalto University, Espoo, Finland
e-mail: mari.lundstrom@aalto.fi

V. Agarwal
Metso-Outotec, Helsinki, Finland

C. Anderson et al. (eds.), *Ni-Co 2021: The 5th International Symposium on Nickel and Cobalt*, The Minerals, Metals & Materials Series,
https://doi.org/10.1007/978-3-030-65647-8_5

that are utilized in the present LIBs, there may be a shift toward more Ni-heavy batteries than Co-heavy batteries. Generally, in batteries, there exists a cathode and anode, separated by membrane soaked in electrolyte that facilitates the transfer of ions. The separator membrane, usually made of polyethylene, polypropylene or a combination of both [5], prevents the anode and cathode and their current collectors from short-circuiting. In LIBs, the cathode current collector is made of metallic Al and anode Cu [6]. In addition to Ni, Mn, Li and Co, other major compounds present in LIBs are Fe (steel casings), volatile electrolytes (e.g. organic carbonates) and graphite (anode) [7].

Many methods have been investigated in recycling of these batteries, from pyrometallurgical pre-treatment to mechanical treatments of various sorts [6, 8–10]. Usually, these process steps must be followed up by hydrometallurgical treatment methods upon which the process specific details and aims of these pre-treatment steps have a huge influence. How well are the metallic compounds separated from the active materials? Has graphite been removed prior to the hydrometallurgical treatment? Have the water-soluble compounds been removed? Leaching with H_2SO_4 is one of the most investigated processing routes, after which the recovery of metals by various precipitation and solvent extraction techniques is a necessity. Basic ammonia leaching has also been investigated for selective extraction of Ni, Li and Co. However, acidic recovery of ammonium-based double salts has not yet been covered in the battery recycling literature. It is known that ammonium ions can be utilized in the recovery of Ni from acidic sulfate solutions [11], and that it can be done relatively selectively over certain elements such as Fe and Mn. Under acidic conditions in sulfate media, ammonium salts undergo reaction that results in a formation of Tutton's salts— a group of double salts—of which one of the more famous is Mohr's salt, i.e. ferrous ammonium sulfate. Reaction Eq. 1 shows the stoichiometry of the reaction between bivalent metal salts and ammonium sulfate in acidic aqueous solution:

$$MSO_4 + (NH_4)_2SO_4 + 6H_2O \rightarrow M(NH_4)_2(SO_4)_2 \cdot 6H_2O \tag{1}$$

Ni-based double salt has a solubility of 6.86 g in 100 g of water at $T = 20$ °C [12]. Co-based double salt, in comparison to Ni, has far greater solubility [13, 14]. Based on this information, a limited, but significant separation of Ni and Co from other elements present in LIBs could potentially be achieved early in the hydrometallurgical purification process of pregnant leach solution. In this paper, the effect of T, quantity of added $(NH_4)_2SO_4$ and pH was investigated on the precipitation of these double salts.

Materials and Methods

This section presents materials, methods, and analyzing methods used in experiments.

Chemicals and Equipment

Synthetic solutions simulating PLS obtained from leaching of spent Li-ion batteries [15, 16] were prepared in volumetric flask by dissolving different metal sulfate salts in 1 M H_2SO_4 (95–97%, EMSURE, Germany) $NiSO_4 \cdot 6\ H_2O$ (>98%, SIGMA-ALDRICH), $CoSO_4 \cdot 7\ H_2O$ (98%, Alfa Aesar,), $MnSO_4 \cdot H_2O$ (100.4%, VWR Chemicals) and $Li_2SO_4 \cdot H_2O$ (>99%, Sigma-Aldrich) in required quantities to obtain nickel rich composition presented in Table 1. The ideal compositions were to mimic the results of Peng et al. [15, 16]. As a precipitate $(NH_4)_2SO_4$ (99%, Riedel-de Haën) was used. The amounts utilized here were chosen based on literature survey. It is expected that if S/L = 1/10 were to be utilized with impure black mass obtained from an industrial recycler, the solution compositions would be akin to what is shown in Table 1.

For variation in *pH*, five additional different synthetic solutions were prepared, for both Co and Ni rich compositions, by first dissolving required amount of metal salts (Table 1) in 60 mL of deionized water in volumetric flask (pH 4.5), and then adjusting the pH using 2 M H_2SO_4, and for higher pH 2 M NaOH (100%, Tamro) with pH meter (Mettler Toledo, SevenCompact S210) attached to a pH electrode (Mettler-Toledo, Expert Pro ISM). The reason separate solutions were prepared was to better control the metal concentrations and sulfates. Alternatively, the pH of stock solutions (Table 1) could have been also adjusted. Final volume of 100 mL was adjusted by adding deionized water.

Table 1 Concentration of synthetic solution used in the study as stock solution (not including the pH study), simulating the PLS of Peng et al. [15]

	Ni (g/L)	Mn (g/L)	Li (g/L)	Co (g/L)
Ideal (Ni-rich)	24	2	4	3
Ideal (Co-rich)	3	2	4	24
Ni-rich	22.873	2.067	3.883	3.047
Co-rich	2.558	1.884	3.816	22.882

Precipitation Experiments

Precipitation experiments were performed in glass beakers (100 mL), under magnetic stirring (IKA, RT10) with stirring rate adjusted to 300 rpm. Volume of synthetic solution in each experiment was 50 mL before the addition of solid $(NH_4)_2SO_4$. Precipitation experiments were performed for 120 min each. The volume ranges after each experiment were estimated from different parameter extremes: the experiments with highest added $(NH_4)_2SO_4$ and highest temperature.

Three experimental parameters and their different levels were investigated: stoichiometry, pH and temperature presented in Table 2. The experiments were performed as single parameter variation experiments, and the constant values were chosen based on literature survey. For instance, in stoichiometry, it is necessary to achieve a supersaturation that enables the precipitation of the desired elements. The stoichiometry was calculated based on Eq. 2:

$$m_{(NH4)_2SO_4} = \mathrm{x_n} \cdot V \cdot \mathrm{M}_{(NH4)_2SO_4} \cdot \left(\frac{c_{Li}}{M_{Li}} + \frac{c_{Ni}}{M_{Ni}} + \frac{c_{Mn}}{M_{Mn}} + \frac{c_{Co}}{M_{Co}}\right) \tag{2}$$

where x_n stands for the molar stoichiometry value, V the initial volume of the solution (L), that was in this situation 50 mL, M_x the molar mass (g/mol) and c_x the wanted concentration of the agent (g/L). It should be noted that in future experiments, as a monovalent cation, Li^+ should not be included in these calculations, resulting in exacerbated over stoichiometric addition. The concentrations used as a basis for the amount of the added $(NH_4)_2SO_4$ in each experiment can be found in Table 1. In this manner, it should be certain that the level of supersaturation is exceeded in a manner necessary for precipitation to occur. Furthermore, it allows the calculation of ammonium and sulfate concentrations in the solution within the limits of error incurred from impurities in the reagents.

In precipitation experiments, the temperature and stirring speed was first adjusted to the desired values, then the precipitating chemical ($(NH_4)_2SO_4$) was added, in a stoichiometrically required quantity (Table 2). For the duration of the experiment, the glass beakers were covered with parafilm and were allowed to react for 120 min under stirring. After the end of the experiment, beakers were taken aside and the precipitates were let to settle for 5 min, after which *ca.* a 5 mL solution sample was collected for AAS analysis. The sample was first compressed through a syringe filter (polyethylene sulfone membranes, −0.45 μm) and diluted 10 times with blank

Table 2 Experimental parameters and their respective levels. Star marks the constant level when any one parameter is adjusted

Parameters	Levels					
x_n	0.5	1	1.5	2*	–	–
pH	0*	1	2	3	4	5
T (°C)	30*	40	50	60	–	–

solution (2% HNO_3). This was done in order to avoid further crystallization after sampling. Rest of the solutions containing the precipitates were filtered with a glass filtration assembly. Vacuum filtration was utilized with Whatman 50 grade filter papers. The filtered sediment was collected and dried for 24 h at 60 °C and then stored in bottles for the later microanalysis. Sediment was not washed in order to avoid dissolving it in water and to avoid the production of sulfates with alcohols.

Solution Analysis

After precipitation, the previously collected solution samples were further diluted in preparation of AAS analysis, and analyzed with AAS (Thermo Scientific, iCE 3000 Series). The solution analysis was utilized in determining the precipitation and co-precipitation of elements. The recovery efficiency was calculated based on Eq. 3:

$$\text{Recovery}\% = 100 - \frac{c_{Mf} \cdot V_f}{c_{Mi} \cdot V_i} \cdot 100 \tag{3}$$

Because of the way the $(NH_4)_2SO_4$ was added as a solid, the initial and final volume of the experiment would have uncertainty. Evaporation was prevented during experiments by covering the beakers. Largest initial volume change was estimated by mixing 50 mL of 1 M H_2SO_4 with the highest amount of $(NH_4)_2SO_4$ that was required in the experiments. V_f was calculated as follows:

$$V_f = \left(0.5652 \cdot m_{(NH_4)_2SO_4} + V_i\right) - \left(\frac{6 \cdot n_{H_2O}}{n_{M(NH_4)_2(SO_4)_2 \cdot 6H_2O}} \cdot m_{ppt}\right) - \left(0.5652 \cdot \left(\frac{n_{(NH_4)_2SO_4}}{n_{M(NH_4)_2(SO_4)_2 \cdot 6H_2O}} \cdot m_{ppt}\right) + 50\right) \tag{4}$$

where $m_{(NH4)2SO4}$ is the mass of $(NH_4)_2SO_4$ added in the experiment, V_i initial volume without $(NH_4)_2SO_4$, m_{ppt} is the mass of recovered dry precipitate. 0.5652 is the slope of volume change that was measured when adding different quantities of $(NH_4)_2SO_4$ into 1 M H_2SO_4 solution at room temperature whose y intersect is V_i.

Results and Discussion

In this study, the precipitation and co-precipitation of precipitates of form $M(NH_4)_2(SO_4)_2 \cdot xH_2O$ were studied. Solid $(NH_4)_2SO_4$ was added into simulated pregnant leach solution (PLS) containing Ni, Co, Mn and Li whose quantities corresponded to a leach solution originating from leaching of spent batteries.

Precipitation Results

Firstly, the paper investigates the effect of adding different amount of ammonium sulfate to solution, Fig. 1. The purpose of the investigation is to observe whether selectivity can be achieved, and whether there is co-precipitation. In the stoichiometric investigation, shown in Fig. 1A, Ni and Co are at recovery efficiencies of 97 and 79%, respectively, and the recovery trend appears increasing over the whole range of 0.5–2 ratio (as per Eq. (2)). It is known that Ni and Co may form mixed double salt crystals [17] which is what likely occurred in present study as well and a later crystallographic analysis should confirm the hypothesis. Slivnik *et al.* studied how the crystals would form and at which elemental ratios within the crystals when equimolar quantities of different elements were present in the common solution [18]. They showed that preferentiality of Ni and Co over Mn and Fe could be achieved to some degree. This is even more applicable to the present study where the situation is not equimolar but in favor of Ni and Co, to which Fe and Mn are relatively small components in comparison. The method of recovering nickel as an ammonium-based Tutton's salt is not new, and has been applied in prior literature [11] and is also a known phenomenon that needs to be avoided in Ni production processes involving NH_4^+ ions in sulfate solutions.

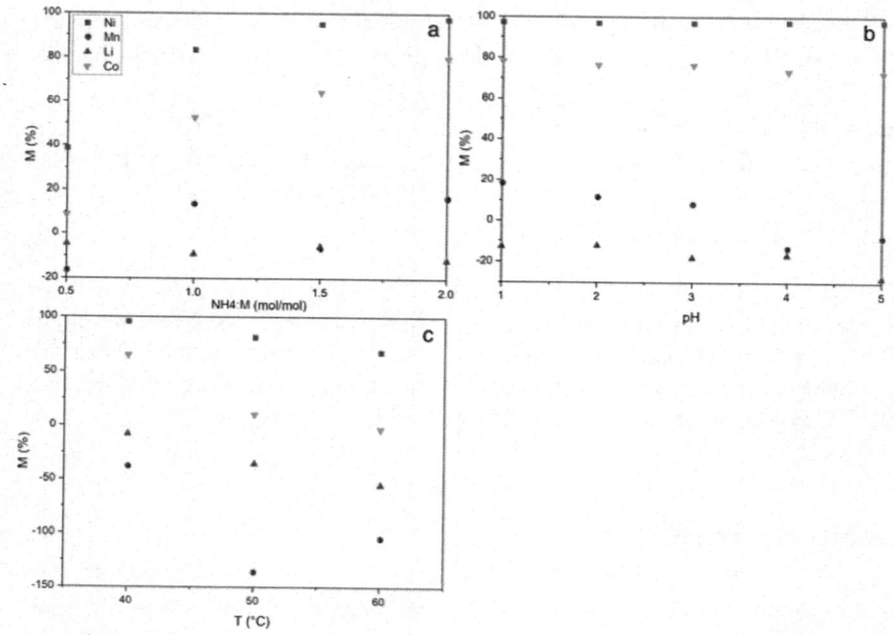

Fig. 1 The results of precipitation from Ni-rich sulfate solutions: **a** as a function of stoichiometric ratio (0.5–2.0), **b** as a function of pH (1–5) and **c** as a function of temperature (40–60 °C). (Color figure online)

Promisingly, low Mn coprecipitation was apparent in all experiments. However, for Mn, the AAS analyses were unreliable and the experiments require further replication and refinement of analysis. The recovered crystals will be later analyzed for the elemental contents. Unlike the erratic Mn results, the measured Li content appears relatively stable and repeatable, however the recovery was in all cases negative which may indicate faulty baseline analysis. It needs to be noted that whatever error that may be present in Ni and Co won't be as readily apparent as in the case of Li and Mn due to the fact that they were relatively well extracted. High decrease in Ni and Co content would also reduce the relative error that could be present in the AAS analyses of Mn and Li. Of note was the result that a higher *T* (>40 °C) resulted in separation of Ni and Co, indicating that the solubility of Co double salt was potentially dramatically higher in the higher temperature than that of the Ni. However, due to the erratic results in Li and Mn, it would be prudent to not draw any significant conclusions from the data presented herein. Experiments must be replicated and the results are verified. Control analyses need to be performed in order to guarantee repeatability of the analyses Fig. 1B.

Similar to trends seen in Fig. 1, the Co rich solutions yielded comparable results with similar uncertainty (Fig. 2). Significant uncertainty is still present in Mn results. Stoichiometric experiments give again similar indications that some selectivity may

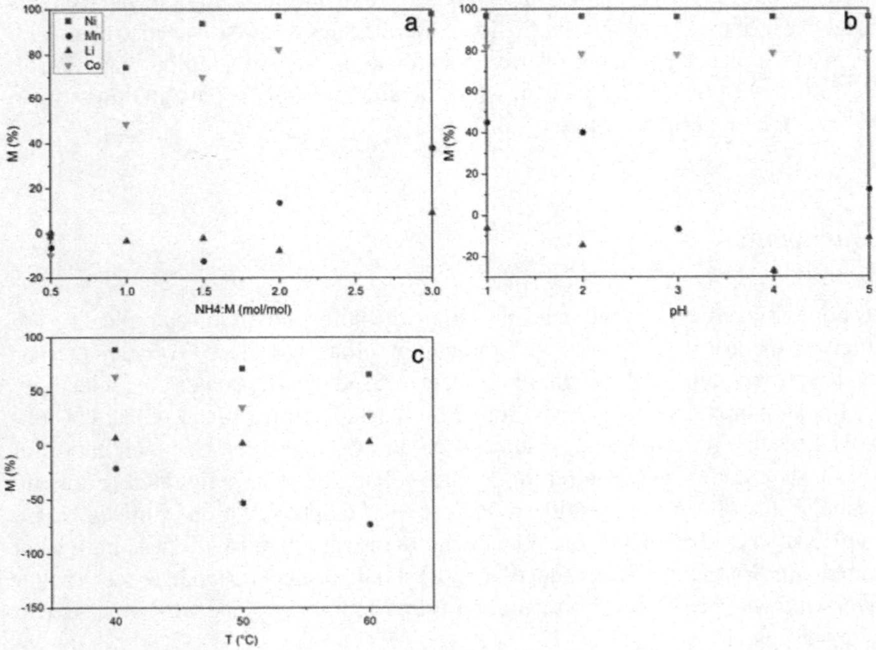

Fig. 2 The results of precipitation from Co-rich sulfate solutions: **a** as a function of stoichiometric ratio (0.5–3.0), **b** as a function of pH (1–5) and **c** as a function of temperature (40–60 °C). (Color figure online)

be obtained over Mn in favor of Ni and Co. Interestingly, Mn recoveries appear to decrease as a function of pH, somewhat similar to what was seen in Fig. 1B. This warrants further experiments in order to obtain a more reliable confirmation of the results.

Other flaws in the study were identified in the validation of results. Ammonium sulfate should be dissolved in a solution matrix akin to where the metal salts were dissolved, and only then mixed with the solution. Because solid ammonium sulfate was utilized in this study, the volume will change as the introduction of ammonium cations will result in change of electrolyte density. It was measured that at stoichiometric ratio of 2 (the highest amount added), the volume changed from the initial 50 mL to 62 mL. Once the precipitation occurs, the density and the volume of the solution would both change as well. The precipitates retain crystal water and also 2 mol of NH_4^+ per metal cation, causing changes to the aforementioned solution properties. Because of this, the results between different experiments are not immediately relatable, for instance, the experiments with pH variations would have alternating sulfate content incomparable to other experiments. Variable volume and density will cause changes in the activities of dissolved species. The extremely anomalous results caused by unreliable baseline might be also because of the used $MnSO_4$: At later inspection, the Mn in storage container appeared clumped which could indicate water absorption.

The results herein highlight the potential to further develop a flowsheet that targets the recovery of majority of Ni and Co first. Several issues would still need to be solved. For instance, the requirement of over-stoichiometric reagent addition would result in raffinate with high NH_4^+ content which could potentially behave in downstream processing in undesirable ways.

Conclusions

Experiments were performed in acidic sulfate medium with dissolved typical Li-ion battery elements Li, Ni, Co and Mn. Ammonium sulfate was added in order to investigate the co-precipitation behavior of the elements and investigate possible selectivity as a function of stoichiometric addition (0.5–3), temperature (40–60 °C) and pH (1–5). The results indicated precipitation of Ni and Co and some co-precipitation of Mn. Further studies are required to confirm reliability of the presented results, in particular that of Mn. The results are indicative of co-precipitation of mixed Ni-Co double salt crystals which is possible due to the isostructurality of these individual double salts. Further work as a function of pH should be undertaken in order to better understand whether Ni and Co separation from Mn improves and what is the extent of improvement.

Acknowledgements This research was funded by BATCircle (XYZ). This work also made use of RAWMatters Research Infrastructure, funded by Academy of Finland. We would like to thank Hannu Revitzer for performing the AAS analyses.

References

1. Zhang T, He Y, Ge L, Fu R, Zhang X, Huang Y (2013) Characteristics of wet and dry crushing methods in the recycling process of spent lithium-ion batteries. J Power Sources 240:766–771. https://doi.org/10.1016/j.jpowsour.2013.05.009
2. Pietrelli L, Bellomo B, Fontana D, Montereali M (2005) Characterization and leaching of NiCd and NiMH spent batteries for the recovery of metals. Spec Issue—VARIREI 2003 25(2):221–226. https://doi.org/10.1016/j.wasman.2004.12.013
3. Cobalt Institute, Production and supply. https://www.cobaltinstitute.org/production-and-supply.html
4. Nickel Institute, About nickel. https://www.nickelinstitute.org/about-nickel#04-first-use-nickel
5. Chagnes A, Swiatowska J (2015) Lithium process chemistry: resources, extraction, batteries, and recycling. Elsevier Science
6. Harper G et al (2019) Recycling lithium-ion batteries from electric vehicles. Nature 575(7781):75–86
7. Zhang X, Xie Y, Lin X, Li H, Cao H (2013) An overview on the processes and technologies for recycling cathodic active materials from spent lithium-ion batteries. J Mater Cycles Waste Manag 15(4):420–430. https://doi.org/10.1007/s10163-013-0140-y
8. Lv W, Wang Z, Cao H, Sun Y, Zhang Y, Sun Z (2018) A critical review and analysis on the recycling of spent lithium-ion batteries. ACS Sustain Chem Eng 6(2):1504–1521. https://doi.org/10.1021/acssuschemeng.7b03811
9. Fan E et al (2020) Sustainable recycling technology for Li-Ion batteries and beyond: challenges and future prospects. Chem Rev
10. Innocenzi V, Ippolito NM, De Michelis I, Prisciandaro M, Medici F, Vegliò F (2017) A review of the processes and lab-scale techniques for the treatment of spent rechargeable NiMH batteries. J Power Sources 362(Supplement C):202–218. https://doi.org/10.1016/j.jpowsour.2017.07.034
11. Barbaroux R, Plasari E, Mercier G, Simonnot MO, Morel JL, Blais JF (2012) A new process for nickel ammonium disulfate production from ash of the hyperaccumulating plant Alyssum murale. Sci Total Environ 423:111–119. https://doi.org/10.1016/j.scitotenv.2012.01.063
12. Mullin JW, Osman MM (1967) Diffusivity, density, viscosity, and refractive index of nickel ammonium sulfate aqueous solutions. J Chem Eng Data 12(4):516–517
13. Manomenova VL, Rudneva EB, Komornikov VA, Lyasnikova MS, Vasilyeva NA, Voloshin AE (2020) The ammonium cobalt sulfate hexahydrate (ACSH) crystal growth from aqueous solutions and some properties of solutions and crystals. J Cryst Growth 532:125416. https://doi.org/10.1016/j.jcrysgro.2019.125416
14. Grebenev VV, Grigor'eva MS, Voloshin AE (2010) Formation of solution inclusions in bicrystals of potassium-cobalt/potassium-nickel and ammonium-cobalt/ammonium-nickel sulfates. Crystallogr Rep 55(5):887–891
15. Peng C, Liu F, Aji AT, Wilson BP, Lundström M (2019) Extraction of Li and Co from industrially produced Li-ion battery waste—using the reductive power of waste itself. Waste Manag 95:604–611. https://doi.org/10.1016/j.wasman.2019.06.048
16. Peng C, Hamuyuni J, Wilson BP, Lundström M (2018) Selective reductive leaching of cobalt and lithium from industrially crushed waste Li-ion batteries in sulfuric acid system. Waste Manag 76:582. https://doi.org/10.1016/j.wasman.2018.02.052
17. Su G, Zhuang X, He Y, Zheng G (2008) A new crystal of ammonium cobalt nickel sulfate hexahydrate for UV light band-pass filter. Opt Mater (Amst) 30(6):916–919. https://doi.org/10.1016/j.optmat.2007.04.002
18. Slivnik J, Rahten A, Gantar D (1985) Study of Tutton's Salts (NH 4) M (SO_4) 2 $6H_2O$ with mixed metal ions in the structure. Croat Chem Acta 58(3):289–294

BATCircle—Towards CO_2 Low Battery Recycling

M. Lundström, A. Porvali, H. Elomaa, M. Rinne, P. Hannula, and P. Kauranen

Abstract Finland-based Circular Ecosystem of Battery Metals (BATCircle, 22 M€) aims at management of materials and value addition during the entire lifetime of Li-ion batteries. The consortium led by Aalto University focuses on enhancing battery metals production from both primary and secondary sources. Along with BATCircle, Finland leads Batteries Europe (ETIP) WG2, where several gaps in the current battery recycling R&D topics have been identified. One of the topics relates to the environmental impacts of emerging battery recycling technologies. The current study (a) summarizes a novel approach for the synergistic hydrometallurgical recycling of LIBs together with NiMH battery waste and (b) highlights the importance of simulation-based life cycle assessment as a tool to provide early stage environmental indicator values for development stage processes.

Keywords Recycling · Battery ecosystem · Process modelling · LCA

Introduction

Finland has initiated a ca. 22 M€ Finland-based Circular Ecosystem of Battery Metals (BATCircle) aiming at value addition during the entire lifetime of the Li-ion batteries and the management of materials. The consortium, led by Aalto University, consists of 23 companies, six research organizations, and two cities. During the previous year, several big investments in battery chemicals production were already launched in Finland. In BATCircle, the academic research has been conducted in five work packages: Sustainable primary resources, Value addition in metals production [1], Recycling of Batteries [2, 3] and Tailored precursors and active electrode

---

M. Lundström (✉) · A. Porvali · M. Rinne · P. Hannula · P. Kauranen
School of Chemical Engineering, Aalto University, Vuorimiehentie 2, P.O. Box 16200, FI-00076 Aalto, Finland
e-mail: mari.lundstrom@aalto.fi

H. Elomaa
Finland Outotec Research Center (Finland) Oy, Kuparitie 10, PO Box 69, Pori, Finland

C. Anderson et al. (eds.), *Ni-Co 2021: The 5th International Symposium on Nickel and Cobalt*, The Minerals, Metals & Materials Series,
https://doi.org/10.1007/978-3-030-65647-8_6

materials. Also, business potential research [4, 5], exploring the untapped business potential in the battery ecosystem and the system-level influence is conducted.

At the European level, the expected forecast for the battery market growth is ca. 250 billion EUR by 2025. European Commission invited Finland (Outotec and Aalto) to coordinate the Strategic Energy Technology Plan (SET-Plan) Action 7 sub-working group on recycling of batteries in 2018. In 2019, the SET-Plan activity was merged with a new European Technology and Innovation Platform (ETIP) named Batteries Europe. During the ongoing work in European level, several gaps in the battery recycling R&D projects have been identified. There are numerous innovations conducted of battery recycling. However, only few organizations focus on holistic processing of batteries, and consequently a true evaluation of the environmental impact of novel development stage processes is missing in the open literature. Therefore, the current paper presents the methodology of simulation-based life cycle assessment and demonstrates its efficacy by one suggested and published holistic recycling route.

Synergistic LIB and NiMH Recycling Process

In order to support development of sustainable recycling processes for batteries, the suggested new process concepts need to be assessed also from environmental point of view. In this study a process model as well as simulation based life cycle assessment is demonstrated for a synergistic recycling process of NiMH and LIB batteries, suggested by Liu et al. [6].

Process Based Simulation-LCA Methodology

Life-cycle assessment is by far the most used method to evaluate the environmental impacts of a product or a process. The method is standardized in ISO 14040 and consists of goal and scope definition, life-cycle inventory (LCI) analysis, impact assessment (LCIA) and interpretation [7]. The method produces quantitative data on the environmental performance of the process in the form of environmental indicators such as global warming potential (GWP, unit kg CO_2-eq), which measures the heat trapped in the atmosphere by the emitted greenhouse gases (GHGs), and terrestrial acidification potential (AP, unit kg SO_2-eq), which refers to the total emissions contributing to acid rain.

Despite the standardization of LCA methodology, inconsistencies between different studies remain. Simulation based LCA methodology has been presented by Reuter (1998) [8] in order to provide accurate and detailed data about metallurgical processes. The methodology combines a rigorous metallurgical process simulation to standardized LCA method, which enables a comprehensive way of investigating the process performance, optimization and environmental impacts. The synergistic advantage of process simulation and LCA helps to generate more detailed LCI inventories and open the "black-box" of metallurgical processes [9]. It is applicable to existing as well as development stage processes and provides early stage information of environmental impacts. Process simulation is a crucial part of any process design and development studies and can be updated through the whole pre-feasibility stage of processes [10, 11]. In this study, the hydrometallurgical battery recycling process is investigated as full process concept by simulation with HSC-Sim 10 followed by LCA in GaBi software.

Process

The investigated process is based on synergistic leaching of LIB and NiMH waste in sulfuric acid media and is described in detail elsewhere by Liu et al. [11]. In short, the process takes an advantage of the reductive nature of the NiMH battery waste, specifically hydride-forming metal alloys on NiMH anode as reductant for highly oxidized LIB phases such $LiCoO_2$ (LCO) or $LiNi_xMn_yCo_{1\text{-}y\text{-}x}O_2$ (NMC) materials. After leaching, REEs are precipitated as double sulfates [12], the solubility decreasing with the introduction of monovalent cations, in particular Na ions. Following the precipitation, the PLS is redirected to Mn removal by oxidation. $KMnO_4$ is utilized at pH = 1.5 in precipitating the dissolved Mn as MnO_2. Ni and Co are selectively extracted by D2EHPA and Cyanex 272, respectively. The loaded organic phase is then stripped by contacting it with 0.5 M H_2SO_4 solution. The stripped Ni and Co are fed to a crystallizer where Co and Ni sulfates are produced. Finally, the final raffinate from the solvent extraction is mixed with a source of phosphates with stoichiometric ratio of 0.9 to 1 (PO_4:3Li). The solution pH is increased to pH = 11 where most of the phosphate reside as free anions, and Li ions proceed to form insoluble Li_3PO_4 [13, 14]. The filtrate is recycled through a Na_2SO_4 crystallizer: i.e., a part of the waste can be again utilized as a precipitating agent in the REE precipitation.

LIB and NiMH Compositions

The compositions of the NiMH and LIB batteries were the following, Table 1. NiMH active materials are known to consist of AB_5 type alloys where A = (La, Ce, Pr, Nd, Y) and B = (Ni, Co, Mn, Al, Zn) as supported by prior analysis [15, 16]. In this study AB_5 was assumed as $LaNi_5$ while factoring in the fact that Co, Zn, Al and Mn are part of the anode B_5.

Lithium ion battery active materials were assumed to be either from $LiCoO_2$ (LCO), $LiMn_2O_4$ (LMO) or $LiNiO_2$ (LNO). It needs to be noted that this is simplification of mixed NMC oxide, only done for the sake of simulation. For example, LNO was never a commercialized cathode material. In terms of acid consumption this is less important as their valence and oxidation states dictate the reagent consumption,

Table 1 The raw material analysis from Liu et al. [11], with supplementary concentrations for Zn and Al which is commonly known to be present

Element (mg/kg)	Li	Co	Ni	Cu	Mn	Fe	La
LIB	39.7	207.9	29.9	4	18.7	3.7	0
NiMH	0	55.9	432.5	1	30.2	3.9	88.2
Element (mg/kg)	Ce	Pr	Nd	K	Na	Zn	Al
LIB	0	0	0	0	0	0	0
NiMH	79.7	12.2	37.2	0.2	0.1	16	11.3

not the state of the matter, although state of the matter contributes strongly to differences in shifts in enthalpy and whether a reaction is able to occur. The treatment of Zn and Al was ignored, similar to the original publication [6].

Results and Discussion

The process flowsheet, suggested by Liu et al. [6], and the technical system boundary used in the LCA study are shown in Fig. 1. Experimentally optimized parameters were used for leaching and double sulfate precipitation (Table 2), whereas the other units were modelled based on known industrially relevant processes and their conditions. The LCI was compiled from the mass and energy balance, and also electricity consumption for the necessary equipment was estimated. The geographical boundary in the study was Europe and the upstream processes were selected accordingly: e.g. the average European electricity production mix was used for power consumption. The functional units were 1 kg of nickel in nickel sulfate and 1 kg of cobalt in cobalt sulfate. No byproduct allocation was conducted.

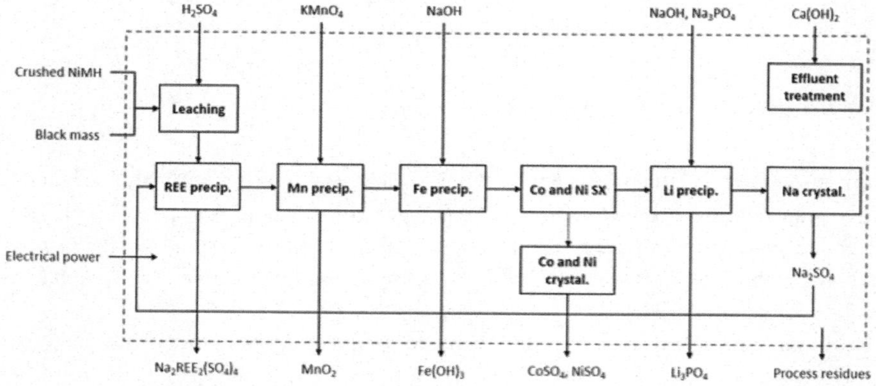

Fig. 1 Simplified flowsheet, technical system boundary highlighted with a dashed line. (Color figure online)

Table 2 Process parameters used in the leaching and double sulfate precipitation steps

Parameter	Value	Unit
NiMH/LIB ratio	0.75	–
H_2SO_4 in leaching	123	g/L
S/L in leaching	78	g/L
Temperature in leaching	75	°C
pH after leaching	0	–
Na/REE ratio in REE precipitation	8	–

The calculated GWP values for nickel and cobalt along with the contribution percentages of upstream processes are shown in Fig. 2. It was estimated that the GWP values were 14.7 kg CO_2-eq/kg for nickel and 26.0 kg CO_2-eq/kg for cobalt. The discrepancy between the metals is explained by the difference in nickel and cobalt content in the raw material mix, and the GWP is consequently affected by raw material grades.

Most of the environmental burden of the investigated process is attributed to NaOH consumption, Fig. 2. NaOH was used as a neutralizing chemical and to bring Na ions for the circulation of Na_2SO_4. However, lime milk ($Ca(OH)_2$) is more commonly used as a neutralizing chemical and it is possible that it could decrease the GWP value of the process. However, this would also necessitate gypsum removal, increase the solid waste burden as well as potentially cause adsorption of valuables on gypsum. It should also be stressed that the GWP is affected by the electricity production mix: operating the plant in fossil reliant countries, leads to higher GHG emissions than in countries with renewable-based electricity, such as Norway.

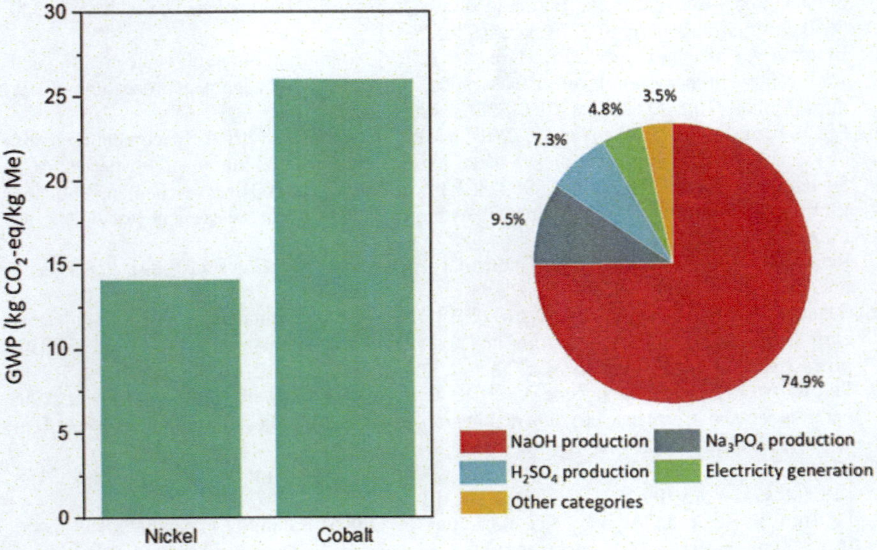

Fig. 2 GWP values calculated for 1 kg of nickel and cobalt in $MeSO_4*xH_2O$ and the contribution (%) of upstream processes to the value. (Color figure online)

Conclusions

The global warming potential values were estimated for nickel and cobalt recoveries from mixed Li-ion and NiMH battery waste by synergistic leaching using simulation-based life-cycle assessment method. The GWP values were determined to be 14.7 kg CO_2-eq/kg for Ni and 26.0 kg CO_2-eq/kg for Co. It also was also demonstrated that the method is a powerful tool in estimating the environmental impacts of emerging processes with no available industry data, and the method can be used as a guiding tool in decision-making when upscaling new process concepts.

Acknowledgements This work has been financed by BATCircle project and "*RawMatTERS Finland Infrastructure*" (RAMI). The provider of the raw material, AkkuSer Ltd, is greatly acknowledged by the authors.

References

1. Peng C, Chang C, Wang Z, Wilson BP, Lundström M Recovery of high-purity MnO_2 from the acid leaching solution of spent Li-ion batteries. JOM, 1–10. https://doi.org/10.1007/s11837-019-03785-1
2. Porvali A, Chernyaev A, Shukla S, Lundström M (2019) Lithium ion battery active material dissolution kinetics in Fe(II)/Fe(III) catalyzed Cu-H_2SO_4 leaching system. Sep Purif Technol, 116305. https://doi.org/10.1016/j.seppur.2019.116305
3. Agarwal V, Khalid MK, Porvali A, Wilson BP, Lundström M Recycling of spent NiMH batteries: integration of battery leach solution into primary Ni production using solvent extraction. Sustain Mater Technol, e00121. https://doi.org/10.1016/j.susmat.2019.e00121
4. Seppälä T, Hakanen E, Lähteenmäki I, Mattila J, Niemi R (2019) The resource dependency of data: a prospective on data sharing in supply chains. SSRN. https://ssrn.com/abstract=3493650 or http://dx.doi.org/10.2139/ssrn.3493650
5. Eloranta V, Hakanen E, Töytäri P, Turunen T (2019) Aligning multilateral value creation and value capture in ecosystem-level business models. Academy of management proceedings 2019. https://doi.org/10.5465/ambpp.2019.18966abstract
6. Liu F, Peng C, Porvali A, Wang Z, Wilson BP, Lundström M (2019) Synergistic recovery of valuable metals from spent nickel–metal hydride batteries and lithium-ion batteries. ACS Sustain Chem Eng, acssuschemeng.9b02863. https://doi.org/10.1021/acssuschemeng.9b02863
7. ISO 14040:2006 (2006) Environmental management-life cycle assessment-principles and framework, pp 20
8. Reuter MA (1998) The simulation of industrial ecosystems. Miner Eng 11(10):891–918. https://doi.org/10.1016/S0892-6875(98)00078-8
9. Segura-Salazar J, Lima FM, Tavares LM (2019) Life cycle assessment in the minerals industry: current practice, harmonization efforts, and potential improvement through the integration with process simulation. J CLEPRO 232:174–192. https://doi.org/10.1016/j.clepro.2019.05.318
10. Pell R, Wall F, Yan X, Li J, Zeng X (2019) Temporally explicit life cycle assessment as an environmental performance decision making tool in rare earth project development. Miner Eng 135:64–73. https://doi.org/10.1016/j.mineng.2019.02.043
11. Wall F, Rollat A, Pell RS (2017) Responsible sourcing of critical metals. Elements 13(5):313–318. https://doi.org/10.2138/gelements.13.5.313
12. Lokshin EP, Tareeva OA, Ivlev KG, Kashulina TG (2005) Solubility of double alkali metal (Na, K) rare-earth (La, Ce) sulfates in sulfuric-phosphoric acid solutions at 20 C. Russ J Appl Chem 78(7):1058–1063. https://doi.org/10.1007/s11167-005-0449-y

13. Song Y-J (2018) Recovery of lithium as Li3PO4 from waste water in a LIB recycling process. Korean J Met Mater 56(10):755–762. https://doi.org/10.3365/KJMM.2018.56.10.755
14. Pinna EG, Ruiz MC, Ojeda MW, Rodriguez MH (2017) Cathodes of spent Li-ion batteries: dissolution with phosphoric acid and recovery of lithium and cobalt from leach liquors. Hydrometallurgy 167:66–71. https://doi.org/10.1016/j.hydromet.2016.10.024
15. Porvali A, Ojanen S, Wilson BP, Serna-Guerrero R, Lundström M (2020) Nickel metal hydride battery waste: mechano-hydrometallurgical experimental study on recycling aspects. J Sustain Metall. https://doi.org/10.1007/s40831-019-00258-2
16. Porvali A, Wilson BP, Lundström M (2017) Lanthanide-alkali double sulfate precipitation from strong sulfuric acid NiMH battery waste leachate. Waste Manag. https://doi.org/10.1016/j.wasman.2017.10.031

Selective Sulfidation and Electrowinning of Nickel and Cobalt for Lithium Ion Battery Recycling

Caspar Stinn and Antoine Allanore

Abstract Processes for recycling lithium ion batteries (LIB), in particular complex chemistries such as those containing nickel-manganese-cobalt oxide (NMC) cathodes are hindered by tradeoffs between capital cost, process sustainability, and materials recovery. Most metal separations in primary and secondary production of critical elements rely on anion exchange chemistries. Herein, we explore the application of a novel oxide-sulfide anion exchange methodology to facilitate LIB recycling. Beginning with selective sulfidation of NMC cathode oxides, we demonstrate that lithium may be stabilized as a sulfate, manganese as an oxysulfide, and nickel and cobalt as sulfides from the mixed metal feed, potentially facilitating isolation of lithium via leaching and nickel-cobalt via flotation. Following, we explore molten sulfide electrolysis as a method of process intensification, combining separation and reduction into a single unit operation for difficult to separate metals such as cobalt and nickel. We demonstrate selective reduction of cobalt from mixed nickel-cobalt sulfide, as produced in selective sulfidation of waste NMC cathodes, using a barium-lanthanum sulfide supporting electrolyte. Our preliminary results suggest that selective sulfidation as a pretreatment for selective molten sulfide electrolysis is a promising avenue for separation of critical elements from complicated materials feeds, such as those found in lithium ion battery recycling streams.

Keywords Nickel · Cobalt · Lithium-ion battery · NMC cathode · Sulfidation · Electrolysis · Recycling · Molten sulfide · Sustainability

Introduction

Increasing demand for electric vehicles has the potential to strain geopolitically sensitive supply chains of battery-grade materials, such as cobalt, motivating efforts to develop recycling technologies to recycle and reuse strategic elements [1]. While a

C. Stinn · A. Allanore (✉)
Department of Materials Science and Engineering, Massachusetts Institute of Technology, 77 Massachusetts Ave, Cambridge, MA 02139, USA
e-mail: allanore@mit.edu

C. Anderson et al. (eds.), *Ni-Co 2021: The 5th International Symposium on Nickel and Cobalt*, The Minerals, Metals & Materials Series,
https://doi.org/10.1007/978-3-030-65647-8_7

variety of pyrometallurgical, hydrometallurgical, and direct recycling processes have been considered for recovery of critical battery materials, present methodologies are plagued by tradeoffs between capital cost, process sustainability, and materials recovery, with additional complications arising from non-standardized chemistries that exhibit inherent, thermodynamic difficulties in separation [2, 3]. For recycling of lithium ion batteries (LIB) containing nickel-manganese-cobalt-based (NMC) cathodes, challenges arise from the fact that nickel, manganese, cobalt, and lithium within the cathode exist as mixed-metal oxide compounds and solid solutions [4, 5]; thus separation of lithium, nickel, manganese, and cobalt presently requires chemical methods to isolate individual elements. Chemical separation of metallic elements is often rooted in anion exchange, where the anion of a metal compound, or in the case of liquid-liquid hydrometallurgy the anionic solvating or chelating species, is exchanged with another to facilitate separation based on property differences between the phases containing different anions. Conventional primary and secondary processing technologies for battery metals including matte smelting [6], carbochlorination [7], and solvent extraction [8, 9] can all be viewed in the framework of anion exchange.

Herein, we consider an oxide-sulfide-based anion exchange chemistry and its potential to handle the multi-element nature of LIB recycling streams. Oxide-sulfide-based anion exchange for battery elements has previously been considered in the context of selective sulfidation of nickel oxides to sulfides [10] and selective sulfation of nickel and cobalt oxides to sulfates [11, 12]. Sulfidation supports separation of oxides from sulfides via froth flotation [10] and sulfation supports separation of oxides from sulfates via leaching [11, 12]. Meanwhile, sulfide chemistry allows for economically competitive [13], environmentally sustainable [14], and selective [15] metal reduction via molten sulfide electrolysis. Through demonstration of selective sulfidation and selective molten sulfide electrolytic reduction of NMC battery cathode metals, we establish the foundation of a sulfide-based processing route for lithium ion battery recycling.

Thermodynamic Basis for Selective Sulfidation and Molten Sulfide Electrolytic Reduction

For the generic sulfidation reaction of a metal (M) oxide (Eq. 1), through the law of mass action and the van't Hoff equation, a critical sulfur to sulfur dioxide partial pressure ratio, $[P_{S2}/P_{SO2}]_{crit}$, can be defined for thermodynamic spontaneity of Eq. 1 at equilibrium (Eq. 2), where R is the gas constant, T is the absolute temperature, a is the compound activity, P is the gas partial pressure, and β, γ, δ, ε are stoichiometric coefficients:

$$M_\beta O_\gamma + \frac{2\beta\varepsilon + \gamma\delta}{4\delta} S_2 = \frac{\beta}{\delta} M_\delta S_\varepsilon + \frac{\gamma}{2} SO_2 \tag{1}$$

$$\left[\frac{P_{S_2}^{\frac{2\beta\varepsilon+\gamma\delta}{4\delta}}}{P_{SO_2}^{\frac{\gamma}{2}}}\right]_{crit} = e^{\frac{\Delta G_r^\circ}{RT}} \frac{a_{M_\delta S_\varepsilon}^{\frac{\beta}{\delta}}}{a_{M_\beta O_\gamma}} \tag{2}$$

Similar relations to Eq. 2 can be defined for the reactions between the metal, oxides, oxysulfides, sulfates, and sulfides. Assuming the metal and its compounds are non-volatile, for a gas phase consisting only of oxygen and sulfur containing species at a given temperature and total pressure, P_{SO2} is a monotonic function of PS_2, thereby defining the P_{S2}/P_{SO2} ratio as a monotonic function of PS_2. When the condensed metal compound reactant and product phases are soluble, their activities at a given P_{S2}/P_{SO2} ratio may be related to one another through the Gibbs-Duhem equation. Assuming insoluble reactant and product metal compounds at unit activity, Eq. 2 converges to a Kellogg formalism. The stability of cobalt and its oxides and sulfides at 1000 °C as a function of gas atmosphere is depicted in the Kellogg diagram of Fig. 1a. Following a similar formalism for nickel, manganese, and lithium, a sulfidation series may be defined for sulfidation of nickel, manganese, cobalt, and lithium oxides (Fig. 1b). While actual nickel-manganese-cobalt oxide (NMC) lithium ion battery (LIB) cathodes are composed of multi-metal oxide compounds and solid solutions, the activity ratio of pure metal oxides within single phases of NMC cathodes is far outweighed by their respective P_{S2}/P_{SO2} ratios differences. This suggests that a sulfidation series composed of pure end-member oxides is a suitable first-pass indication of sulfidation selectivity. As predicted in Fig. 1b, during sulfidation of a NMC cathode, lithium may be stabilized as a sulfate, while nickel, cobalt, and manganese may be individually stabilized as either oxides or sulfides. Such outcome opens recovery methods for NMC cathode elements including leaching of lithium sulfate and froth flotation to separate nickel, cobalt, and manganese. This may be facilitated by oxide-sulfide anion exchange.

Molten sulfide electrolysis is a potential process intensification to integrate separation and reduction of mixed metal sulfides resulting from selective sulfidation. For the generic metal sulfidation reaction (Eq. 3), the corresponding decomposition potential (ΔE) may be found using the Nernst equation (Eq. 4), where $\Delta G_r°$ is the standard Gibbs energy, z is the number of electrons involved in the reaction (for 1 mol of S_2, $z = 4$), F is the Faraday constant, and β, γ are stoichiometric coefficients.

$$\frac{2\beta}{\gamma} M + S_2 = \frac{2}{\gamma} M_\beta S_\gamma \tag{3}$$

$$\Delta E = \frac{-\Delta G_r^\circ}{zF} - \frac{RT}{zF} \ln\left(\frac{a_{M_\beta S_\gamma}^{\frac{2}{\gamma}}}{P_{S_2} a_M^{\frac{2\beta}{\gamma}}}\right) \tag{4}$$

Following an Ellingham formalism for pure species at unit activity under a pure sulfur atmosphere, a reduction series may be tabulated for NMC cathode endmember sulfides, as well as supporting electrolytes [14–16] for molten sulfide electrolysis (Fig. 1c). While the solution thermodynamics of relevant molten sulfides are under

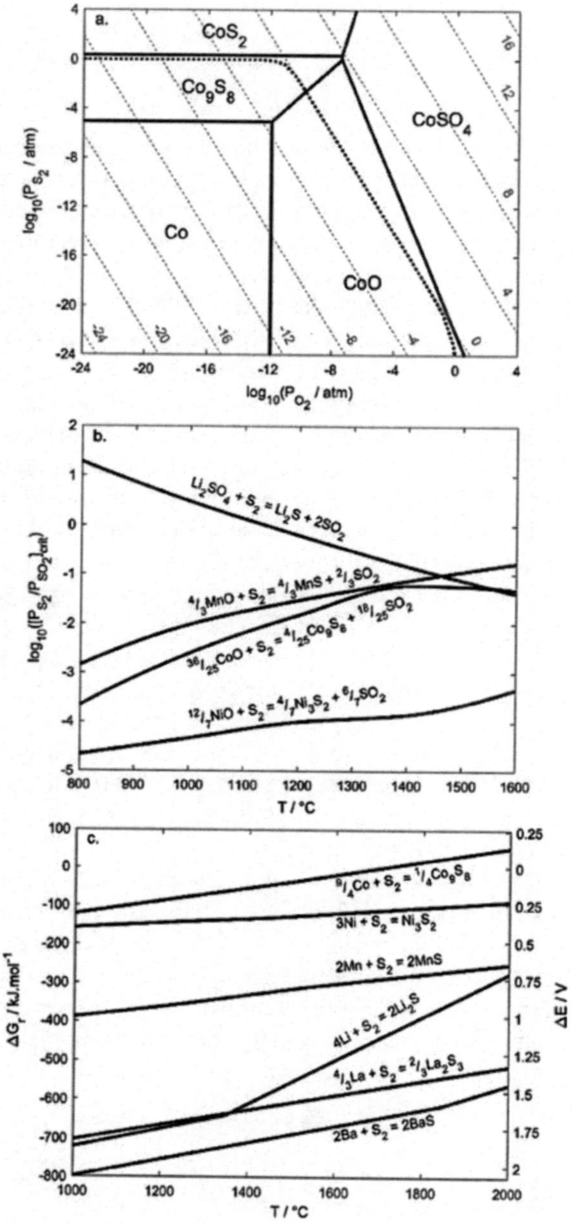

Fig. 1 Oxide-sulfide-based anion exchange chemistry offers selective separation opportunities for battery metals. **a**. A Kellogg formalism, shown here for cobalt at 1000 °C, can be used to determine a critical gas atmosphere for sulfidation to occur. **b**. Comparison of critical gas ratios for sulfidation allows for construction of a sulfidation series, shown here for battery metals. **c**. Ellingham comparison of the reduction selectivity of battery metal sulfides and sulfide supporting electrolytes, suggesting that molten sulfide electrolysis may be selective for critical battery elements

development, review of the decomposition voltages between pure species suggests that molten sulfide electrolysis at 1500 °C is a promising approach for selective reduction of nickel, manganese, cobalt, and lithium sulfides using a barium-lanthanum sulfide supporting electrolyte, due to the large relative difference in decomposition potential between species. In the following sections, we demonstrate selective sulfidation of NMC cathode oxides and selective molten sulfide electrolysis, constituting an integrated sulfidation-electrolytic NMC cathode recycling pathway.

Selective Sulfidation of Nickel-Manganese-Cobalt Oxide Cathodes

Experimental

Simulated nickel-manganese-cobalt (NMC) oxide cathode material ($LiNi_{1/3}Mn_{1/3}Co_{1/3}O_2$, 98% purity, Sigma Aldrich) was used as the oxide precursor for selective sulfidation experiments. While industrially H_2S and CS_2 are commonly employed for sulfidation of oxides, their use for complete sulfidation of mixed transition metal oxide catalysts [17] calls into question the suitability of H_2S and CS_2 for selective sulfidation. Therefore, vaporized elemental sulfur (S_X, 99.5%, purity, Acros Organics) served as the sulfidizing agent. At temperatures above 800 °C, sulfur gas is predominantly diatomic [18]. Selective sulfidation of metals from the mixed Li-Ni-Mn-Co oxide was conducted in a packed bed reactor, consisting of a 20 mm OD, 14 mm ID crucible fabricated in-house from machinable alumina (Al_2O_3, 96% purity, Rescor 960, Cotronics Corp), with seven 1 mm diameter holes spaced 2 mm apart drilled through the bottom of the crucible to support sulfur gas flow to the oxide. NMC oxide powder was loaded into the alumina crucible at a porosity of 90%, at a sulfidation charge of 2 g of oxide. Care was taken to avoid loss of the oxide precursor through the bottom crucible holes. The alumina crucible was held in the hot zone of a vertical tube furnace (SS15R-2.50X6V- 1Z, Mellen) within an alumina tube (25 mm OD, 21 mm ID). Leveraging the thermal gradient of the tube furnace, sulfur gas was boiled from a 316 stainless steel crucible held below the alumina crucible, with sulfur gas transported to the oxide-containing crucible via an argon (Ar, 99.95% purity, Air Gas) carrier gas flowing at 1000 sccm. Sulfidation conversion was monitored by measuring the rate of SO_2 generation via IR gas analysis (IR208, Infrared Industries) of the product gas stream. Sulfur gas flowrate was similarly monitored using an H_2S tracer calibrated to sulfur evaporation rate. The sulfidized NMC was mounted in epoxy, cross sectioned, polished, and then imaged optically and via scanning electron microscope (SEM, JEOL JSM-6610LV, JEOL Ltd.) equipped with an energy dispersion spectroscopy analyzer (EDS, Sirius SD detector, SGX Sensortech Ltd.). To determine the extent of sulfidation of lithium oxide from the NMC, x-ray diffraction analysis (XRD) was conducted on crushed sulfidation product.

Results and Discussion

Sulfidation of 2 g of nickel-manganese-cobalt oxide (NMC) lithium ion battery cathode was conducted following the methodology above, at 1000 °C and a P_{S2}/P_{SO2} ratio of approximately 10, P_{S2} of approximately 0.1 atm. As determined from the rate of oxygen liberation in the form of SO_2, the reaction rate became kinetically negligible after about 50 min. This was taken to be the point at which the NMC reached a quasi-equilibrium with the S_2/SO_2 atmosphere, upon which the sulfur flow was stopped and the furnace cooled. Under such conditions, nickel, cobalt, and manganese were thermodynamically predicted to form sulfides, while lithium was predicted to form a sulfate (Fig. 1b). As observed through energy dispersion spectroscopy (EDS), nickel and cobalt from the NMC oxide were indeed sulfidized (Fig. 2). Surprisingly, manganese from the NMC oxide did not fully sulfidized, forming instead a separate oxysulfide phase. At 1000 °C, nickel sulfide and cobalt sulfide are reported to form a fully miscible liquid [19]. While nickel and cobalt sulfides did melt, they were observed in this study to phase separate into distinct cobalt-rich and nickel-rich liquid sulfides. The nickel-rich sulfide, cobalt-rich sulfide, and manganese oxysulfide phases coalesced into single phase regions on the order of 100 μm to 1 mm in size during the sulfidation reaction. While the distribution of lithium was unobservable post sulfidation with EDS, x-ray diffraction (XRD) revealed that lithium existed largely as a sulfate.

Starting from synthetic NMC oxide cathode material, the sulfidation treatment resulted in the selective sulfidation of nickel and cobalt to sulfides, manganese to oxysulfide, and lithium to sulfate. Furthermore, not only was sulfidation conducted selectively, but the resultant nickel-rich sulfide, cobalt-rich sulfide, and manganese oxysulfide coalesced into large phases >100 μm, large enough to support practical liberation of nickel and cobalt from manganese via comminution alone [20]. The differing hydrophobicity of sulfides and oxysulfides suggests that nickel and cobalt may therefore be isolated from manganese using flotation [21]. Meanwhile the soluble nature of alkali metal sulfates in water, in contrast to the largely insoluble behavior of transition metal oxides, oxysulfides, and sulfides, suggests that lithium can possibly be selectively leached and recovered from selectively sulfidized lithium ion battery cathodes. Indeed, selective sulfidation of lithium-nickel-manganese-cobalt oxide may facilitate the isolation of nickel-cobalt, manganese, and lithium, a promising avenue for recycling of lithium ion batteries. For isolation of cobalt from nickel, some preliminary separation is possible during selective sulfidation due to the immiscibility of the cobalt-rich and nickel-sulfide liquids. Following, selective molten sulfide electrolysis was explored as a complementary avenue to integrate cobalt and nickel separation and reduction into a single processing step.

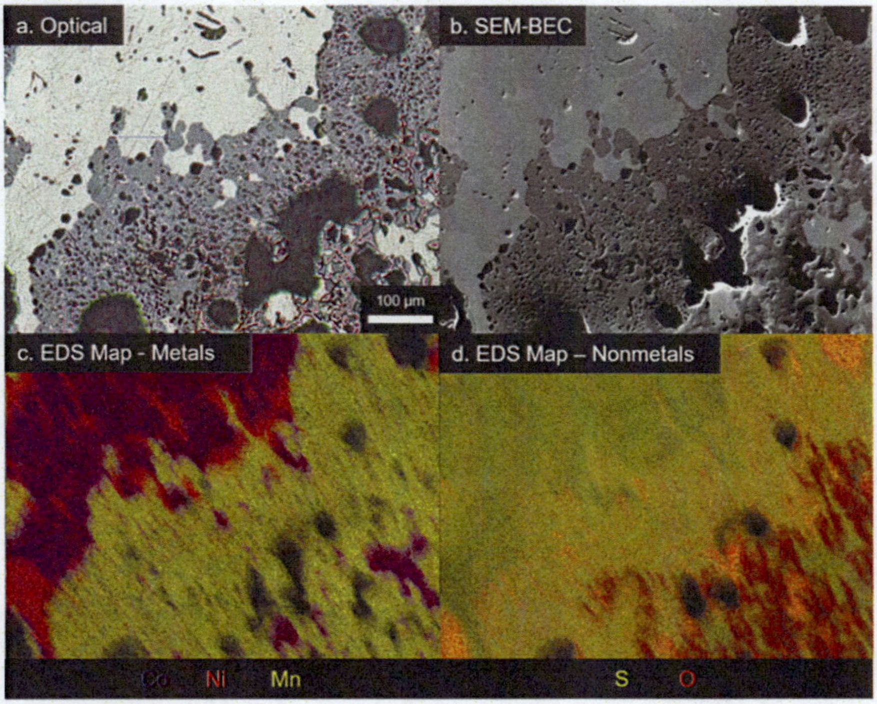

Fig. 2 Lithium, nickel, manganese, and cobalt were selectivity sulfidized from synthetic nickel-manganese-cobalt oxide cathode material. Optical microscopy (**a**), SEM microscopy (**b**), and EDS mapping (**c, d**) reveal that upon selective sulfidation, a cobalt-rich sulfide (**c**, purple, $Co_{0.67}Ni_{0.33}S$), nickel-rich sulfide (**c**, pink, $Ni_{0.75}Co_{0.25}S$), and manganese oxysulfide (**c**, yellow, $MnO_{0.2}S_{0.8}$) phases formed. These phases coalesced to sizes between 100 µm and 1 mm, suggesting that nickel, manganese, and cobalt may be liberated via comminution and physical separation. Minimal inclusion between manganese and nickel-cobalt is observed. While lithium is indiscernible on the EDS maps, XRD reveals lithium to exist as a sulfate, suggesting oxygen-rich regions in (**c, d**) lacking Ni, Mn, or Co may correspond to lithium sulfate. Due to differences in aqueous solubilities between sulfides, oxysulfides, and sulfates, lithium may be potentially recovered from sulfidized metals via leaching. (Color figure online)

Selective Electrowinning of Nickel and Cobalt from Molten Sulfides

Experimental

For selective electrowinning of nickel and cobalt from liquid sulfides, the electrolyte consisted of nickel sulfide (Ni_3S_2, 99.7% metals basis purity, Sigma Aldrich) cobalt sulfide (CoS, 99.5% metals basis purity, Fisher Scientific), barium sulfide (BaS, 99.7% metals basis purity, Alfa Aesar), and lanthanum sulfide (La_2S_3, 99.5% metals basis purity, Strem Chemical) at a composition of $(Ni_3S_2)_7(CoS)_{21}(BaS)_{54}(La_2S_3)_{15}$.

The $(BaS)_{54}(La_2S_3)_{15}$ supporting electrolyte has recently been shown to increase the ionic transference number of molten transition metal sulfides [15], making them suitable for electrolytic decomposition. While the phase stability of some transition metal sulfides in the BaS-La_2S_3 supporting electrolyte is known [15, 16], the liquidus behavior of mixed Ni_3S_2, CoS, BaS, and La_2S_3 remains unclear. Therefore, a sample of the electrolyte was held at 1500 °C for one hour to confirm that the electrolyte formed a stable liquid at temperatures necessary for liquid nickel and cobalt deposition. To understand the electrochemical behavior of the electrolyte, open circuit potential (OCP), potentiostatic electrochemical impedance spectroscopy (EIS), and cyclic voltammetry (CV) were performed on the liquid $(Ni_3S_2)(CoS)_3(BaS)_{54}(La_2S_3)_{15}$ at 1500 °C using a potentiostat/galvanostat (Reference 3000, Gamry) with a three electrode cell constructed in-house from an alumina crucible, with cell design and electrode orientation described elsewhere [14, 15]. Galvanostatic electrolysis was also conducted at an anodic current density of 1.5 A/cm^2 using the same cell, reconfigured with two electrodes similarly described elsewhere [14, 15]. All electrodes were machined in-house from EDM graphite (Isostatically Pressed, EC-12/ AC-12, Tokai Carbon). The cell was operated within the hot zone of a tube furnace (SS15R-2.50X6V- 1Z, Mellen) inside of an alumina tube of 64 mm OD, 58 mm ID, under inert atmosphere with an argon (Ar, 99.95% purity, Air Gas) flowrate of 400 sccm. Post-electrolysis, the electrolyte was mounted in epoxy, cross sectioned, polished, and then imaged optically and via scanning electron microscope (SEM, JEOL JSM-6610LV, JEOL Ltd.) equipped with an energy dispersion spectroscopy analyzer (EDS, Sirius SD detector, SGX Sensortech Ltd.).

Results and Discussion

Due to the unexplored nature of the $(Ni_3S_2)_7(CoS)_{21}(BaS)_{54}(La_2S_3)_{15}$ electrolyte, the liquid phase stability was tested by holding a sample of the electrolyte in an alumina crucible at the electrolysis operating temperature of 1500 °C for one hour. Once cooled, the sample was observed to have melted and wetted the alumina, with less than 1% mass loss observed. Following, the electrolytic nature of the molten sulfide was explored using a three electrode cell with graphite working, counter, and quasi-reference electrodes as described previously. Before electrochemical measurements, the melt was held at the operating temperature of 1500 °C for one hour. The electrodes were confirmed to be in contact with the electrolyte by measuring the open circuit potential (OCP). The cell resistance between the working and quasi-reference electrodes (WE, quasi-RE) with a 4 cm separation was found through electrochemical impedance spectroscopy (EIS) to be 0.5 Ω. Cyclic voltammetry was conducted at a scan rate of 5 mV per second with a potential sweep between +1.5 V to −1.5 V versus the quasi-RE (Fig. 3a). Reported potentials depicted in voltammetric data have been corrected post-measurement by 80% of the EIS-measured solution resistance between the graphite WE and quasi-RE. At very positive potentials, the anodic current showed noise that is characteristic of fluctuating cell resistance, typically due

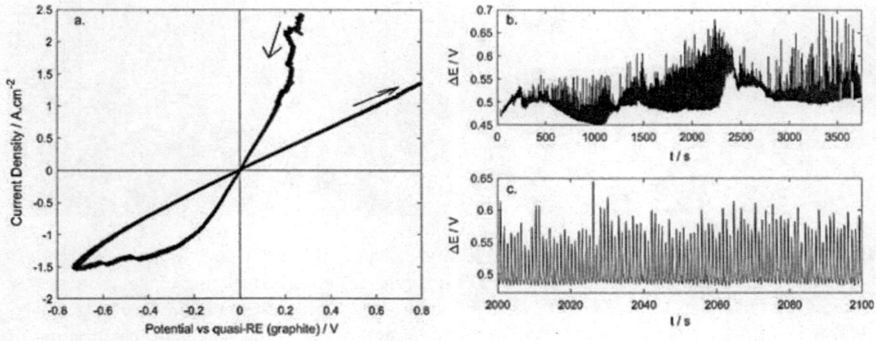

Fig. 3 Cyclic voltammetry and galvanostatic electrolysis suggest the presence of faradaic events corresponding to metal reduction and gas evolution. **a**. During cyclic voltammetry analysis of the molten $(Ni_3S_2)(CoS)_3(BaS)_{54}(La_2S_3)_{15}$ at 1500 °C at a scan rate of 5 mV/s, fluctuations during high anodic current densities suggest sulfur gas evolution, while peaks at high cathodic current densities suggest nickel and cobalt metal reduction shows a potential difference of approximately 140 mV. Reported potentials depicted in voltammetric data have been corrected post-measurement by 80% of the EIS-measured solution resistance (0.5 Ω) between the graphite working and quasi-reference electrodes. **b**. Bulk galvanostatic electrolysis of $(Ni_3S_2)(CoS)_3(BaS)_{54}(La_2S_3)_{15}$ at current densities observed to facilitate faradic events revealed recurring fluctuations in cell potential versus time, attributed changing solution resistance due to sulfur gas evolution at the anode. **c**. Fluctuations during bulk galvanostatic electrolysis of $(Ni_3S_2)(CoS)_3(BaS)_{54}(La_2S_3)_{15}$ occurred with a period of approximately 1 s

to gas evolution [14, 15, 22]. As the potential sweep moved to negative values, two current peaks are observed at current densities between −1.3 and −1.5 A/cm^2, likely corresponding to nickel sulfide and cobalt sulfide reduction. The potential difference between the two peaks was observed to be approximately 140 mV. Galvanostatic electrolysis was conducted for 3750 s at an anodic current density of 1.5 A/cm^2, corresponding to the upper bound of current density at which the two faradic events were observed during CV (Fig. 3b). As shown in Fig. 3c, the applied voltage shows rapid fluctuations with a period on the order of a second. This is attributed to changing cell resistance during electrolysis due to evolution of sulfur gas bubbles on the anode [14, 15, 22].

Post electrolysis, the electrolyte was mounted in epoxy and examined under optical and scanning electron microscopes (Fig. 4). The electrolyte was observed to wet both the alumina crucible and the graphite electrodes. Metal droplets on the order of 10–100 μm were observed in a cloud near the graphite cathode, decreasing in frequency toward the graphite anode. EDS analysis revealed the metal to have a composition of approximately 64 at.% Co, 34 at.% Ni, 2 at.% S, illustrating that cobalt was selectively reduced from mixed, equimolar (metal basis), cobalt-nickel sulfide via molten sulfide electrolysis in the barium sulfide/lanthanum sulfide supporting electrolyte. Cobalt-rich sulfide phases of approximately the same cobalt-nickel ratio, free of lanthanum and barium sulfide, were observed surrounding the reduced metal, suggesting that post-electrolysis, the reduced metal may have back-reacted with dissolved sulfur gas in the system. The presence of residual sulfur in the system

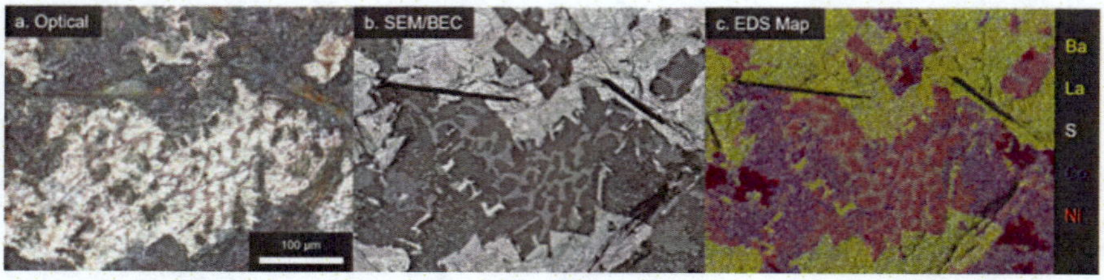

Fig. 4 Molten sulfide electrolytes support selective electrowinning of cobalt from nickel. Optical microscopy (**a**), SEM microscopy (**b**), and EDS mapping (**c**) reveal cobalt and nickel reduction at a composition of 64 at.% Co, 34 at.% Ni, and 2 at.% S (dark blue) from equimolar (metal basis) Ni-Co-S in a Ba-La-S supporting electrolyte. Sulfide regions of similar metal composition to the reduced metal (light blue) surround the cobalt-nickel alloy, suggesting that back-reaction of cobalt-nickel to sulfide may have occurred post-electrolysis during cooling and solidification. (Color figure online)

is of no surprise, considering the solubility for sulfur gas previously observed in liquid barium-lanthanum sulfide-containing systems [22]. The successful reduction of cobalt and nickel metal from the molten sulfide at current densities observed in CV to correspond with faradic events suggests the two peaks 140 mV apart observed in Fig. 3a correspond to cobalt and nickel reduction. This 140 mV difference in decomposition potential of nickel sulfide and cobalt sulfide in the barium-lanthanum sulfide suggests that molten sulfides are candidate electrolytes for selective reduction of difficult to separate metals.

Conclusions

Herein, we demonstrate that for simulated nickel-manganese-cobalt oxide cathode material, selective sulfation results in the formation of lithium sulfate, manganese oxysulfide, and cobalt-nickel sulfide. During the selective sulfidation process, manganese oxysulfide and nickel-cobalt sulfides each coalesced into single-phase regions on the order of 100 µm—1 mm in size, large enough for practical recovery via comminution and physical separation. The existence of lithium as a sulfate suggests that it may be recovered from the other insoluble metal compounds via leaching. For isolation and reduction of cobalt from the mixed nickel-cobalt sulfides, we demonstrate selective molten sulfide electrolysis at 1500 °C, employing barium-lanthanum sulfide as a supporting electrolyte. Many promising electrochemical engineering avenues exist to optimize selective electrowinning of nickel and cobalt from molten sulfides, including the employment of liquid cathodes and the optimization of sulfide electrolytes and operating conditions. These process parameters may be informed and optimized through studies of thermodynamics, mass transport, and electrochemical kinetics in molten sulfide systems. Even now, crude, un-optimized selective sulfidation and molten sulfide electrolysis present a promising avenue for lithium ion battery recycling, particularly for complicated waste streams containing nickel-manganese-cobalt oxide cathodes. Selective sulfidation as demonstrated herein suggests that oxide-sulfide anion exchange chemistry is a tenable tool for challenging metal separations. Meanwhile selective molten sulfide electrolysis facilitates process intensification by combining separation and reduction operations for difficult to separate metals, such as nickel and cobalt, into a single unit operation.

References

1. Olivetti EA, Ceder G, Gaustad GG, Fu X (2017) Lithium-Ion battery supply chain considerations: analysis of potential bottlenecks in critical metals. Joule 1:229–243
2. Harper G et al (2019) Recycling lithium-ion batteries from electric vehicles. Nature 575:75–86
3. Ciez RE, Whitacre JF (2019) Examining different recycling processes for lithium-ion batteries. Nat Sustain 2:148–156
4. Wang M, Navrotsky A (2004) Enthalpy of formation of $LiNiO_2$, $LiCoO_2$ and their solid solution, LiNi1-xCoxO$_2$. Solid State Ion 166:167–173
5. Chang K, Hallstedt B, Music D (2011) Thermodynamic and electrochemical properties of the Li−Co−O and Li−Ni−O systems. Chem Mater 24:97–105
6. Piskunen P et al (2018) Precious metal distributions in direct nickel matte smelting with Low-Cu mattes. Metall Mater Trans B Process Metall Mater Process Sci 49:98–112
7. Gaballah I, Djona M (1995) Recovery of Co, Ni, Mo, and V from unroasted spent hydrorefining catalysts by selective chlorination. Metall Mater Trans B 26B:41–50
8. Zante G et al (2020) Solvent extraction fractionation of manganese, cobalt, nickel and lithium using ionic liquids and deep eutectic solvents. Miner Eng 156:106512
9. Dhiman S, Gupta B (2019) Partition studies on cobalt and recycling of valuable metals from waste Li-ion batteries via solvent extraction and chemical precipitation. J Clean Prod 225:820–832
10. Harris CT, Peacey JG, Pickles CA (2013) Selective sulphidation and flotation of nickel from a nickeliferous laterite ore. Miner Eng 54:21–31
11. Korkmaz K, Alemrajabi M, Rasmuson Å, Forsberg K (2018) Recoveries of valuable metals from spent nickel metal hydride vehicle batteries via sulfation, selective roasting, and water leaching. J Sustain Metall 4:313–325
12. Shi J et al (2019) Sulfation roasting mechanism for spent lithium-ion battery metal oxides under SO_2-O_2-Ar Atmosphere. JOM 71:4473–4481
13. Stinn C, Allanore A (2020) Estimating the capital costs of electrowinning processes. Electrochem Soc Interface 29:44–49
14. Sokhanvaran S, Lee S-K, Lambotte G, Allanore A (2016) Electrochemistry of molten sulfides: copper extraction from BaS-Cu2S. J Electrochem Soc 163:D115–D120
15. Sahu SK, Chmielowiec B, Allanore A (2017) Electrolytic extraction of copper, molybdenum and rhenium from molten sulfide electrolyte. Electrochim Acta 243:382–389
16. Stinn C, Nose K, Okabe T, Allanore A (2017) Experimentally determined phase diagram for the barium sulfide-copper(i) sulfide system above 873 K (600 C). Metall Mater Trans B 48:2922–2929
17. Afanasiev P et al (2000) Preparation of the mixed sulfide Nb2Mo3S10 catalyst from the mixed oxide precursor. Catal Lett 64:59–63
18. Meyer B (1976) Elemental sulfur. Chem Rev 76:367–388
19. Jacob KT (1980) Isothermal section of the Ni-Co-S phase diagram at 1273 K. Metall Trans B 11B:640–643
20. Gallios GP, Kyzas GZ, Matis KA (2019) Flotation in the 2010s: focus on mineral processing. Adv Low-Cost Sep Tech Interface Sci 30:43–68. (Elsevier B.V., 2019)

21. Kyzas GZ, Lazaridis NK, Matis KA (2019) Flotation: recent innovations in an interesting and effective separation process. Adv Low-Cost Sep Tech Interface Sci 30:15–42. (Elsevier B.V., 2019)
22. Chmielowiec BJ (2019) Electrochemical engineering considerations for gas evolution in molten sulfide electrolytes. Massachusetts Institute of Technology

Additive Manufacturing of 3D Microlattice Lithium-Ion Battery Electrodes: A Review

Modupeola Dada and Patricia Popoola

Abstract Lithium-ion batteries are one of the most desirable energy storage systems which consist of two electrodes and an electrolyte. These batteries are currently restricted to consumer electronics attributed to the low performance of the cathode and/or anode electrode, and the limitations of the electrode's manufacturing process. Li-ion battery electrodes are widely manufactured using the slurry casting technique. However, the organic solvent used in slurry casting increases the manufacturing costs while electrodes fabricated can only be used for portable electronics and not for large hybrid automobiles. Additive manufacturing is a solvent-free technique which prints large complex geometries, therefore, producing porous electrodes through interdigitated geometries and more recently through three-dimensional micro-lattice structures. This manufacturing process can fabricate Ni-Co based high entropy oxide electrodes which are deposited directly with uniform distribution of particles. Electrodes fabricated with 3D micro-lattice structures are optimal, having controlled porosity which improves the overall Li-ion battery capacity.

Keywords Li-ion batteries · Electrodes · Slurry casting · Additive manufacturing · Energy storage systems · High entropy oxides

Introduction

The evolution from environmentally unfriendly fossil fuels to renewable energy supply was significant in mitigating the climate changes caused by the emission of CO_2 on a long-term basis [1]. Höök and Tang [2] did a thorough review of the relationship between the climate effect of fossil fuels and its depletion. The authors concluded that the limitation to the accessibility to fossil fuels will also limit the

M. Dada (✉) · P. Popoola
Chemical, Metallurgical and Materials Engineering, Tshwane University of Technology, Pretoria, South Africa
e-mail: dadadupeola@gmail.com

P. Popoola
e-mail: PopoolaAPI@tut.ac.za

C. Anderson et al. (eds.), *Ni-Co 2021: The 5th International Symposium on Nickel and Cobalt*, The Minerals, Metals & Materials Series,
https://doi.org/10.1007/978-3-030-65647-8_8

ability of mankind to affect climate changes. Nonetheless, meeting the need for both grid storage systems and cleaner sources of energy requires the expansion of alternative methods such as solar photovoltaics, hydroelectric, natural gas, wind and new battery technologies [3]. Johansson et al. [4] assessed the economic and technical potential of using renewable energy in producing electricity. According to Cho et al. [5], batteries as an electrochemical storage system is one of the most preferred storage devices which convert chemical energy into electricity. This is attributed to their efficiency, discharge time, scalability and weight. They consist of negative and positive electrodes soaked with a dissociated salt solution as an electrolyte [6]. There are lithium-ion and lithium-sulfur batteries; the latter is cheap and has theoretical specific energy about ten times that of the lithium-ion; however, lithium-sulfur batteries have an intermediate discharge of lithium polysulfide which dissolves and damages the electrodes [7]. Cerdas et al. [8] also reported that the lithium-sulphur batteries have a relatively low volumetric energy density which limits its applications in electric vehicles. Lithium-ion batteries, on the other hand, are versatile and self-discharge at a very low rate. They are referred to as rocking chair batteries and are the most desirable energy storage systems for rechargeable portable applications such as mobile phones, laptops and cameras. Vetter et al. [9] stated that lithium-ion is the most promising battery for stationary energy storage systems and a power source for hybrid electric vehicles because the battery is introduced into a two-dimensional layer of TiS_2 and graphite hosts which increase its capacity for long travel distances [10].

Nonetheless, lithium-ion batteries are susceptible to supply and resource risks, there is a great need to increase their energy densities thus, they are restricted to consumer electronics and are currently not widely used for large hybrid and electric vehicles [11]. This restriction is also attributed to the low performance of the cathode/anode material; the volume expansion experienced during cycling and the limitations of the electrode's manufacturing processes [12]. Therefore, without sacrificing the battery's efficiency, the energy density needs to be enhanced to expand the application of Li-ion batteries, and this can be achieved through the manufacturing technique. In literature, commercial Li-ion battery electrodes are manufactured using the slurry casting technique [13, 14]. This conventional method of fabrication involves the mixing of the electrode material with a solvent to create a slurry. Nonetheless, the organic solvent used in slurry casting increases the manufacturing costs and the electrodes fabricated can only be used for portable electronics. Choi et al. [15] explained that the poor interfacial strength between the electrolyte and the active materials limits the battery performance of slurry cast composite electrodes.

Additive manufacturing is a dry powder and solvent-free technique which prints complex geometries, therefore, producing high-density porous electrodes through interdigitated geometries and more recently through three-dimensional micro-lattice structures [16]. This manufacturing process can fabricate electrode materials which are dry mixed and deposited directly with a uniform distribution of particles [17]. Electrodes fabricated with 3D micro-lattice structures are optimal, having controlled porosity which improves the overall Li-ion battery capacity and the charge or discharge rate of the battery. Lithium through this structure utilizes the electrodes by penetrating through the electrode volume, increasing its energy storage capacity.

Energy Storage Classifications

In recent times, systems such as computer innovations, vehicles and mobile phone technologies have their primary source of power supply as electrical energy. Other sources of energy include gas, Hydro River and dam, biogas, coal and lignite, yet static electricity is the most preferred for its availability, reliability and flexibility [18]. Static electricity occurs temporarily when objects are rubbed together; however, these can cause repulsive and attractive forces. According to Guney et al. [19], the fluctuations in electricity demand energy storage systems which can continuously separate positive and negative charges to extract energy for more potential applications, consequently, the major shortcomings of electricity usage are electrical storage systems. A secondary battery characterized by an open-circuit voltage, charge capacity, shelf life, cycle life, service life, energy and power density uses chemical energy to separate positive and negative charges to generate power and is currently the most efficient storage system amongst other electrical storage systems like pumped hydro systems, compressed air, flywheel and superconducting magnetic systems [20].

Li-Ion Battery

The transport industry with extreme economic impact comprises the maritime, aviation and land transport. Nonetheless, land transportation contributing to over 60% of the total output, having millions of vehicles on the road means these vehicles also consume billions of litres of fuel. Road transportation, therefore, is the largest source of methane, nitrous oxide, carbon dioxide, water vapour and ozone which makes up greenhouse gases through the heavy consumption of fossil fuels. This fuel not only steadily increases in the cost, but it also has several environmental challenges detrimental to humanity [21].

Therefore, targeted research activities on innovative alternatives to fuel consumption and solutions to air pollution have made the demand for electric, hybrid and plug-in hybrid vehicles increase significantly. Hybrid and plug-in hybrid vehicles have two storage units; fuel and electricity, electric vehicles only run on batteries and/or grid electricity. These vehicles have high efficiency, low noise, excellent drive control, fault tolerance and acceleration with the electromotor's ability to function as a generator. However, these vehicles are challenged by their travel range and this can be attributed to their storage capacity [22].

The battery pack in most electric vehicles is a significant powerhouse characterized by a high power and energy density, safety and low cost while experiencing several cycles in the lifespan of the vehicle. The battery comprises electrochemical cells; positive and negative electrodes connected by an electrolyte which converts chemical energy to electricity. In recent times, Lead-acid, Nickel-Metal Hydride, Lithium-Sulphur, Molten Salt and Lithium-ion batteries have been used for electric vehicles; however, most of these batteries have very low specific energy and are engineered for short distances. Although the consumption of energy by the vehicle depends on the road conditions, the vehicle body shape, size and auxiliary systems; potential for longer travel ranges are met using the Lithium-ion battery [23].

The lithium-ion battery is found in most consumer electronics, however, in recent times, it is used in automobiles and the Chevrolet volt automobile is being run using the lithium battery because the battery has potential for applications that require lightweight, high energy densities and operating voltage [24]. They have high reversibility and discharge rates with no memory effects. They can have either high energy density or high power density, but not both. The energy density shows the quantity of energy in litres or kilogram, while the power density shows how the energy is released through the discharge per time in litres or kilogram. For automobiles, a high energy density Lithium-ion battery is required because it guarantees longer travel distances. However, the Lithium-ion cells determine how operational the battery will be. Like most batteries, the Lithium-ion battery comprises electrolytic cells that have these components; an electrolyte, two electrodes and a separator. The electrodes consist of an anode made up of metal oxides like Ni-Co based high entropy oxide or graphite and a cathode made up of any composite material like lithium metal oxide ($LiCoO_2$, $LiMn_2O_4$, and $LiFePO_4$) [25].

The progress in the innovation of high entropy alloys using the concept of high entropy stabilization has led to the development of oxide systems. High entropy oxides characterized by their entropy stabilization effect are compounds consisting of several metal cations which form a single-phase structure with excellent properties especially for energy storage applications. Their compositional system consisting of at least five cations in equal or near-equal concentrations invariably leads to a single-phase fluorite, perovskite or rock-salt structures with high Lithium-ion conductivity [26].

Wang et al. [27] explored the potential applications of $Co_{0.2}Cu_{0.2}Mg_{0.2}Ni_{0.2}Zn_{0.2}$ high entropy oxide cell for the anode and $LiNi_{1/3}Co_{1/3}Mn_{1/3}O_2$ high entropy oxide for the cathode. The authors reported that the oxides showed high power and specific energy densities with stable cycles. Qiu et al. [28] studied the $Mg_{0.2}Co_{0.2}$ $Ni_{0.2}Cu_{0.2}$ $Zn_{0.2}O$ high entropy oxide as Lithium-ion battery anode material and they reported that the oxide had superior stable cycles with high specific energy density with stable microstructure and surface morphology. Lökçü et al. [29] fabricated $(MgCoNiZn)_{1-x}$ Li_xO high entropy oxide to investigate their electrochemical performance as lithium-ion anode battery materials with ($x = 0.05$,0.25, 0.35 and 0.15). They recommended the use of high entropy oxides; however, the report stated that the synthesis of high entropy oxides with high oxygen vacancies has the potential for enhanced electrochemical performance.

There are rare-earth, mixed and transition-metal based high entropy oxides. Transition-metal based oxides especially the Ni-Co based on high entropy oxides are widely used for lithium-ion batteries because they can interpolate lithium ions without changing the structure. They display enhanced characteristics; specific capacity attributed to Nickel and kinetics attributed to cobalt. They have excellent electrical properties and they are stable with the electrolyte. Their conversion reaction in the anode makes the process of multi-electron redox, thus ensuring the potential for enlarged capacity. Wang et al. [30] synthesized high entropy oxides and they described how the lithiated rock-salt structure of the oxides had great potential for positive and negative electrodes. In another study, Wang et al. [31] demonstrated how the entropy stabilization of Ni-Co based high entropy oxides leads to the formation of lithiated materials which facilitates oxyfluorides that enhances the energy storage capacity of the material and the cycling performances. These oxides ensure safety by lowering the risk of short-circuiting attributed to their capacity to get to high voltages compared to graphite anodes. For this reason, Lithium transition-metal Ni-Co high entropy oxides are also being studied for their potential as cathode materials [32]. However, these oxides have been reported to lack the needed efficiency attributed to the processing route [33].

Other types of batteries are:

Battery	Advantages	Disadvantages
Lead-Acid	Low cost of maintenance, readily available, durable, technologically advanced	Consumes too much space, difficulty in recycling and/or disposal, low energy density, low storage capacity
Zinc Air Cells	Low cost, very high energy density, environmentally friendly	Low storage capacity, Large charging cycles required
Nickel Metal Hydride	Good lifespan, High energy density	Unstable cycles, heavy, self-discharges, efficiency is weather dependent
Sodium Nickel Chloride	Has good series resistance, Technologically advanced, good energy density, weather independent, large charge cycles, environmentally friendly	Low power density, high costs, poor storage capacity, limited in usage
Sodium Sulphur	Cost-effective, availability of raw materials	Environmentally unfriendly, insufficient power electrolyte, self-discharges

Processing Routes of Lithium-Ion Electrodes

Lithium-ion electrodes are characterized by their electrochemical performance, crystallinity, particle size, surface area and morphology and this is greatly influenced by their fabrication techniques.

Slurry Casting

Commercial battery electrodes for lithium-ion are extensively fabricated by the slurry casting method. The electrodes consist of an anode and cathode made up of electrochemically active materials, polymer binder and conductive additive. The slurry is derived from mixing the electrode materials with a solvent. The solvent used is dependent on the binder, which influences how the binder is uniformly dispersed or dissolved [15]. The mixture of the solvent and the binder is coated to a current collector using a slot-die; afterwards, the electrodes are dried using ovens that help increase porosity by evaporating the solvent with higher speed. This fabrication process which can be aqueous-based or solvent-based is meant to guarantee excellent mechanical strength and conductivity. However, N-methyl-2-pyrrolidine used in the solvent-based slurry casting is expensive and the material can be toxic. More so, the rheological properties needed in aqueous slurries cannot be obtained without additional additives to get uniformity in the system, and these additives complicate the whole fabrication process [34]. Additives are responsible for increased surface tension which decreases the wettability of the slurry. The increased surface tension makes drying the slurries very difficult and sometimes the electrodes may have defects while drying. Some researchers have reported using co-solvents to reduce the surface tension; however, this didn't change the need for a solvent recovery during fabrication of the electrodes [35, 36].

Electrostatic Spray Deposition

This fabrication technique is a non-vacuum, cost-effective method that has a high deposition efficiency and low deposition temperature. Electrodes fabricated using this technique is fractal-like, sponge-line porous and dense. The electrochemical performance of electrodes can be optimized by the control of electrodes structuring through a reduction in the ion diffusion pass way, volume change buffering and the increment in the surface area between the electrolyte and the electrode [37]. The electrostatic spray deposition process is assembled with a nozzle which is connected to a syringe, and through the syringe pump, the precursor solution is supplied. A high voltage DC power supply which is used to generate an electrostatic force is placed between a substrate and the nozzle. The atomized liquid droplets are charged to form an aerosol which is deposited on the substrate to design the film electrode. The films fabricated using this technique compared to other deposition processes show promising properties for lithium-ion battery applications, yet they require too much energy and production time. Consequently, there is a need for alternative manufacturing processes in the fabrication of lithium-ion electrodes [38].

Additive Manufacturing

To eliminate the cost of solvents, the removal and recovery solvent process and a reduction in manufacturing time, a solvent-free manufacturing route is a preferred alternative to fabricating electrodes used for lithium battery applications. The revolution of technological advances in the fabrication of electrodes brought about three-dimensional printing processes which are dry processes of producing electrodes with complex structures, lightweight, time and cost-effective [16]. Thus, using any of its classifications; extrusion, sheet lamination, direct laser deposition, material jetting, binder jetting or powder bed fusion, the principle is the same. A computer-aided design constructs the material in an orderly layered fashion with 15–500 micron thickness which affords minimal to zero post-processing. These give additive manufacturing greater advantages over other fabrication routes and it has been reported to give properties to an electrode that shows better performance in batteries [39].

Pulsed laser deposition is a versatile additive manufacturing technique which uses an energy source located outside an ultrahigh vacuum chamber in preparing film electrodes through the ablation of targets at 45° illuminated by a pulsed laser beam. The ions and atoms ablated from the targets are deposited unto a substrate which is attached to the surface parallel to the target's surface at a 2–10 cm distance [40]. The dynamic nature of the process parameters makes this technique versatile in the preparation of electrodes, oxides, carbides, polymers and nitrides. Nonetheless, this technique cannot be used for large-scale production of electrodes due to its porosity [41].

Three-dimensional lattice structures are innovative ways of fabricating electrodes that are porous, lightweight without sacrificing the stability of the material. They produce geometrically high complex parts using a laser with minimal waste or material consumption. Micro lasers sintering of lattice electrode structures guarantee no design limitations. They fabricate random systematic open-cell lattice structures with very high resolution, reduced relative density, enhanced mechanical strength with an ordered pore distribution and interconnection [41]. Maloney et al. [42] used 3D micro-lattice structures in fabricating a heat exchanger for aerospace and automotive applications. Wu et al. [43] reviewed the potential of micro-lattice structures in fabricating photon devices. Micro-lattice structures possess heat transfer proficiencies, vibration control and energy absorption capabilities, hence, they are widely used for several applications; however, recently, they have been applied to the fabrication of lithium-ion electrodes using stereolithography. Nonetheless, the use of selective laser melting (SLM), a rapid prototyping additive manufacturing process in fabricating 3D micro-lattice lithium-ion high entropy oxide electrodes is limited.

Selective laser melting is a powder bed fusion process via a laser beam using a computer-aided design data input. This process fabricates micro-lattice structures with excellent properties and structure performance [44]. Ni-Co based high entropy oxides can be fabricated by electroplating the oxides on a polymer-based micro-lattice structure. To avoid imperfections, the process parameters such as scan speed, hatch speed, laser power, scanning direction, hatch overlap and hatch style are optimized.

It is as versatile as the pulsed laser beam process, producing a wide range of oxide electrodes. Campanelli et al. [45] fabricated Ti6Al4V lattice components using SLM and they reported the alloy having enhanced properties. Li et al. used finite element stimulation to study the deformation behaviour of stainless steel micro-lattice structures fabricated using SLM and they also recorded having enhanced the performance of the steel. Therefore, 3D micro-lattice structures via selective laser melting have great potential in fabricating high entropy oxides for lithium-ion battery electrodes applications with improved storage performance.

Conclusion

This study presents high entropy oxides as potential materials for lithium-ion battery electrode applications and additive manufacturing of solvent-free high entropy oxide electrodes with 3D micro-lattice structures via selective laser melting as a potential manufacturing route for large energy storage systems in hybrid/electric automobiles.

Acknowledgements The authors appreciate the National Laser Center of the Council for Scientific and Research (CSIR) and Tshwane University of Technology, Pretoria, South Africa.

References

1. Johnsson F, Kjärstad J, Rootzén J (2019) The threat to climate change mitigation posed by the abundance of fossil fuels. Clim Polic 19:258–274
2. Höök M, Tang X (2013) Depletion of fossil fuels and anthropogenic climate change—a review. Energy Polic 52:797–809
3. Panwar N, Kaushik S, Kothari S (2011) Role of renewable energy sources in environmental protection: a review. Renew Sustain Energy Rev 15:1513–1524
4. Johansson TB, Kelly H, Reddy Ak, Williams RH (1993) Renewable energy: sources for fuels and electricity
5. Cho J, Jeong S, Kim Y (2015) Commercial and research battery technologies for electrical energy storage applications. Prog Energy Combust Sci 48:84–101
6. Oudalov A, Chartouni D, Ohler C, Linhofer G (2006) Value analysis of battery energy storage applications in power systems. in 2006 IEEE PES Power Systems Conference and Exposition, pp 2206–2211
7. Qiu X, Hua Q, Zheng L, Dai Z (2020) Study of the discharge/charge process of lithium–sulfur batteries by electrochemical impedance spectroscopy. RSC Adv 10:5283–5293
8. Cerdas F, Titscher P, Bognar N, Schmuch R, Winter M, Kwade A et al (2018) Exploring the effect of increased energy density on the environmental impacts of traction batteries: a comparison of energy optimized lithium-ion and lithium-sulfur batteries for mobility applications. Energies 11:150
9. Vetter J, Novák P, Wagner MR, Veit C, Möller K-C, Besenhard J et al (2005) Ageing mechanisms in lithium-ion batteries. J Power Sour 147:269–281
10. Li M, Lu J, Chen Z, Amine K (2018) 30 years of lithium-ion batteries. Adv Mater 30:1800561
11. Nayak PK, Yang L, Brehm W, Adelhelm P (2018) From lithium-ion to sodium-ion batteries: advantages, challenges, and surprises. Angew Chem Int Edn 57:102–120

12. Schipper F, Erickson EM, Erk C, Shin J-Y, Chesneau FF, Aurbach D (2016) Recent advances and remaining challenges for lithium ion battery cathodes. J Electrochem Soc 164:A6220
13. Klavetter KC, Wood SM, Lin Y-M, Snider JL, Davy NC, Am Chockla et al (2013) A high-rate germanium-particle slurry cast Li-ion anode with high Coulombic efficiency and long cycle life. J Power Sour 238:123–136
14. Ludwig B, Zheng Z, Shou W, Wang Y, Pan H (2016) Solvent-free manufacturing of electrodes for lithium-ion batteries. Sci Rep 6:23150
15. Choi S-J, Choi S-H, Bui AD, Lee Y-J, Lee S-M, Shin H-C et al (2018) LiI-doped sulfide solid electrolyte: enabling a high-capacity slurry-cast electrode by low-temperature post-sintering for practical all-solid-state lithium batteries. ACS Appl Mater Interfaces 10:31404–31412
16. Pang Y, Cao Y, Chu Y, Liu M, Snyder K, MacKenzie D et al (2020) Additive manufacturing of batteries. Adv Funct Mater 30:1906244
17. Azhari A, Marzbanrad E, Yilman D, Toyserkani E, Pope MA (2017) Binder-jet powder-bed additive manufacturing (3D printing) of thick graphene-based electrodes. Carbon 119:257–266
18. Krishan O, Suhag S (2019) An updated review of energy storage systems: classification and applications in distributed generation power systems incorporating renewable energy resources. Int J Energy Res 43:6171–6210
19. Guney MS, Tepe Y (2017) Classification and assessment of energy storage systems. Renew Sustain Energy Rev 75:1187–1197
20. Budde-Meiwes H, Drillkens J, Lunz B, Muennix J, Rothgang S, Kowal J et al (2013) A review of current automotive battery technology and future prospects. Proc Inst Mech Eng Part D J Automob Eng 227:761–776
21. Corbett JJ, Winebrake J (2008) The impacts of globalisation on international maritime transport activity. In: Global forum on transport and environment in a globalising world, pp 1–31
22. Nordelöf A, Messagie M, Tillman A-M, Söderman ML, Van Mierlo J (2014) Environmental impacts of hybrid, plug-in hybrid, and battery electric vehicles—what can we learn from life cycle assessment? Int J Life Cycle Assess 19:1866–1890
23. Ding Y, Cano ZP, Yu A, Lu J, Chen Z (2019) Automotive Li-ion batteries: current status and future perspectives. Electrochem Energy Rev 2:1–28
24. Zeng X, Li M, Abd El-Hady D, Alshitari W, Al-Bogami AS, Lu J et al (2019) Commercialization of lithium battery technologies for electric vehicles. Adv Energy Mater 9:1900161
25. Harper G, Sommerville R, Kendrick E, Driscoll L, Slater P, Stolkin R et al (2019) Recycling lithium-ion batteries from electric vehicles. Nature 575:75–86
26. Sarkar A, Velasco L, Wang D, Wang Q, Talasila G, de Biasi L et al (2018) High entropy oxides for reversible energy storage. Nature Commun 9:1–9
27. Wang Q, Sarkar A, Li Z, Lu Y, Velasco L, Bhattacharya SS et al (2019) High entropy oxides as anode material for Li-ion battery applications: a practical approach. Electrochem Commun 100:121–125
28. Qiu N, Chen H, Yang Z, Sun S, Wang Y, Cui Y (2019) A high entropy oxide (Mg0. 2Co0. 2Ni0. 2Cu0. 2Zn0. 2O) with superior lithium storage performance. J Alloys Comps 777:767–774
29. Lökçü E, Toparli C, Anik M (2020) Electrochemical performance of (MgCoNiZn) 1-xLixO high entropy oxides in lithium ion batteries. ACS Appl Mater Interfaces
30. Wang J, Stenzel D, Azmi R, Najib S, Wang K, Jeong J et al (2020) Spinel to rock-salt transformation in high entropy oxides with Li incorporation. Electrochem 1:60–74
31. Wang Q, Sarkar A, Wang D, Velasco L, Azmi R, Bhattacharya SS et al (2019) Multi-anionic and-cationic compounds: new high entropy materials for advanced Li-ion batteries. Energy Environ Sci 12:2433–2442
32. Zheng Y, Yi Y, Fan M, Liu H, Li X, Zhang R et al (2019) A high-entropy metal oxide as chemical anchor of polysulfide for lithium-sulfur batteries. Energy Storage Mater 23:678–683
33. Sarkar A, Wang Q, Schiele A, Chellali MR, Bhattacharya SS, Wang D et al (2019) High-entropy oxides: fundamental aspects and electrochemical properties. Adv Mater 31:1806236
34. Bauer W, Nötzel D (2014) Rheological properties and stability of NMP based cathode slurries for lithium ion batteries. Ceram Int 40:4591–4598

35. Li J, Lu Y, Yang T, Ge D, Wood III DL, Li Z (2010) Water-based electrode manufacturing and direct recycling of lithium-ion battery electrodes—a green and sustainable manufacturing system. iScience, 101081
36. Li J, Daniel C, An SJ, Wood D (2016) Evaluation residual moisture in lithium-ion battery electrodes and its effect on electrode performance. MRS Adv 1:1029–1035
37. ávan der Put PJ (1996) Morphology control of thin LiCoO 2 films fabricated using the electrostatic spray deposition (ESD) technique. J Mater Chem 6:765–771
38. Chen C, Kelder E, Jak M, Schoonman J (1996) Electrostatic spray deposition of thin layers of cathode materials for lithium battery. Solid State Ion 86:1301–1306
39. Cobb CL, Ho CC (2016) Additive manufacturing: rethinking battery design. Electrochem Soc Interface 25:75
40. Trzciński K, Szkoda M, Gazda M, Karczewski J, Cenian A, Grigorian GM et al (2020) Pulsed laser deposition of bismuth vanadate thin films—the effect of oxygen pressure on the morphology, composition, and photoelectrochemical performance. Materials 13:1360
41. Schälicke G, Landwehr I, Dinter A, Pettinger K-H, Haselrieder W, Kwade A (2020) Solvent-free manufacturing of electrodes for lithium-ion batteries via electrostatic coating. Energy Technol 8:1900309
42. Maloney KJ, Fink KD, Schaedler TA, Kolodziejska JA, Jacobsen AJ, Roper CS (2012) Multi-functional heat exchangers derived from three-dimensional micro-lattice structures. Int J Heat Mass Transf 55:2486–2493
43. Wu S, Serbin J, Gu M (2006) Two-photon polymerisation for three-dimensional micro-fabrication. J Photochem Photobiol A Chemi 181:1–11
44. Schleifenbaum H, Meiners W, Wissenbach K, Hinke C (2010) Individualized production by means of high power selective laser melting. CIRP J Manuf Sci Technol 2:161–169
45. Campanelli SL, Contuzzi N, Ludovico AD, Caiazzo F, Cardaropoli F, Sergi V (2014) Manufacturing and characterization of Ti6Al4V lattice components manufactured by selective laser melting. Materials 7:4803–4822

A Strategy for Acid-Free Waste Lithium Battery Processing

Mark L. Strauss, Luis Diaz Aldana, Mary Case, and Tedd Lister

Abstract The current methods for the extraction of cobalt, lithium, nickel, and manganese from waste lithium-ion batteries require reagents such as HCl, $H2SO_4$, HNO_3 and excess of a reductants such as of hydrogen peroxide. This work provides a new strategy for metal recovery and impurity removal without the use of mineral acids, bases or discrete reductants. This study is an alternative strategy using unique pyrometallurgical method that may be more environmentally sustainable because reagent consumption may be less, fewer toxic reagents are utilized, and fewer unit operations are employed. The effects of temperature, time, and reagent concentration were studied upon the recovery of cobalt, lithium, nickel, iron, and aluminum.

Keywords Critical materials · Recycling · Lithium-ion battery · Acid-free

Introduction

To prepare for a circular economy, the electricity grid requires critical materials to form the backbone of the operations for backup energy systems, wind-turbines, electric motors, and batteries—which are not available by primary production in the United States [1]. Therefore, developing a local source of these metals is important to minimize supply constraints. Recycling lithium-ion batteries, magnets, waste fluorescent scrap could be a source of these critical materials. However, the current industrial methods that recover critical materials from electronic waste (E-waste) streams such as lithium batteries require excess quantities of expensive, toxic reagents, generate non-reusable waste, and require several unit operations in order to recover metals. The demand for lithium used in batteries for electric vehicle cars is expected to drive the production fivefold to 1.3 million metric tons of lithium carbonate equivalent (LCE) in 2025 [2]. The world demand for cobalt is expected to rise to 325,000

M. L. Strauss (✉) · L. D. Aldana · M. Case · T. Lister
Idaho National Laboratory, Biological and Chemical Processing Department, Idaho Falls, Idaho 83415, USA
e-mail: mark.strauss@inl.gov

C. Anderson et al. (eds.), *Ni-Co 2021: The 5th International Symposium on Nickel and Cobalt*, The Minerals, Metals & Materials Series,
https://doi.org/10.1007/978-3-030-65647-8_9

metric tons by 2030 and new resources such as waste lithium battery cathode scrap are a logical source for current and future cobalt supply [3].

The development of a fast, cost-efficient, and more environmentally benign extraction processes for lithium and cobalt recovery from waste lithium-ion battery (LIB) cathodes is a major challenge for present and future lithium battery recycling operations. From the literature, the challenges of the common methods for LIB leaching include long leaching time, low recoveries, reducing agents for cobalt (III) reduction, high waste generation, and low pulp densities[4]. Larouche et al. 2020 charts the various methods for LIB cathode extraction from previous studies as well as the pros and cons of each method [4]. In general, the methods requiring mineral acids such as sulfuric acid require up to 50% excess iron (II) sulfate or hydrogen peroxide reductants [5]. Adding iron impurities to the leach solution adds complexity during purification. Diaz-Aldana et al. 2020 reported an electrochemical approach which used much lower than stochiometric quantities of iron sulfate, where iron (II) is continuously regenerated at the cathode [6]. Alternatively, American Manganese has patented a unique process using sulfuric acid and rechargeable sulfur dioxide as the reductant with success at the pilot scale [7, 8]. Alternatively, organic acids combined with a reductant have been shown to recover metals from LIB cathodes—albeit with low pulp densities and long leaching times [9]. One favorable example is methanesulfonic acid has had high recoveries and high pulp densities [10]. Other methods such as deep-eutectic solvents have shown success, but with low pulp densities [11].

Originally, solid-state chlorination was developed to extract rare earth ore in China [12]. In addition, solid-state chlorination has been demonstrated to be effective at rare earth extraction from waste fluorescent lamps, Nd-Fe-B hard drive scrap, and samarium cobalt magnet swarf [13–17]. Depending on the study, there are several possible overall reactions by which critical materials are recovered. Lorenz et al. 2019 & 2020 suggests that decomposing ammonium chloride into hydrogen chloride gas (between 225 and 325 °C), over a long period of time, reacts with rare earths (RE) to form rare earth chlorides as in Eqs. (1) and (2) from either NdFeB or SmCo magnets [16, 17]. The optimized condition for RE recovery NdFeB chlorination was 146 min at 285 °C, NH_4Cl:NdFeB weight ratio of 3:1 and for SmCo was 76 min at 325° C, NH_4Cl:magnet weight = 3:1.

$$NH_4Cl \rightarrow HCl_{gas} + NH_{4gas} \tag{1}$$

$$RE_{(s)} + 3HCl_{(g)} \rightarrow RECl_{3(s)} + \frac{3}{2}H_2 \tag{2}$$

A similar equation from Lorenz et al. 2015 was proposed for rare earth oxides in waste fluorescent lamps was optimized for a slightly different temperature and time (85 min, 285 °C, mass ratio 3:1) [13].

$$Eu_2O_{3(s)} + 6HCl_{(g)} \rightarrow 2EuCl_{3(s)} + 3H_2O \tag{3}$$

However, a different overall equation is indicated for recovering metals from LIB cathodes. Qu et al. suggests, based on the Ellingham Diagram, HSC optimized reaction conditions, that most dominant reaction in regard to lithium-ion battery cathodes is the following equation in (4) [18].

$$NH_4Cl + \frac{1}{3}LiCoO_2 + \frac{2}{3}O_2 \rightarrow \frac{1}{3}CoCl_2 + \frac{1}{3}LiCl + \frac{1}{2}N_2 + 2H_2O \quad (4)$$

In Eq. (4), nitrogen is oxidized, and cobalt and oxygen are reduced. In essence, ammonium chloride acts as a solid-state leachant and reductant. In addition, the recommended ammonium chloride to lithium cobalt oxide weight ratio is 2:1 (1.6:1 is stoichiometric) and the recovery of lithium is 99% and cobalt is 90% at 300 ºC. This information suggests a 22% excess ammonium chloride to optimize metal recovery. The paper suggests, based on the overall data, that the mechanism is solid–solid reaction rather than a solid–gas reaction [18]. The pretreatment phase involves planetary ball milling for 30 min. Finally, these halides are easily dissolved in water and recovered.

Finally, the economics of solid-state chlorination are favorable. A total reagent cost comparison of ammonium chloride leaching versus hydrochloric acid leaching was made for NdFeB magnets [16]. At the time of the article, ammonium chloride costed 133 euros per tonne while 35% hydrochloric acid was 215 euros per ton [16]. As a comparison, to leach, one tonne of magnet scrap requires a significantly greater quantity of material hydrochloric acid than ammonium chloride chlorination. As a result, reagent costs are 51% more in the traditional leaching process for NdFeB magnets versus solid-state chlorination [16]. Similarly, a comparison was made with regard to sulfuric acid SmCo magnet leaching—which does not consider additional processing agents needed to precipitate the double sulfate salt. Actually, traditional sulfuric acid leaching was cheaper in this case [17]. However, the ammonia generated from ammonium chloride decomposition could be captured as a 25% ammonia high purity saleable product and sold at 352 euros per tonne [17]. The value of high purity ammonia causes the overall reagent cost including ammonia sales to less than zero, but absent producing a saleable ammonia product, the economics for ammonium chloride leaching are less favorable than traditional acid leaching [17]. Although a similar economic comparison was not made with regard to lithium cobalt oxide peroxide-assisted leaching versus solid-state chlorination, the reagent cost comparison could give a similar result [16, 18]. Similarly, all three studies demonstrated that a continuous rotary kiln could be used to scale up the processes. In addition, it is possible that disposal cost may be higher for traditional leaching due to the higher costs of disposal of unreacted and unrecovered hydrochloric acid.

Ultimately, there is information to demonstrate that the solid-state chlorination of LIB cathodes could be fast (20 min), more economic (less reagent costs), more benign (less toxic reagents), have high recoveries (99% Li, 90% Co), high pulp densities (chlorides have high solubility), and requires no additional reductants versus standard leaching techniques. As a result, this method should be studied more at depth with more LIB cathode types, and a complete economic analysis should be conducted in order to compare leaching costs.

References

1. Bauer D, David D, Mckittrick M (2011) U.S. Department of energy critical materials strategy. https://doi.org/10.1017/CBO9781107415324.004
2. Ellsmoor J (2019) Electric vehicles are driving demand for lithium—with environmental consequences. Forbes
3. Administrator (2012) Analyzing China's heavy rare earth element reserves (Part 1): ion-absorbed clays & illegal mining. 2. https://investorintel.com/technology-metals-intel/analyzing-chinas-heavy-rare-earth-element-reserves-part-1-ion-absorbed-clays-illegal-mining/
4. Larouche F et al (2020) Progress and status of hydrometallurgical and direct recycling of Li-ion batteries and beyond. Materials (Basel) 13
5. Wohlgemuth D, Schneider M, Spielau R, Willems J, Steinbild M (2017) Method for the hydrometallurgical recovery of lithium, nickel and cobalt from the lithium transtion metal oxde-containing fraction of used galvanic cells, vol 2
6. Diaz LA et al (2020) Electrochemical-assisted leaching of active materials from lithium ion batteries. Resour Conserv Recycl 161:104900
7. Warkentin DD, Crystal N (2019) Processing of cobaltous sulphate/dithionate liquors derived from cobalt resource, vol 1
8. American Manganese (2019) American manganese Inc. achieves improved pilot plant results of 99.93% purity from recovered NMC Cathode material. https://americanmanganeseinc.com/american-manganese-inc-achieves-improved-pilot-plant-results-of-99-93-purity-from-recovered-nmc-cathode-material/
9. Li L et al (2013) Recovery of metals from spent lithium-ion batteries with organic acids as leaching reagents and environmental assessment. J Power Sources 233:180–189
10. Wang B et al (2019) Recycling LiCoO2 with methanesulfonic acid for regeneration of lithium-ion battery electrode materials. J Power Sources 436:226828
11. Peeters N, Binnemans K, Riaño S (2020) Solvometallurgical recovery of cobalt from lithium-ion battery cathode materials using deep-eutectic solvents. Green Chem 22:4210–4221
12. Shi WZ, Zhu GC, Hua J (2003) Recovery of RE from Baotou rare earth concentrate with chlorination roasting. Trans Nonferrous Met Soc China (English Ed, 438–442)
13. Lorenz T, Golon K, Fröhlich P, Bertau M (2015) Recycling of rare earth Hg-containing fluorscent lamp scrap by solid state chlorination. Chemie Ing Tech 87:1373–1382
14. Lorenz T, Golon K, Fröhlich P, Bertau M (2015) Rückgewinnung seltener erden aus quecksilberbelasteten leuchtstoffen mittels feststoffchlorierung. Chemie Ing Tech 87:1373–1382
15. Lorenz T, Bertau M (2019) The SepSELSA-Project—how rare earths are recovered from fluorescence lamps. Chemkon 26:72–77
16. Lorenz T, Bertau M (2019) Recycling of rare earth elements from FeNdB-magnets via solid-state chlorination. J Clean Prod 215:131–143
17. Lorenz T, Bertau M (2020) Recycling of rare earth elements from SmCo5-magnets via solid-state chlorination. J Clean Prod 246:118980
18. Qu X et al (2020) Recovery of LiCoO 2 from spent lithium-ion batteries through a low-temperature ammonium chloride roasting approach: thermodynamics and reaction mechanisms. ACS Sustain Chem Eng. https://doi.org/10.1021/acssuschemeng.0c01205

The Role of Nickel in Batteries

Ken Rudisuela

Abstract This paper covers a short history of the use of nickel in batteries from invention and leading up to advanced state-of-the-art Li-ion, an overview of the technology and the advantages that nickel brings. A discussion on relative performance of competing technologies including safety issues is included. Opportunities for cost reduction of Li-ion batteries with the use of nickel enabling the electric vehicle market will be discussed. Finally, information is provided on the availability of nickel for batteries, life cycle, and recyclability of nickel-based Li-ion batteries is provided.

Keywords Nickel in batteries · Li-ion batteries · Life cycle of batteries

History of Nickel in Batteries

Italian physicist Alessandro Volta is credited with inventing the first electrochemical battery in 1799 using zinc, copper and sulfuric acid. One hundred years later in 1899, the Swedish scientist Waldemar Jungner invented the Ni-Cd battery, which was the first portable rechargeable battery.

The Ni-Cd battery uses nickel oxide hydroxide as a cathode, metallic cadmium as an anode and potassium hydroxide as an electrolyte. The nominal voltage is 1.2 V which is lower than the 1.5 V of the standard non-rechargeable alkaline battery. At the time, Ni-Cd became popular due to its light weight, high power and significantly higher energy density than the only other commercial rechargeable battery, being the Pb-acid. The main drawbacks of Ni-Cd were high self-discharge rates and the use of the toxic and carcinogenic cadmium. Even with these drawbacks, the Ni-Cd battery enjoyed great commercial success for over ninety years in lighting, portable devices and electronics, emergency back-up power and aerospace.

In the late 1980s, the Ni-Cd battery was starting to be replaced by another nickel technology being Ni-metal hydride (NiMH). This new type of battery uses the same

K. Rudisuela (✉)
Nickel Institute, Toronto, Canada
e-mail: krudisuela@nickelinstitute.org

C. Anderson et al. (eds.), *Ni-Co 2021: The 5th International Symposium on Nickel and Cobalt*, The Minerals, Metals & Materials Series,
https://doi.org/10.1007/978-3-030-65647-8_10

nickel oxide hydroxide cathode as the Ni-Cd battery, but the toxic cadmium anode is replaced by a hydrogen-absorbing alloy made up of rare earth elements such as cerium, lanthanum and neodymium. Potassium hydroxide is the electrolyte and the nominal voltage of 1.2 V is the same as Ni-Cd.

The technical advantages that NiMH has over Ni-Cd is a 40% increase in energy density, better cycle life and limited memory effect. This relatively, environmentally friendly technology, also replaced the Ni-Cd battery in Europe through order of the Battery Directive.

NiMH had great commercial success in not only replacing Ni-Cd in nearly all applications, but it was also the battery technology that started the electrification of the automobile. Nearly all first-generation hybrid electric vehicles used NiMH as did many first-generation EVs. GM, Ford, Toyota and Honda were early adaptors of NiMH due to its reliability and low-temperature performance albeit cycle life and power capability kept the automotive industry looking for something better.

After only a few years of NiMH dominance, it started losing market share to the new Li-ion technology, which has higher voltage, higher energy density and longer cycle life.

Li-Ion, the Beginning

Today, Li-ion is the dominate battery technology in almost every portable application and even in stationary energy storage.

Li-ion started in the late 1970s when Prof Stan Whittingham of Binghamton University, New York, USA, discovered that lithium ions could be inserted reversibly, without chemical bonding, into small pockets within a TiS_2 structure, becoming the first functional rechargeable lithium battery.

In the early 1980s, Prof John Goodenough while at Oxford University discovered that using transition metal oxides, instead of TiS_2 resulted in a significantly higher voltage and therefore very useful energy density.

At the same time, Akira Yoshino, while working at Asahi Kasei discovered that layered carbon could be highly reversible and hence a substitute for lithium metal.

Collectively, these three pioneers were awarded the Noble Prize in Chemistry for their contribution to the development of Li-ion technology.

Since the beginning in the early 1990s when Li-ion was first commercialized, there has been significant advancements in the technology to increase energy density, cycle life and safety with multiple generations of technology occurring.

Three generations of lithium technology have already occurred and two more will occur in the future.

Gen1: Li-ion (LCO/HC) enabled communication with small portable devices such as phones.

Gen2: (NMC/graphite) with higher energy density and cycle-life enabled portable information, laptop computers could travel and operate all day without having to be plugged in.

Gen3: (advanced high nickel) advancements in safety and thermal management have enabled transportation and up to 500 km on a single charge
Gen4: (solid-state) with no liquid electrolyte enables the safe use of lithium metal as the anode, significantly increasing energy density.
Gen5: (Li-S and Li-air) theoretically, the highest energy density possible.

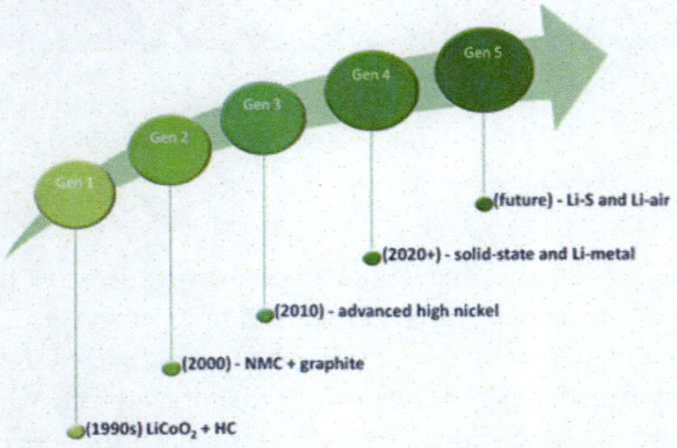

Nickel and Li-Ion

The Li-ion battery was first commercialized by Sony in 1991.

A Li-ion battery fundamentally consists of a graphite anode, an ionically conductive electrolyte and a transition metal oxide cathode. The anode and cathode being electronically separated from each other by a porous membrane.

The transition metal cathodes possess a molecular structure such that they are able to store lithium ions within their matrix. These structures can be layered as in NiO_2 and CoO_2, or spinel as in Mn_2O_4, or olivine as in $FePO_2$. On charge, lithium ions move from the cathode to the anode, through the electrolyte and during discharge the reverse happens.

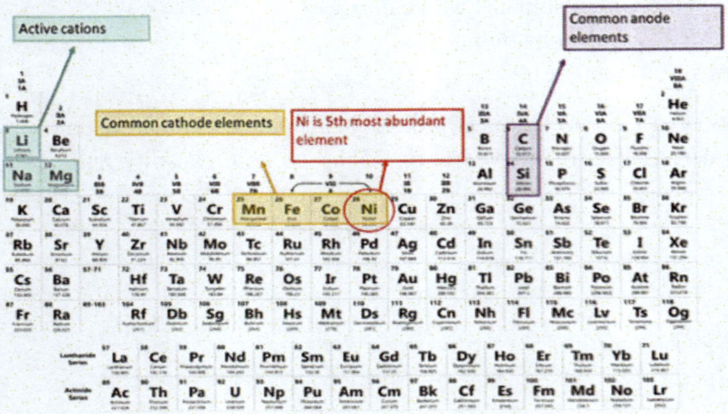

Nickel is the 5th most abundant element in the earth's makeup and results in the highest energy-dense layered oxide.

Of the three main layered transition metal oxides, nickel oxide has the highest energy density.

Cathode chemistry	Practical capacity (mAh/g)	Vol energy density (mAh/cc)	Safety	Cost
$LiNiO_2$	220	1,056	poor	fair
$LiCoO_2$	160	808	fair	high
$LiMnO_2$	110	462	good	low

Nickel oxide is the material of choice due to its high energy density, but poor thermal stability led to cobalt oxide being commercialized first.

Although $LiCoO_2$ was the first transition metal oxide to be commercialized there was a strong desire to move away from using cobalt due to its toxicity, high cost and uncertainty of supply. This desire to reduce cobalt as well as the need to increase energy density, led to the discovery of partially substituting one transition metal oxide (MO_2) with another. This maintains the overall layered structure but optimizes the properties of nickel, (capacity), manganese, (safety), and cobalt (rate capability), as mixed oxide materials.

Michael Thackeray and separately Prof Jeff Dahn were the first to report on and patent this mixed oxide technology.

Over time there have been many different percent combinations of the mixed oxide technology designated as $Li(Ni_xMn_yCo_z)O_2$, where $x + y+z = 1$. Along with additional substitution elements for optimizing performance attributes other competitive technologies have been developed, they include:

Lithium Nickel Cobalt Aluminum (NCA cathode)
Lithium Cobalt Oxide (LCO cathode), consumer electronics
Lithium Ferro Phosphate (LFP cathode)
Lithium Manganese Oxide spinal (LMO cathode)
Lithium Sulfur (Li-S, Sulfur cathode)
Lithium Titanate (LTO anode)

Energy density of any given chemistry is dependent upon its characteristic voltage discharge curve derived from the integration of the specific capacity and voltage profile.

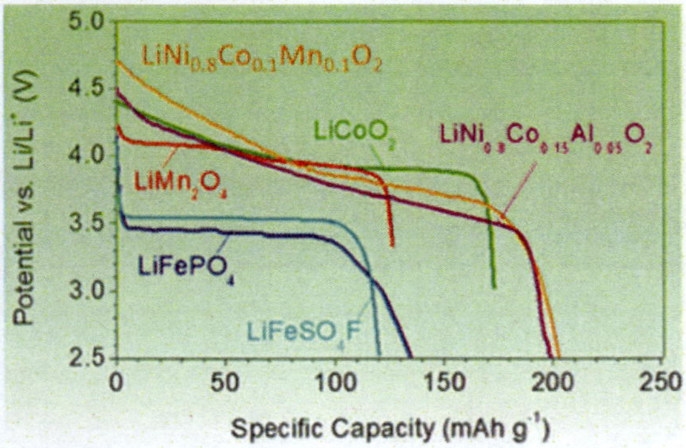

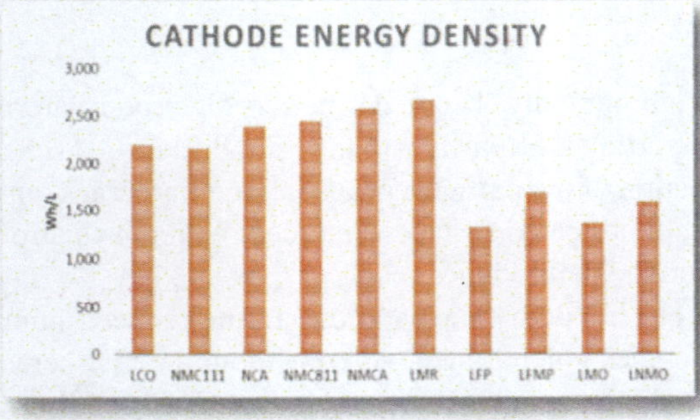

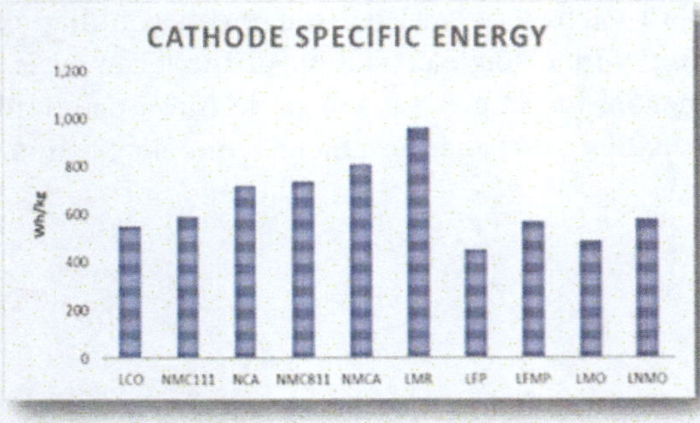

Development of the various technologies has been driven by goals to increase both gravimetric and volumetric energy density, as power density and also to broaden the temperature range in which the battery would effectively operate. In addition was the need to improve safety and to increase the cycle life and durability under real life conditions. Finally, for commercialization, the cost had to be driven down.

Ref: Modified from Nitta, N., Wu, F., Lee, J. T., & Yushin, G. (2015). Li-ion battery materials: present and future. *Materials today*, *18*(5), 252-264.

LFP and LMO chemistries are less expensive than those using nickel, but their inherent much lower energy density prohibits mass adoption in many markets, including automotive. Lithium sulfur chemistries suffer from poor cycle life and lithium titanate is both low in energy density and very expensive.

Increasing the nickel content has been the constant in Li-ion development because it satisfies more of the criteria than any other technology. Not only increased performance attributes such as energy density, power and run time but also higher nickel content result in a lower cost due to reducing the amount of cobalt in the battery. Over time the amount of nickel in commercial Li-ion batteries has increased from 33% to 50% to 80% by weight. For this to happen, the thermal dynamic stability and durability of the high nickel technology had to be improved. Several solutions have been successfully used to strengthen the high nickel structure through surface coatings, cell core technology and passivating electrolyte additives.

Single Crystal

Electro-mechanical degradation is commonly observed in battery electrode materials, often prepared as polycrystalline particles. Expansion of the cathode during discharge creates high forces at the intersection of the particles propagating microcracks in the polycrystalline matrix, exposing unprotected surfaces that react with the electrolyte causing the impedance to rise and the cell to fade.

Large single crystals have more freedom to expand due to the reduced number of particle interfaces and hence cycles much longer. Also, without the microcracks, the high nickel material is thermal dynamically more stable due to the much lower surface area and potential for reduction.

The single crystal, high nickel Li-ion battery, has led to the so-called million-mile battery. With the structural stability of the single crystal, this battery can cycle for more than 3,000 full cycles, potential for an electric vehicle to travel one million miles before replacement or for auxiliary use such as vehicle to grid application.

Cost

For mass acceptance of EVs, their cost must come down more in line with equivalent internal combustion engine (ICE) cars. Currently, the purchase price of an EV, without subsidies, is 30% to 50% higher than and ICE. The single largest cost of an EV is the Li-ion battery and hence cost reduction in materials, design, processes and manufacturing are constantly being addressed.

Availability of Nickel

Nickel is considered a critical raw material for Li-ion batteries used in EVs. With the potential of a 30% CAGR of EV sales over the next decade and beyond, the demand for nickel in the form of nickel sulphate is high. Although nickel sulphate, used in Li-ion batteries, is most commonly derived from high-grade sources, it can be derived in many different ways from multiple sources. Both lateritic and sulfidic ores can and are used as input materials for Li-ion batteries.

Currently, known resources amount to 300 million tons of nickel, 180 million tons from laterites and 120 million tons from sulfides. Mining companies continuously convert resources into reserves, which form the economically viable extraction, currently at 89 million tons. Approximately two million tons of nickel are mined each year with approximately 5% currently going into Li-ion batteries.

Life Cycle of Batteries

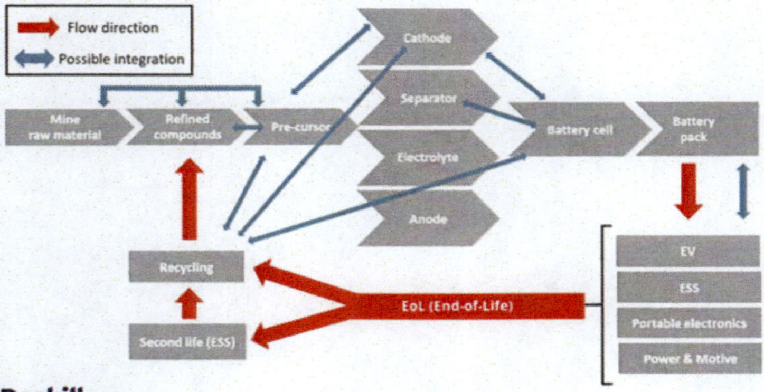

Roskill

From: Lithium-ion Batteries: Outlook to 2029
Source: Roskill
Created for: [User] / Nickel Institute
IP Address: 24.36.229.52

Nickel has a high economic value and is a main driver for the recycling of EV batteries. Nickel is recycled at high efficiency (> 95%) with either standard hydro or pyrometallurgical processes.

The high economic value of nickel will ensure that batteries are collected at the end of life and are not landfilled reducing the overall carbon footprint of nickel-containing batteries.

The Effect of Cu, Al and Fe Impurities on Leaching Efficiency of Two Lithium-Ion Battery Waste Fractions

A. Chernyaev, J. Partinen, and M. Lundström

Abstract It has been established that high extraction of cobalt in LIB waste leaching is achieved with an addition of reducing agents, such as H_2O_2. However, battery current collectors (Al and Cu) and casings (Fe) can also act as reducing agents in leaching. The target of this work was to investigate the extractions of Co, Mn and Ni from <1250 μm and < 500 μm fractions of waste battery concentrate in sulfuric acid leaching ($T = 60$ °C, [H2SO_4] = 2 M, and 200 g/L). It was found that 90% of Co and Mn and all of Ni could be extracted from <1250 μm fraction in the absence of externally added reductants, whereas in the case of < 500 μm fraction, the extraction was 14–40%-units lower. The results suggest that extensive sieving of the concentrate can be avoided, but the residual current collectors present in the coarse fraction can be utilized as reducing agents in leaching, and the use of external reduction chemicals can be minimized.

Keywords Waste battery · Recycling · Reductive leaching · Hydrometallurgy

Introduction

Efficient leaching and extraction of metals from $LiCoO_2$ require a reductant with a lower standard electrode potential than that of $LiCoO_2$ (Eq. 1). One of the most widely researched reducing agents for this purpose is hydrogen peroxide [1–3]. The major advantage of hydrogen peroxide is that it introduces no new elements or compounds into the system, as O_2 and H_2O are its only decomposition products (Eq. 2). However, H_2O_2 is also a strong oxidant (Eq. 3), which causes some of it to be consumed in reactions involving oxidation of other metals present in a battery waste concentrate, such as Cu, Al, and Fe [4, 5]. Moreover, hydrogen peroxide decomposes in sulfuric acid at elevated temperatures, which further increases its consumption [6].

Many alternatives to H2O2 have been researched, including sodium bisulfite, glucose, ascorbic acid, citric acid [7], and nickel-metal hydride battery waste [8].

A. Chernyaev · J. Partinen · M. Lundström (✉)
Aalto University, Espoo, Finland
e-mail: mari.lundstrom@aalto.fi

C. Anderson et al. (eds.), *Ni-Co 2021: The 5th International Symposium on Nickel and Cobalt*, The Minerals, Metals & Materials Series,
https://doi.org/10.1007/978-3-030-65647-8_11

The high standard reduction potential of $LiCoO_2$ (2.13 V vs. SHE) indicates that also several metals in elemental form already present in the battery waste could be used as reductants in leaching of the cathode material. Peng et al. [9] demonstrated that the overflow fraction of the crushed $LiCoO_2$ battery waste consisting largely of Cu and Al foils can be used to improve the extraction of Co and Li during leaching. It has also been shown that aluminum foil particles of the battery waste concentrate contain up to 11 wt% of Co on their surfaces [5], which indicates that the total extraction of cobalt could alternatively be increased by feeding the foils to the leaching process rather than separating them from the concentrate before leaching.

$$LiCoO_{2(s)} + 4H^+ + 2e^- \rightarrow Li^+ + Co^{2+} + 2H_2O \quad E_{eq} = 2.131 \text{ V vs. SHE} \quad (1)$$

$$O_{2(g)} + 2H^+ + 2e^- \rightarrow H_2O_{2(aq.)} \quad E_{eq} = 0.584 \text{ V vs. SHE} \quad (2)$$

$$H_2O_{2(aq.)} + 2H^+ + 2e^- \rightarrow 2H_2O \quad E_{eq} = 1.817 \text{ V vs. SHE} \quad (3)$$

Previous research has shown that copper, iron, and aluminum particles present in the battery waste can increase the final extraction of Co from $LiCoO_2$ in the sulfate system [1, 10]. In the present work, the leaching performances of two fractions of battery waste concentrate—as received fraction (<1250 μm) and sieved underflow (<500 μm)—were investigated. Besides Co, extraction of Ni and Mn was observed.

Materials and Methods

The waste battery concentrate used in the experiments was acquired from an industrial battery-recycling company. The concentrate comprised mostly of batteries from mobile devices. The concentrate was prepared by the initial pre-sorting of the batteries, followed by two-step crushing, magnetic separation, and sieving [11]. The concentrate was rich in cobalt (26 wt%), having also minor fractions of impurities such as Ni, Mn, Cu, Fe, and Al.

The concentration of the main elements in the battery waste concentrate is presented in Table 1. The < 1250 μm fraction was sieved with a 500 μm mesh screen for five minutes using a vibratory sieve in order to remove a part of Al and Cu. Also, some iron and plastics were removed in sieving, and the underflow (<500 μm) was used as the feed material in leaching.

Table 1 Average metal content (mg/g) in the investigated waste battery concentrate fractions

Fraction	Ni	Mn	Co	Al	Fe	Cu
<1250 μm	22.8	26.3	259	34.1	5.9	40.1
<500 μm	25	27.6	260	23.1	4.7	15.4

The lixiviant (2 M H2SO_4) was prepared from sulfuric acid (95–97%, Merck Emsure). Nitric acid (0.3 M) prepared from concentrated HNO_3 (65%, Merck Emsure) was used for sample dilution.

Extraction of metals from the battery concentrate and the sieved underflow fraction (<500 μm) was investigated at $T = 60$ °C, initial $[H_2SO_4] = 2$ M, agitation speed of 350 rpm, and $S/L = 200$ g/L, over a leaching time of three hours. The experiments were carried out in a 1L glass jacketed reactor equipped with a 4-blade impeller agitator (90 mm diameter), and a reflux condenser. The metal extractions Y_{Me} (in %) of each sample were calculated using Eq. 4.

$$Y_{Me} = \frac{C_{Me} \cdot V_L}{x_0 \cdot m_0} \cdot 100\% \tag{4}$$

where C_{Me} is the concentration of the metal (g/L) in solution, V_L is the volume of the leach solution (L), x_0 is the fraction of metal in the concentrate (%), and m_0 is the initial weight of the concentrate (g).

Results and Discussion

Two fractions of the battery waste concentrate (<1250 μm and <500 μm) were leached separately and the metal extractions were compared. The XRD analysis carried out for <500 μm fraction revealed the presence of graphite, $LiCoO_2$, $LiMn_2O_4$, and nickel oxide, possibly in the form of $LiNiO_2$. According to Fig. 1, the extraction data of <500 μm leaching suggests that the dissolution of Co may progress through a different reaction route compared to Ni and Mn, since the dissolution of $LiCoO_2$ ceases after $t = 0.5$ h, while Ni and Mn-based oxides continue to dissolve throughout the entire leaching time. This could be associated with the formation of an intermediate phase in the case of $LiCoO_2$, such as Co_3O_4 [12, 13], thus hindering the charge transfer between the $LiCoO_2$ particle and the lixiviant.

The highest leaching rate for the <1250 μm fraction was observed during the first hour of the experiment, whereas the leaching kinetics of the <500 μm fraction declined after the first 30 min. The final extraction of Co (90% vs. 53%), Ni (100% vs. 90%) and Mn (90% vs. 70%) was found to be higher when leaching of < 1250 μm fraction compared to <500 μm (Fig. 1). Such significant differences in the kinetics and final extractions are suggested to be attributed to the varying contents of the impurities in the fractions, the amounts being significantly higher in <1250 μm fraction (Table 1), i.e. 4 g vs. 1.5 g for Cu, 3.4 g and 2.3 g for Al, as well as 6 g vs. 4.7 g for Fe.

It has been previously demonstrated that copper, iron and aluminum-rich sieve overflow acts as a reductant towards $LiCoO_2$ [3]. The dissolved iron alone is known to catalyze the electron transfer between copper and $LiCoO_2$ (Eq. 5) [10]. Additionally, aluminum could contribute to the reductive leaching directly or via intermediate

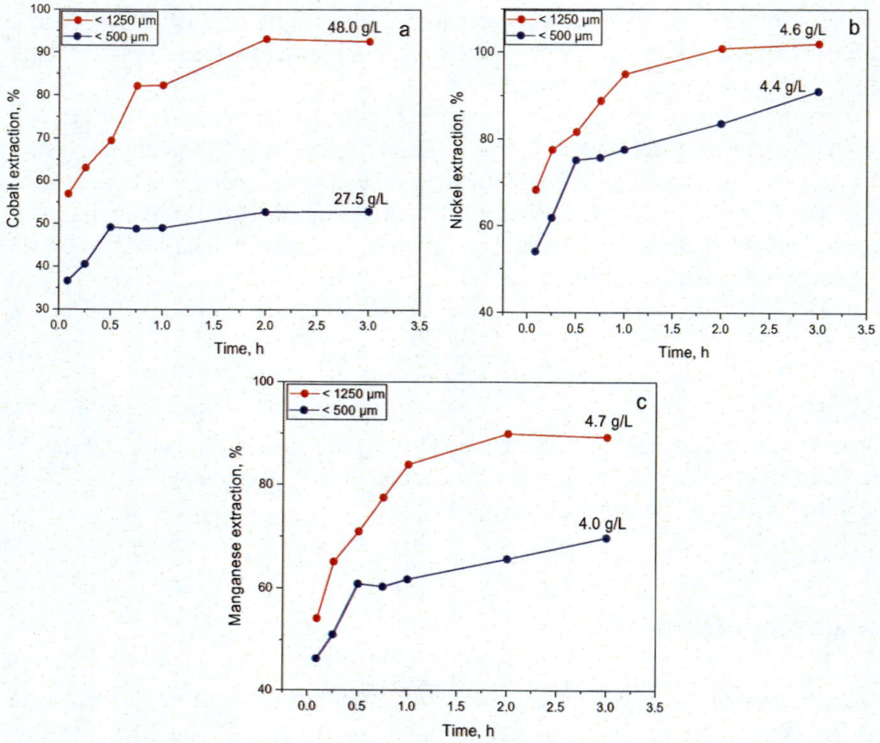

Fig. 1 Extraction and final concentration of **a** cobalt, **b** nickel and **c** manganese. T = 60 °C, 200 g/L of solids, [H2SO4] = 2 M. (Color figure online)

reactions. In the case of <1250 μm fraction, the metallic elements, specifically copper, continues to react even after t = 1 h, whereas in <500 μm fraction, the dissolution of Al, Cu and Fe has almost ceased after t = 0.5 h (Fig. 2). The greater amount of aluminum dissolved in the leaching of the coarse fraction (Fig. 2a) corresponds to the higher content in < 1250 μm fraction. All iron (Fig. 2b) and copper (Fig. 2c) were also shown to dissolve throughout the leaching. Higher extraction of valuable metals from the fraction with a larger particle size is associated with the higher amount of copper available for electron transfer to $LiCoO_2$, Mn_2O_4 and nickel oxide with the aid of a larger amount of iron to improve the kinetics (1.2 g/L vs. 0.8 g/L, Fig. 2B).

$$2LiCoO_2 + Cu + 4H^+ \rightarrow 2Li^+ + Cu^{2+} + 2Co^{2+} + 2H_2O \quad \Delta G = -37.8\,\text{kcal} \quad (5)$$

For the investigated concentrate, the cathode metal extraction of 90% or higher was achieved with the aid of the impurities present in the <1250 μm fraction feed. The content of aluminum was 3.4 g, copper 4 g and iron 0.6 g per 100 g of the concentrate, which contains 26 g of Co, 2.3 g of Ni and 2.6 g of Mn. The findings demonstrate that the presence of metallic impurities in the concentrate, namely Al, Cu and Fe, benefits

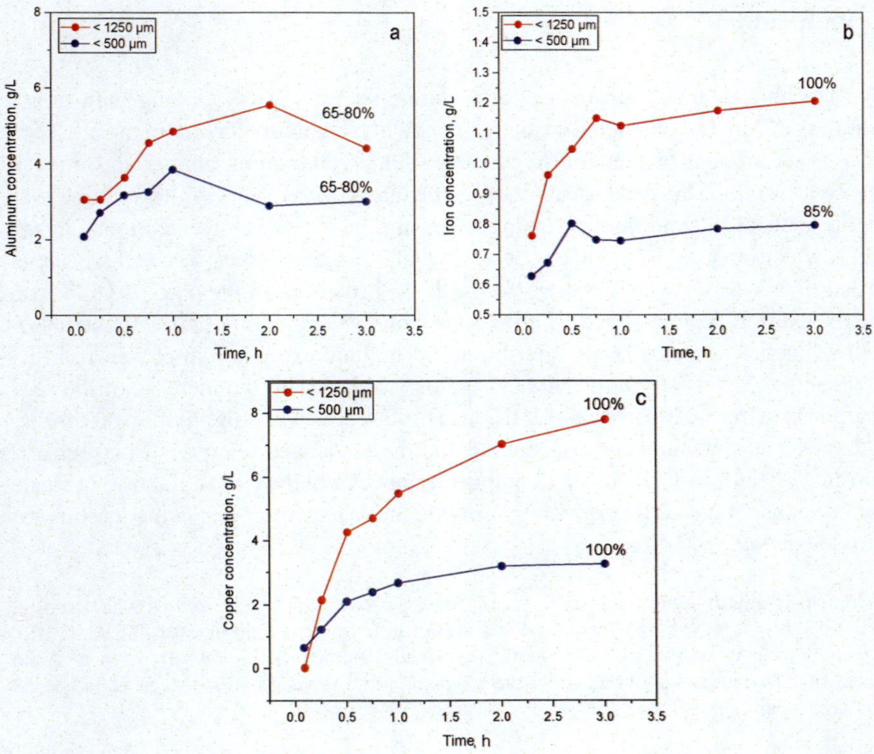

Fig. 2 Concentration and final extraction of **a** aluminum, **b** iron and **c** copper. T = 60 °C, 200 g/L of solids, [$H2SO_4$] = 2 M. (Color figure online)

the leaching efficiency. However, the role of each current collector material and other metallic impurities and their efficiencies as reducing agents needs to be investigated separately. If the use of the metallic impurities in the waste battery concentrate is proven effective in an industrial scale leaching, the use of other reducing agents, such as H2O2, could be minimized or even replaced. A part of the waste current collectors and metal casings could potentially be used directly in leaching as additives, thus the need to recycle these waste components separately would be minimized. However, the use of the metallic reductants in leaching would affect the solution purification process, as the concentration of these metals in the final leach solution would increase, e.g. 5 g/L of Al, 1.2 g/L of Fe and 8 g/L of Cu (Fig. 2). Of these elements, the removal of aluminum in the neutralization stage poses significant challenges due to the formation of gel-like precipitates which result in poor filtration efficiency and a tendency for Ni and Co co-precipitation.

On the other hand, feeding a coarser fraction to the leaching process is an attractive alternative, as it requires fewer sieving steps. Besides, cobalt loss with aluminum current collectors in sieving overflow would be decreased, and less hydrogen peroxide would be required in the leaching stage.

Conclusions

The leaching of the waste battery concentrate rich in $LiCoO_2$, along with minor amounts of Mn_2O_4 and nickel oxide, was shown to be more efficient in the presence of a larger amount of metallic impurities, such as aluminum, copper and iron, in the concentrate. The yield of all cathode metals (Co, Ni, Mn) in the solution was higher with <1250 μm fraction owing to the presence of the abovementioned impurities, which act as reducing agents towards cathode metal oxides. The cathode metal dissolution kinetics were different for each feed fraction. In the case of <500 μm, the dissolution declined after 30 min of leaching, whereas for <1250 μm the dissolution kinetics decreased only after 60 min. The final extraction improved from 53 to 90% for Co, 91–100% for Ni and 70–90% for Mn with the increase of the feed fraction size from <500 μm to <1250 μm. This research work highlights a possibility to bypass the sieving of the concentrate and the associated removal of plastics and current collector foils. Instead, the impurities present in the coarse fraction of waste battery concentrate could potentially provide the reductive power in the leaching of cathode metals.

Acknowledgements This research work has been supported by Revolt project (08_2018_IP167_ReVolt), by Business Finland BatCircle project (grant number 4853/31/2018), and the Academy of Finland's RawMatTERS Finland Infrastructure (RAMI) based at Aalto University. The authors are grateful for the research grant provided by Finnish Steel and Metal Producers' Fund.

References

1. Joulié M, Billy E, Laucournet R, Meyer D (2017) Current collectors as reducing agent to dissolve active materials of positive electrodes from Li-ion battery wastes. Hydrometallurgy 169:426–432. https://doi.org/10.1016/j.hydromet.2017.02.010
2. Meng Q, Zhang Y, Dong P (2018) Use of electrochemical cathode-reduction method for leaching of cobalt from spent lithium-ion batteries. J Clean Prod. https://doi.org/10.1016/j.jclepro.2018.01.101
3. Peng C, Liu F, Aji AT, Wilson BP, Lundström M (2019) Extraction of Li and Co from industrially produced Li-ion battery waste—using the reductive power of waste itself. Waste Manag. https://doi.org/10.1016/j.wasman.2019.06.048
4. Aaltonen M, Peng C, Wilson B, Lundström M (2017) Leaching of metals from spent lithium-ion batteries. Recycling 2:20. https://doi.org/10.3390/recycling2040020
5. Porvali A, Aaltonen M, Ojanen S, Velazquez-Martinez O, Eronen E, Liu F, Wilson BP, Serna-Guerrero R, Lundström M (2019) Mechanical and hydrometallurgical processes in HCl media for the recycling of valuable metals from Li-ion battery waste. Resour Conserv Recycl. https://doi.org/10.1016/j.resconrec.2018.11.023
6. Wu LK, Chen KY, Cheng SY, Lee BS, Shu CM (2008) Thermal decomposition of hydrogen peroxide in the presence of sulfuric acid. J Therm Anal Calorim 115–120. Springer. https://doi.org/10.1007/s10973-007-8829-6
7. Zheng X, Zhu Z, Lin X, Zhang Y, He Y, Cao H, Sun Z (2018) A mini-review on metal recycling from spent lithium ion batteries. Engineering 4:361–370. https://doi.org/10.1016/J.ENG.2018.05.018

8. Liu F, Peng C, Porvali A, Wang Z, Wilson BP, Lundström M (2019) Synergistic recovery of valuable metals from spent nickel-metal hydride batteries and lithium-ion batteries. ACS Sustain Chem Eng. https://doi.org/10.1021/acssuschemeng.9b02863
9. Peng C, Hamuyuni J, Wilson BP, Lundström M (2018) Selective reductive leaching of cobalt and lithium from industrially crushed waste Li-ion batteries in sulfuric acid system. Waste Manag. https://doi.org/10.1016/j.wasman.2018.02.052
10. Porvali A, Chernyaev A, Shukla S, Lundström M (2019) Lithium ion battery active material dissolution kinetics in Fe(II)/Fe(III) catalyzed Cu-H_2SO_4 leaching system. Sep Purif Technol 116305. https://doi.org/10.1016/j.seppur.2019.116305
11. Pudas J, Erkkila A, Viljamaa J (2015) Battery recycling method, US8979006B2
12. Ferreira DA, Prados LMZ, Majuste D, Mansur MB (2009) Hydrometallurgical separation of aluminium, cobalt, copper and lithium from spent Li-ion batteries. J Power Sources 187:238–246. https://doi.org/10.1016/j.jpowsour.2008.10.077
13. Musariri B, Akdogan G, Dorfling C, Bradshaw S (2019) Evaluating organic acids as alternative leaching reagents for metal recovery from lithium ion batteries. Miner Eng 137:108–117. https://doi.org/10.1016/j.mineng.2019.03.027

A Sustainable Oxalate Process for Recovery of Metals from $LiCoO_2$: Experimental and Modeling Study

Ankit Verma, David R. Corbin, and Mark B. Shiflett

Abstract Green and sustainable recycling of lithium-ion batteries is an important problem that requires immediate attention to meet the rapid growth in critical materials demand for the electric vehicle (EV) market. In this work, a holistic study is presented to develop a closed-loop Li and Co metal recovery and separation process for $LiCoO_2$ electrode materials using oxalate chemistry. More than 97% of Co and Li extraction and almost complete separation between these metals were found in the digestion using oxalic acid at a concentration of 0.46 M, a solid-to-liquid ratio of 15 g/L, agitator speed of 600 rpm, and a temperature of 100 °C. An optimum pH of about 2.5 was identified to achieve efficient extraction and separation between Li and Co. A novel oxalic acid regeneration process using ion-exchange resins is proposed to recover oxalate, reduce waste production, and lower the overall cost of the process. The experimental results are supported with a chemical speciation study and a shrinking core model based kinetic study to develop a deeper understanding of the interfacial processes.

Keywords Lithium-ion battery · Oxalate chemistry · Recycling · Shrinking core model

Introduction

Lithium-ion batteries (LiBs) have made a tremendous impact on the portable electronics industry and are about to do the same for electric vehicles (EVs) [1]. Industry experts predict that by 2030 the world will generate two million metric tons of used LiBs every year. Currently, because of various technical and economic factors, less

A. Verma · M. B. Shiflett (✉)
Institute for Sustainable Engineering, University of Kansas, 1530 W. 15th St., Lawrence, KS 66045, USA
e-mail: Mark.B.Shiflett@ku.edu

A. Verma · D. R. Corbin · M. B. Shiflett
Center for Environmentally Beneficial Catalysis, University of Kansas, 1501 Wakarusa Dr., Lawrence, KS 66047, USA

C. Anderson et al. (eds.), *Ni-Co 2021: The 5th International Symposium on Nickel and Cobalt*, The Minerals, Metals & Materials Series,
https://doi.org/10.1007/978-3-030-65647-8_12

than 5% of waste batteries are being recycled. Most of the LiBs that do get recycled currently undergo a high-temperature pyrometallurgical processes that suffers from high operating costs and harmful emissions [2, 3].

A typical LiB consists of a positive electrode (cathode during discharge), a negative electrode (anode during discharge), and an electrolyte along with separators and current collectors. A intercalated lithium-containing transition metal oxide (such as $LiMO_2$) is used as the positive electrode, graphite is used as the negative electrode, and an organic solvent in which a lithium salt is dissolved (such as $LiPF_6$) is used as the electrolyte [4]. Lithium cobalt oxide ($LiCoO_2$) is one of the most common electrode materials, which had been used widely over the past two decades; however, with the fluctuating Co prices, recent efforts have been focused on developing similar battery performance with lower Co content [5, 6]. This has led to the development of various electrode materials with other metals such as Ni, Mn, Al, and Fe. As most of the critical metals are required in the positive electrode, efficient metal recovery has been the focus of several studies [7]. Currently, there are three primary technologies for metals recovery from the spent positive electrode including pyrometallurgy, hydrometallurgy, and biometallurgy. Pyrometallurgy is an energy-intensive process with high operating costs. Biometallurgy involves long reaction time and the development of micro-organisms. Hydrometallurgy, which involves treating electrode material in an acidic environment to leach metals into the aqueous phase, is a promising technology provided that separation efficiency and recycling of reagents is achieved [8–10]. Direct regeneration of the cathode via electrochemical relithiation is another process gaining traction where the layered morphology of the electrode material remains unaffected [9, 11]. In hydrometallurgical treatment, metals are extracted from the layered structure into the aqueous phase, and different separation processes can be used to recover the critical metal products. Hydrometallurgical treatment offers the opportunity to recycle traditional $LiCoO_2$ electrodes by converting them into modern electrodes with lower Co content.

In this work, Li and Co are recovered using a hydrometallurgical process based on oxalic acid ($H_2C_2O_4$) chemistry. Oxalic acid is an organic acid that allows the direct separation of Li and Co via precipitation of cobalt oxalate dihydrate ($CoC_2O_4 \cdot 2H_2O$). This one-step separation provides an economical and environmental-friendly route to recover metals from $LiCoO_2$ [12]. In our previous work, the details have been provided that outline the advantages of oxalic acid and other organic acids over inorganic acids for hydrometallurgical processes [6, 10]. In this work, the separation and recovery of Li and Co from $LiCoO_2$ has been investigated using oxalic acid, and the fundamental kinetic and chemical speciation has been modeled.

Chemical speciation modeling refers to the distribution of elemental metals amongst the chemical species in an aqueous system typically as a function of pH. It is critical for understanding the pH range required for efficient metal separation. In this work, a speciation model to study the effect of pH on complexation and precipitation of Co will be discussed to support the experimental observations [10]. In addition, a constant radius shrinking core model has been developed to better understand the Li and Co extraction kinetics.

Experimental

Materials and Reactor Setup

In this work, commercially available lithium cobalt oxide (99.8 wt%, Sigma-Aldrich) and oxalic acid dihydrate (99.5 wt%, ACROS Organics) were used for the digestion reactions. The concentration of extracted metals in the aqueous phase was measured using inductively coupled plasma optical emission spectroscopy (ICP-OES) in a matrix of 5 wt% HNO_3. To characterize the solid precipitates, powder X-ray diffraction was performed. Additional details for the material characterization can be found in our previous work [10].

A major aspect of this study involves measuring the kinetics of Li extraction into the aqueous phase. Hence, a similar reactor setup discussed in Verma et al. was used with minor modifications to the sampling technique [10]. In this work, the solid and aqueous phases were separated using centrifugation rather than a syringe filter.

Metal Extraction Experiments

All of the metal extraction experiments were carried out in the digestion reactor by heating the oxalic acid solution to the set temperature, and then the required amount of $LiCoO_2$ was added. To obtain the kinetic data, samples were extracted in Falcon® tubes, and the metal extraction process was immediately quenched by submerging the Falcon® tubes in an ice bath. Later, the samples were centrifuged, phases were separated, and the aqueous phase was diluted in 5 wt% HNO_3 to determine the concentration of Li and Co. All the metal extraction experiments were performed at a fixed solid-to-liquid ratio (S/L) of 15 g/L and an agitator speed of 600 rpm.

Results and Discussion

Digestion of $LiCoO_2$ in Oxalic Acid

$LiCoO_2$ digestion in oxalic acid (OA) has been well-studied in the literature [13, 14]. In $LiCoO_2$, Co exists in a + 3 state and is known to form soluble oxalate complexes [15]. Hence, to achieve a direct separation between Co and Li, Co has to be reduced from the + 3 to a + 2 state and precipitated in the form of $CoC_2O_4{\cdot}2H_2O$. Therefore, the concentration of oxalic acid required for complete digestion needs to be optimized. To perform this study, the molar ratio between $LiCoO_2$ and oxalic acid (i.e., Co:OA ratio) becomes critical. The amount of Co present in this study was set by the constant solid-to-liquid ratio (15 g/L). Hence, the amount of oxalic acid was

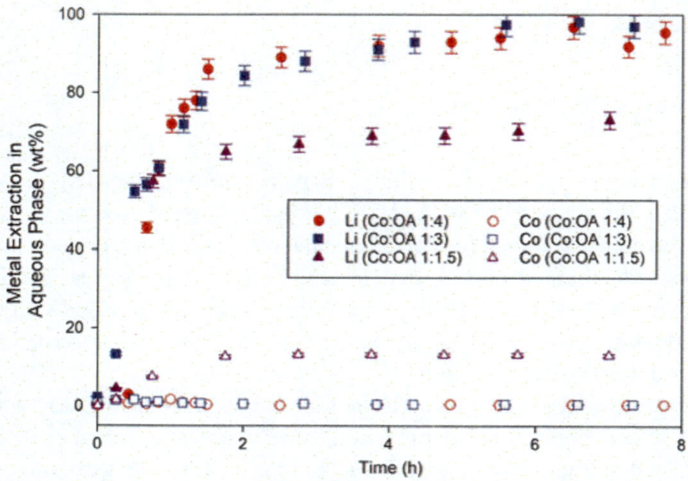

Fig. 1. Concentration of Li and Co as a function of time in the aqueous phase at T = 100 °C, S/L = 15 g/L, and agitator speed = 600 rpm. (Color figure online)

varied, and the results are illustrated in Fig. 1. A Co:OA ratio of 1:3 (0.46 M oxalic acid) to 1:4 (0.61 M oxalic acid) led to complete extraction and separation of Li and Co. A Co:OA ratio of 1:1.5 (0.23 M oxalic acid) led to incomplete Li extraction even after a reaction time of 6 h. This conclusion supports the stoichiometric ratio of $LiCoO_2$ to oxalic acid of 1:2 as shown in Eq. 1.

$$LiCoO_2\,(s) + 2H_2C_2O_4\,(aq) \rightleftharpoons \frac{1}{2}Li_2C_2O_4\,(aq) + CoC_2O_4{\cdot}2H_2O(s) + CO_2\,(aq) \tag{1}$$

An interesting observation was the large amount of Co being extracted (18 wt%) into the aqueous phase with a Co:OA ratio of 1:1.5. To investigate this phenomenon, the pH of the extracted samples was measured (shown in Fig. 2) and indicates that even though oxalic acid was present in a limited amount, a higher pH can lead to the dissolution of insoluble cobalt oxalate dihydrate. The sensitivity of the reaction to pH, including the addition of NaOH, was studied as a function of time (Fig. 3). As expected, with an increase in pH, the concentration of Co in the aqueous phase increased, which led to further study of the Co-oxalate speciation and cobalt chemistry.

Chemical Speciation of Co-oxalate

Oxalic acid is a dibasic acid with a $pKa_1 = 1.23$ and $pKa_2 = 4.19$; therefore, at a pH between 1.23 to 4.19, $HC_2O_4^-$ is the predominant species, whereas at a pH above

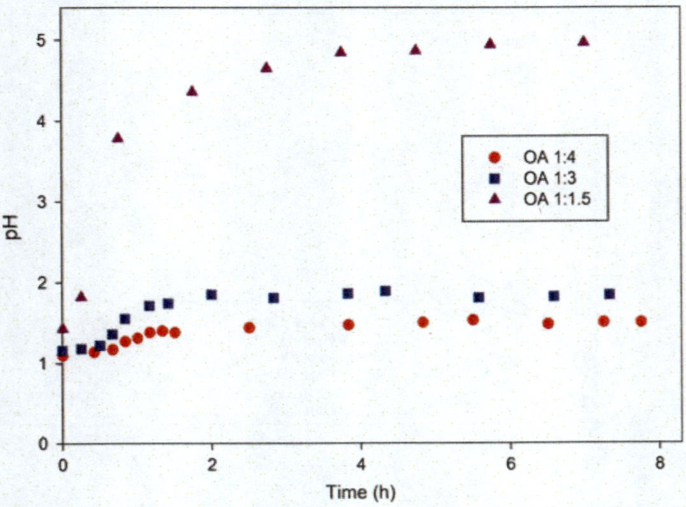

Fig. 2. pH as a function of time in the aqueous phase for different digestion reactions at T = 100 °C, S/L = 15 g/L, and agitator speed = 600 rpm. (Color figure online)

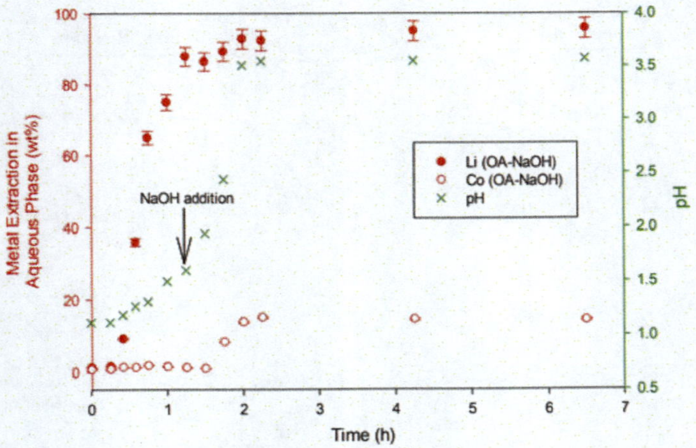

Fig. 3. Effect of NaOH addition on Li and Co metal concentration as a function of time at T = 100 °C, S/L = 15 g/L, and agitator speed = 600 rpm for Co:OA 1:4 case. (Color figure online)

4.19, $C_2O_4^{2-}$ is the predominant species. The absence of dissolved Co in the aqueous phase for the Co:OA 1:3 and 1:4 cases indicates the complexation is mainly due to the presence of the $C_2O_4^{2-}$ anion. Krishnamurthy summarized the possibility of soluble complexes such as $Co(C_2O_4)_2^{2-}$ and $Co(C_2O_4)_3^{4-}$ in the presence of excess $C_2O_4^{2-}$ anion [15]. Meng et al. [16] described a similar observation and provided the equilibrium constants for the Co-oxalate complexation steps, as shown in Eqs. 2–4.

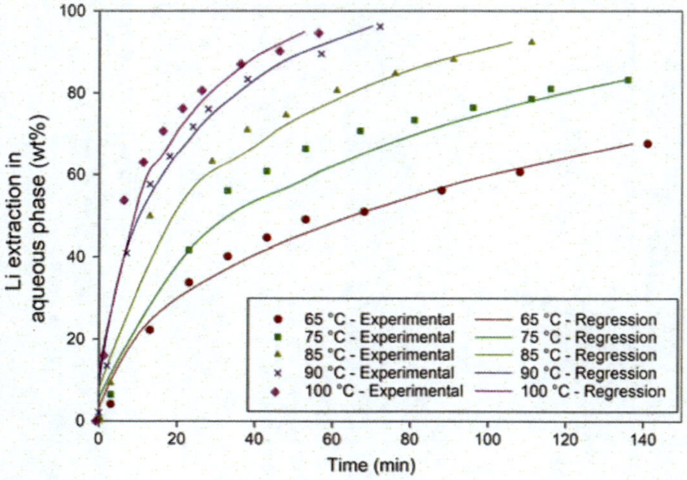

Fig. 4. Concentration of Li as a function of time at various temperatures for Co:OA 1:3 case, S/L = 15 g/L, and agitator speed = 600 rpm. Solid lines represent the spline curves generated from multi-linear regression. (Color figure online)

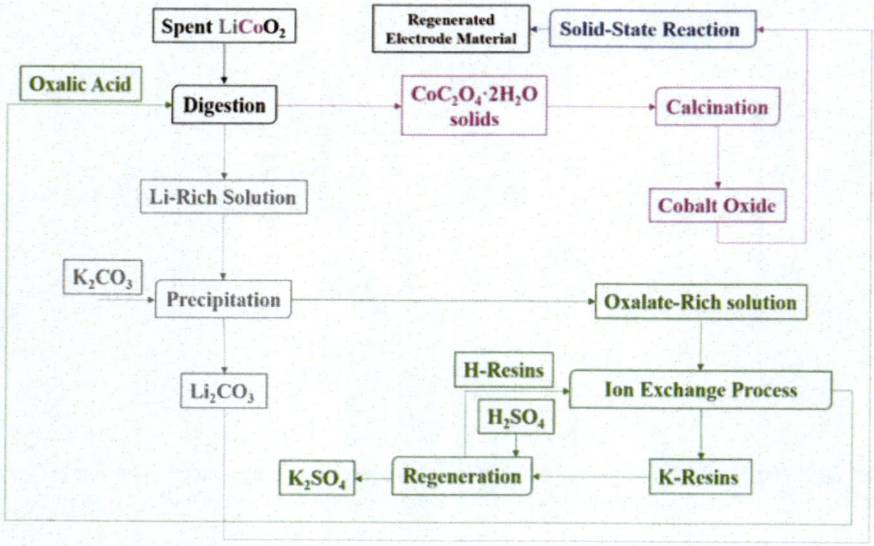

Fig. 5. Flowsheet of the proposed closed-loop process. (Color figure online)

$$Co^{2+} + C_2O_4^{2-} \rightleftharpoons Co(C_2O_4)\, LogK_1 = 7.26 \quad (2)$$

$$Co^{2+} + 2C_2O_4^{2-} \rightleftharpoons Co(C_2O_4)_2^{2-}\, LogK_2 = 6.7 \quad (3)$$

$$Co^{2+} + 3C_2O_4^{2-} \rightleftharpoons Co(C_2O_4)_3^{4-}\ \mathrm{LogK}_3 = 9.7 \quad (4)$$

To understand the potential complexes at these reaction conditions, Visual MINTEQ (version 3.1) was used to generate the chemical speciation of the Co-oxalate system, as shown in Table 1. This analysis was performed by assuming all the Co^{3+} was reduced to Co^{2+} and was available for complexation with the $C_2O_4{}^{2-}$ ligand. Interestingly, Visual MINTEQ showed no significant amount of $Co(C_2O_4)_3^{4-}$ formed, and most of the Co precipitated as either $CoC_2O_4{\cdot}2H_2O$ or was dissolved in the aqueous phase as $Co(C_2O_4)_2{}^{2-}$. In Table 1, under the equimolar condition of Co^{2+} and $C_2O_4{}^{2-}$ (columns 2 and 3), the Co remained as a precipitate over the acidic pH range. Increasing the concentration of $C_2O_4{}^{2-}$ (columns 3–6), the amount of soluble $Co(C_2O_4)_2{}^{2-}$ also increases at higher pH. This result is consistent with the observation of a larger amount of dissolved Co at higher pH. The increase in the dissolved Co concentration at a pH > 4 supports the hypothesis that the $C_2O_4{}^{2-}$ species generated after the loss of two protons from oxalic acid ($H_2C_2O_4$) plays a major role in the dissolution of Co into the aqueous phase. The chemical speciation study is ongoing and additional results will be provided in future publications.

Kinetic Modeling

In order to understand the kinetics of Li extraction, a shrinking core model (SCM) was developed. This model is applicable to solid–liquid reactions and has been used previously for $LiCoO_2$ digestion in inorganic acids; however, this is the first application of the SCM for digestion of $LiCoO_2$ with oxalic acid [17–19]. This reaction system is more interesting and different from other acid leaching systems because of the formation of a solid $CoC_2O_4{\cdot}2H_2O$ product. The particle size was similar (10–15 μm) before and after the digestion. Hence, a shrinking core model for spherical particles of constant size was applied to describe the Li and Co dissolution kinetics. According to the SCM, the overall size of the particle remains constant, but the size of the reactant core ($LiCoO_2$) decreases, and the product layer ($CoC_2O_4{\cdot}2H_2O$) increases. The SCM is presented in detail in references [17, 20].

In order to determine the rate-controlling step, a new technique was used where a summation of the film layer control, diffusion layer control, and chemical reaction control as shown in Eq. 5 were fit to the experimental data using multi-linear regression to determine the parameters (τ_f, τ_p, and τ_r). This method provided values for the fitted parameters and identified the control mechanism.

$$t = \tau_f X + \tau_r\left(1 - (1 - X)^{\frac{1}{3}}\right) + \tau_p\left(1 - 3(1 - X)^{\frac{2}{3}} + 2(1 - X)\right) \quad (5)$$

The parameters τ_f, τ_p, and τ_r in Eq. 5 for the SCM are defined by Levenspiel [20].

The SCM was applied to the Li dissolution kinetics in the aqueous phase at temperatures ranging from 65 to 100 °C. The regressed experimental data using the

Table 1 Chemical speciation of the Co-Oxalate system generated with Visual MINTEQ

pH	Co:OA = 1:1		Co:OA = 1:1.5		Co:OA = 1:3	
	CoC_2O_4(mole %)	$Co(C_2O_4)_2^{2-}$(mole %)	CoC_2O_4(mole %)	$Co(C_2O_4)_2^{2-}$(mole %)	CoC_2O_4(mole %)	$Co(C_2O_4)_2^{2-}$(mole %)
1	91.65	1.26×10^{-4}	97.58	9.63×10^{-4}	98.21	4.36×10^{-3}
2	98.74	2.95×10^{-4}	99.81	1.68×10^{-2}	99.74	9.16×10^{-2}
3	99.69	7.18×10^{-4}	99.79	1.81×10^{-1}	99.01	9.75×10^{-1}
4	99.89	1.68×10^{-3}	98.90	1.09	95.09	4.91
5	99.93	2.49×10^{-3}	98.13	1.86	92.22	7.78
6	99.93	2.66×10^{-3}	98.01	2.00	91.73	8.27
7	99.93	2.69×10^{-3}	97.99	2.01	91.67	8.33

Table 2 SCM parameters for Li dissolution in Eq. 5

Digestion temperature (°C)	Fitting parameters		
	τ_p(min)	τ_r(min)	τ_f(min)
65	581.43	0	0
75	319.16	0	0
85	172.04	0	0
90	96.09	0	0
100	75.78	0	0

SCM is shown in Fig. 4. The Co:OA ratio of 1:3 and the S/L ratio of 15 g/L were kept constant. The fitted parameters (τ_f, τ_p, and τ_r) for each temperature are provided in Table 2. Under the current conditions over the range of temperature studied, the diffusion of oxalic acid (i.e., reactant) through the formation of the product layer (i.e., $CoC_2O_4{\cdot}2H_2O$) was found to be the rate-determining step. To ensure that the agitation speed did not have an effect on the kinetics, the digestion reaction at 90 °C was studied at several speeds (200, 350, 600, and 850 rpm) and similar Li dissolution kinetics were observed. This observation supports the conclusion that the rate of reaction is controlled by the product layer formation as shown in Table 2. Additional studies to improve the Li extraction kinetics are ongoing.

Proposed Closed-Loop Metal Recovery Process

This study is focused on the development of an environmentally responsible and economical closed-loop process for the recovery of Li and Co from $LiCoO_2$ using oxalic acid as a green digestion reagent. Hence, recovery of oxalic acid and no waste generation are critical towards maintaining the sustainability and economics for this chemistry. The closed-loop process is shown in Fig. 5 and provides a method for separating Li and Co into separate phases during the digestion followed by precipitation of the Li from the aqueous phase in the form of Li_2CO_3 using K_2CO_3. The precipitated cobalt oxalate is calcined in a vertical tube furnace to produce cobalt oxide. Additional details on the precipitation and calcination can be found in our previous work [10]. Some oxalate is lost as CO_2 during calcination to form cobalt oxalate; therefore, it is important to recover the majority of the oxalate in the aqueous phase after Li precipitation. Hence, an ion-exchange process using a strong acid cation exchange resin (like Amberlyst® 15) was used to exchange the K^+ ion with H_3O^+ in order to regenerate the oxalic acid. A solid-state reaction between Li_2CO_3 and Co_3O_4 can be performed to synthesize $LiCoO_2$ and regenerate the electrode material [21, 22]. In order to reduce the amount of Co in the recycled electrode material, nickel oxide or manganese oxide can be added to synthesize new electrode materials.

Conclusions

Recovery of Li and Co from spent $LiCoO_2$ electrodes is essential for meeting the rapid increase in the anticipated future LiB market demand. An oxalic acid-based chemistry provides a process for the economic recovery and separation of Li and Co from $LiCoO_2$ in a sustainable and environmental-friendly technology. For an efficient metal extraction and separation using oxalic acid, the acidity of the digestion is an important factor. Experimental observations and preliminary speciation results indicate that the presence of excess $C_2O_4{}^{2-}$ anion (at pH > 4) leads to the formation of soluble complexes (i.e., $Co(C_2O_4)_2^{2-}$). To understand the Li extraction kinetics for future scale-up, a shrinking core model was developed and identified the rate-limiting step as the formation of a solid $CoC_2O_4{\cdot}2H_2O$ layer over the reacting $LiCoO_2$ core, which creates additional resistance for the diffusion and reaction of oxalic acid with the core. Further experiments are underway to improve the extraction and separation efficiency of Li and Co from $LiCoO_2$. In addition, oxalate chemistry can be applied to other metal recovery processes such as Fe and Al from bauxite, and Fe and Ti from ilmenite as well as other critical metals from used NMC cathodes and REEs from coal fly ash.

Conflict of Interest The authors declare that they have no conflict of interest.

References

1. van Schalkwijk W, Scrosati B (2002) Advances in lithium ion batteries introduction. Advances in lithium-ion batteries. Springer, Boston, pp 1–5
2. Gaines L (2014) The future of automotive lithium-ion battery recycling: charting a sustainable course. SM&T 1–2:2–7. https://doi.org/10.1016/j.susmat.2014.10.001
3. Curry C (2017) Lithium-ion battery costs and market. Bloom New Energy Financ 5:4–6
4. Wakihara M, Li G, Ikuta H (2007) Cathode active materials with a three-dimensional spinel framework. In Lithium Ion Batteries; Wiley, Hoboken, pp 26–48
5. Monge M, Gil-Alana LA (2019) Automobile components: lithium and cobalt. Evid Persistence Energy 169:489–495
6. Verma A et al (2019) Recovery of metals from cathode of lithium-ion batteries: a closed-loop approach using oxalate chemistry. Paper presented at the 2019 AIChE Annual Meeting, Orlando, Florida, 10–15 Nov 2019
7. Heelan J, Gratz E, Zheng Z, Wang Q, Chen M, Apelian D, Wang Y (2016) Current and prospective li-ion battery recycling and recovery processes. JOM 68(10):2632–2638. https://doi.org/10.1007/s11837-016-1994-y
8. Zeng X, Li J, Singh N (2014) Recycling of spent lithium-ion battery: a critical review. Crit Rev Env Sci Technol 44(10):1129–1165. https://doi.org/10.1080/10643389.2013.763578
9. Harper G, Sommerville R, Kendrick E, Driscoll L, Slater P, Stolkin R, Anderson P (2019) Recycling lithium-ion batteries from electric vehicles. Nature 575(7781):75–86. https://doi.org/10.1038/s41586-019-1682-5
10. Verma A, Johnson GH, Corbin DR, Shiflett MB (2020) Separation of lithium and cobalt from LiCoO2: a unique critical metals recovery process utilizing oxalate chemistry. ACS Sustain Chem Eng 8(15):6100–6108. https://doi.org/10.1021/acssuschemeng.0c01128

11. Wang T, Luo H, Bai Y, Li J, Belharouak I, Dai S (2020) Direct recycling of spent NCM cathodes through ionothermal lithiation. Adv Energy Mater 10(30):2001204
12. Verma A, Kore R, Corbin DR, Shiflett MB (2019) Metal recovery using oxalate chemistry: a technical review. Ind Eng Chem Res 58(34):15381–15393. https://doi.org/10.1021/acs.iecr.9b02598
13. Sohn J-S, Shin S-M, Yang D-H, Kim S-K, Lee C-K (2006) Comparison of two acidic leaching processes for selecting the effective recycle process of spent lithium ion battery. Geosyst Eng 9(1):1–6. https://doi.org/10.1080/12269328.2006.10541246
14. Zeng X, Li J, Shen B (2015) Novel approach to recover cobalt and lithium from spent lithium-ion battery using oxalic acid. J Hazard Mater 295:112–118. https://doi.org/10.1016/j.jhazmat.2015.02.064
15. Krishnamurty KV, Harris GM (1961) The chemistry of the metal oxalato complexes. Chem Rev 61(3):213–246. https://doi.org/10.1021/cr60211a001
16. Meng Q, Zhang Y, Dong P (2018) A combined process for cobalt recovering and cathode material regeneration from spent LiCoO2 batteries: Process optimization and kinetics aspects. Waste Manage 71:372–380. https://doi.org/10.1016/j.wasman.2017.10.030
17. Takacova Z, Havlik T, Kukurugya F, Orac D (2016) Cobalt and lithium recovery from active mass of spent Li-ion batteries: theoretical and experimental approach. Hydrometallurgy 163:9–17. https://doi.org/10.1016/j.hydromet.2016.03.007
18. Setiawan H, Petrus HTBM, Perdana I (2019) Reaction kinetics modeling for lithium and cobalt recovery from spent lithium-ion batteries using acetic acid. Int J Min Met Mater 26(1):98–107. https://doi.org/10.1007/s12613-019-1713-0
19. Yuliusman Y, Fajaryanto R, Nurqomariah A, Silvia (2018) Acid leaching and kinetics study of cobalt recovery from spent lithium-ion batteries with nitric acid. E3S Web of Conferences 67:03025. https://doi.org/10.1051/e3sconf/20186703025
20. Levenspiel O (1999) Chemical reaction engineering. Wiley, New York
21. Lagergren C, Lundblad A, Bergman B (1994) Synthesis and performance of LiCoO2 cathodes for the molten carbonate fuel cell. J Electrochem Soc 141(11):2959
22. Chen H, Qiu X, Zhu W, Hagenmuller P (2002) Synthesis and high rate properties of nanoparticled lithium cobalt XE "Cobalt" oxides as the cathode material for lithium-ion battery. Electrochem Commun 4(6):488–491

Refining of Mixed Sulphide Precipitate to Produce Battery Grade Metals Using Outotec Pressure Oxidation Process

C. Ecott

Abstract The demand for base metals is accelerating with interest in nickel and cobalt accelerating due to the global demand for electrical vehicles and cobalt featuring on the list of critical raw materials (CRM) for the European Union, that is materials fundamental to Europe's economy, growth and jobs. Terrafame Oy is an established metals producer of nickel, zinc, cobalt and copper in Sotkamo, Finland. The process utilizes bioleaching and precipitation to produce sulphide metals with lower environmental impact and in an energy efficient manner. This paper describes the next stage of metals production at the Terrafame site, a hydrometallurgical refining process utilizing Outotec proprietary equipment to treat the existing NiCo-sulfide intermediate, producing battery grade nickel and cobalt sulphate crystals as products. Focus is put to the oxidative pressure leaching of the NiCo sulphide as well as the purification of pregnant leach solution to achieve battery grade quality.

Keywords Nickel · Cobalt · Battery metals · Pressure leaching

Introduction

Approximately 2–3% of new cars sold globally are electric or hybrid vehicles [6], but this market share is expected to increase whereby 25–30% of new cars worldwide are estimated to be electric or hybrid vehicles by 2030 [2]. Combined with a growing Additionally the share of nickel which is used in battery applications is growing. Annually these combined factors are expected to correspond to an additional 400,000 tonnes of nickel demand for battery applications annually as of 2030 [5].

Terrafame Oy is a multi-metal company producing nickel, zinc, cobalt and copper at its mine and metals processing plant in Sotkamo, Finland (Fig. 1). As one of Europe's largest sulphide nickel resources, the mining area covers over 60 km^2, open

C. Ecott (✉)
Outotec (Finland) Oy, Puolikkotie 10, 02230 Espoo, Finland
e-mail: christopher.ecott@mogroup.com

Terrafame Oy, Malmitie 66, 88120 Tuhkakylä, Finland

C. Anderson et al. (eds.), *Ni-Co 2021: The 5th International Symposium on Nickel and Cobalt*, The Minerals, Metals & Materials Series,
https://doi.org/10.1007/978-3-030-65647-8_13

Fig. 1 Site location of Terrafame's mine and processing facility. (Color figure online)

pit mining is followed by materials handling; crushing and agglomeration; before 2 stages of bio-heap leaching; 13–18 months of primary leaching and a 3–4-year secondary leaching stage. Pregnant leach solution then undergoes a series of metal precipitation stages.

The plant's main product is currently a nickel cobalt mixed sulphide precipitate which is currently sold as intermediates to metal refining companies. Terrafame will advance their position in the value chain by commencing production of Ni and Co chemicals suitable for use in electrical batteries. This mixed sulphide precipitate (MSP) intermediate will directly feed the new battery grade chemical plant which is being constructed in a new facility on the same site. The production targets for this new facility are 170,000 tonnes of $NiSO_4$ and 7,400 tonnes of $CoSO_4$ annually. Approximately 115,000 tonnes of ammonium sulphate will also be produced as a by-product of the process [7]. Construction began in 2019 with production expected to commence in early 2021.

Terrafame Hydrometallurgical Mixed Sulphide Leaching Process

Terrafame's hydrometallurgical leaching process starts with pulping of the NiCo sulphide to a solid density of ~ 200 g/l. The plant capacity was designed to process 35,000 t/a of Ni and 670 t/a or Cobalt.

Slurry is further diluted using recirculating feed from ion removal filter and make up water to achieve solids concentration of 135 g/l before feed to the autoclave. The NiCo Sulphide is leached under pressure oxidation (POX) conditions in a continuously operating, five compartment autoclave. The autoclave retention time is approximately 2.5 h and leaching is conducted at approximately 180 °C with oxygen overpressure of 5–6 bar. Total pressure in the autoclave is approx. 15 bar. Oxygen is injected to each individually agitated compartment, oxidizing sulphidic compounds and ferrous iron to ferric. The pH of the process is typically less than 1. Pressure leaching is followed by a one stage flashing of solution and vent gas scrubbing.

After leaching residual iron is precipitated in atmospheric reactors with ammonia water as a neutralization reagent. After precipitation of iron, solids are separated by thickening and filtration. There is a filtration stage for both overflow and underflow of the thickener. The filtered and washed thickener underflow is discharged to Outotec design battery limit. The overflow nickel and cobalt containing pregnant leach solution (PLS) is pumped from thickener overflow to impurity solvent extraction (SX) then through cobalt solvent extraction circuit where cobalt is extracted from the nickel containing solution. The purified and concentrated cobalt solution goes to $CoSO_4 \cdot 7H_2O$ crystallization. The Co Raffinate (Ni rich solution) continues to nickel solvent extraction which purifies the $NiSO_4$ before transfer to $NiSO_4 \cdot 6H_2O$ crystallization. Raffinate from Nickel SX is fed to ammonium sulphate crystallization stage (Fig. 2).

Pressure Oxidative Leaching

Solids are leached the in acidic oxidative conditions forming dissolved metal sulphate solution of nickel and cobalt as presented in Eqs. 1–5 below. Gypsum leaches according to reaction 3 and re-precipitates until calcium concentration is at saturation point. Leaching of ferrous sulphide (FeS) consumes acid as presented in Eq. 2. Leached iron precipitates partially as hematite producing acid according to reaction 6. Acid is also produced by leaching of elemental sulfur in reaction 5.

$$MeS_{(s)} + 2\,O_{2(g)} \rightarrow Me^{2+}_{(aq)} + SO^{2-}_{4\,(aq)} \tag{1}$$

$$Me = Ni,\ Co,\ Zn,\ Cu,\ Mn$$

$$2\,FeS_{(s)} + 4.5\ O_{2(g)} + 2\,H^{+}_{(aq)} \rightarrow 2\,Fe^{3+}_{(aq)} + 2\,SO^{2-}_{4(aq)} + H_2O_{(l)} \tag{2}$$

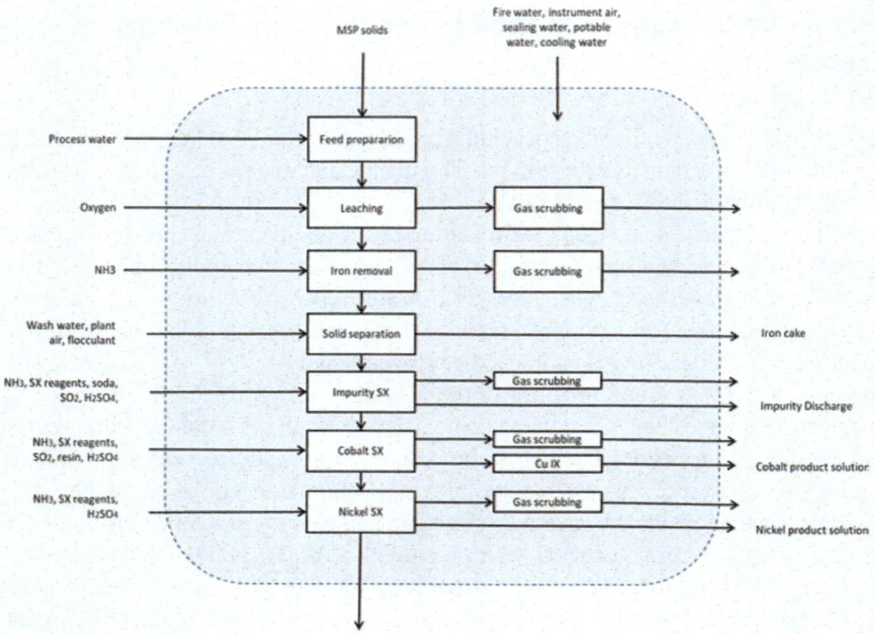

Fig. 2 Battery limit diagram of the pressure oxidation and solvent extraction stages of Ni Cobalt Sulfide Plant

$$CaSO_4 \cdot 2H_2O_{(s)} \leftrightarrow Ca^{2+}_{(aq)} + SO^{2-}_{4\ (aq)} + 2H_2O_{(1)} \tag{3}$$

$$Mg(OH)_{2(s)} + 2\,H^{+}_{(aq)} \rightarrow \ Mg^{2+}_{(aq)} + \ 2\,H_2O_{(l)} \tag{4}$$

$$S_{(s)} + 1.5\ O_{2(g)} + H_2O_{(1)} \rightarrow SO^{2-}_{4\ (aq)} + 2\,H^{+}_{(aq)} \tag{5}$$

$$2\,Fe^{3+}_{(aq)} + \ 3\,H_2O_{(l)} \rightarrow \ Fe_2O_{3(s)} + \ 6\,H^{+}_{(aq)} \tag{6}$$

The final free acid concentration in autoclave is ~15g/l, and the pregnant leach solution (PLS) contains a nickel concentration of ~ 65 g/l, and iron concentration ~0.45 g/l and sulphate concentration of ~130 g/l.

The leaching reactions are exothermal and thus external cooling is required to regulate the autoclave temperature at operating value (Fig. 3). The temperature is maintained using 1–4 cooling elements in each autoclave compartment which continuously circulate cooling water. Each element can be closed and isolated if necessary. In the first two compartments the majority of the reactions take place, as such these have the highest cooling requirement. Part of this cooling is carried out by cool slurry feed to autoclave in the first compartment but additional cooling of total 10 MW is needed in autoclave.

Fig. 3 The POX autoclave during installation. (Color figure online)

Autoclave pressure is controlled by venting gases of unreacted oxygen, nitrogen and steam from the autoclave using ejector venturi gas scrubber prior to atmospheric release. All the autoclave utilities: oxygen, cooling water, sealing water, flushing water and steam for start-up, operate at higher pressure than autoclave operating pressure to prevent slurry blow back from autoclave through oxygen and steam feed lines.

Slurry from the autoclave is discharged by pressure difference to a one stage flashing; allowing for safe controlled decrease of pressure and temperature from the autoclave operating conditions to atmospheric pressure and temperature. The vapor generated in the flashing is fed to venture-type gas scrubber in the gas scrubbing area and the slurry from the flash tank flows by gravity to leach slurry tank which provides a buffer before atmospheric iron removal stages. The sulfuric acid concentration after flashing is ~ 30 g/l, nickel concentration ~ 80 g/l iron concentration of ~0.5 g/l and sulphate concentration ~170 g/l.

Iron Removal

Iron precipitation occurs under atmospheric conditions, the temperature is not regulated by the process but contains latent heat of close to 100 °C after flashing. The remaining iron in solution is precipitated as ferric hydroxide using ammonia water as a neutralization agent to achieve pH 3.5–4. The target iron levels in PLS is less than 60 ppm after the 2-h residence time. Low pressure oxygen can be fed to reactors to oxidize ferrous iron to ferric if needed.

Acid neutralization is presented in Eq. (7). Iron and gypsum precipitate according to Eqs. (8) and (3).

$$H^+_{(aq)} + OH^-_{(aq)} \rightarrow H_2O_{(l)} \quad (7)$$

$$Fe^{3+}_{(aq)} + 3\,H_2O_{(l)} \rightarrow Fe(OH)_{3(s)} + 3\,H^+_{(aq)} \quad (8)$$

The slurry is thickened with flocculant addition to accelerate solid settling before pressure filtration stage, part of the thickener underflow can be recycled back to iron precipitation reactors as a seed material enhancing crystal growth. Since the flow rate of both underflow removal and underflow recycle to reactors is small, continuous slurry circulation is arranged for the underflow to help preventing blockages in the slurry line.

To minimize crud formation in solvent extraction the thickener overflow solution is pumped to the agitated polishing filter feed tank through heat exchanger where it is cooled down to approx. 60 °C. The target solid concentration in the PLS is less than 50 ppm. Due to the small volume of solids the filtration cycle is long 12–24 h. The slurry from the filter cake slurry tank is recycled gradually to autoclave feed tank. The filtered solution is collected in the filtrate tank and diluted to target nickel concentration of ~ 75 g/l. The ammonium concentration is ~7 g/l. From the filtrate tank PLS is pumped to impurity SX.

With high nickel and ammonium concentration there is possibility of formation of nickel-ammonium double salt. As the nickel concentration is high, ammonium concentration must be kept low by controlling ammonium-containing recycle flows to pressure leaching from filter wash water and polishing filter drain slurry. The solubility of double salt is also dependent on temperature and it decreases with decreased temperature, the risk of double salt formation increases in further process steps, iron removal and solvent extraction, where ammonia-based neutralization reagent is used, and temperature is lower. According to experimental data at 60 °C temperature with nickel concentration of 75 g/l double salt is formed if ammonium concentration exceeds 25–30 g/l. If temperature decreases to 20 °C the double salt is formed with as low ammonium concentration as 5 g/l. The risk of ammonium-nickel double salt formation must therefore be kept in mind and the ammonium concentration monitored. If temperature is dropped to 20 °C the double salt will already form with operating concentrations of nickel and ammonium.

Thickener underflow slurry is pumped to the filter feed tank, whose cooling elements cool the slurry down to 60 °C, from here slurry is pumped to Iron residue filter. One pressure filter is used for filtration and washing of iron residue slurry and the filtrate and wash filtrate are collected into separate tanks before the wash filtrate is recycled back to autoclave feed tank. Filtrate is pumped to thickener and continues to polishing filtration with the overflow. Iron residue cake is discharged after drying to Outotec's design battery limit (Fig. 4).

Fig. 4 Installation of filters and reactors in iron removal area. (Color figure online)

Leaching and Iron Removal Gas Scrubbing

Autoclave off-gas and flashed steam are treated in a venturi scrubber, the solids and acid carryover in the off-gas are washed out by spraying water through nozzles to the venture throat and the scrubbed gas is released to atmosphere. The wash water is collected in the internal water reservoir of the scrubber and re-circulated to the nozzles by recycle pumps. Make up water is added to the reservoir to cool the water and dilute impurities in the circulation. Bleed from the wash water circulation is pumped to autoclave feed tank. 25% NaOH-solution can be added to recycle pump suction to neutralize acid that may be built-up in the wash water circulation.

Off-gases from leached slurry tank and iron precipitation reactors are also collected prior to atmospheric release and scrubbed in a spray chamber scrubber. The wash water is bled to iron precipitation reactors when needed to remove condensate, if any, and control impurities in the wash water circulation.

Solvent Extraction

The Terrafame process has three distinct solvent extraction (SX) processes, Impurity SX from removal of metals such as Al, Fe, Zn, Ca, Mn; Cobalt SX producing a purified concentrated cobalt solution for crystallization; Nickle SX producing a purified concentrated Nickel solution for crystallization. The generic SX process is shown in Fig. 5 and the main reactions occur in all solvent extraction stages as shown below in Eqs. 9 and 10 [1, 3]:

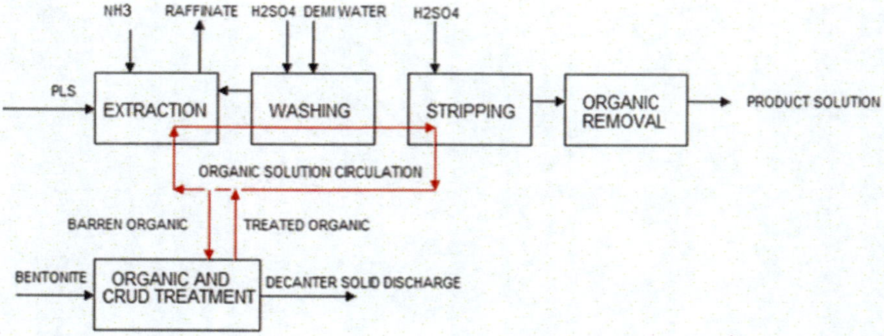

Fig. 5 Generic block diagram of typical solvent extraction process. (Color figure online)

$$\text{Extraction; } Me^{2+}_{(aq)} + SO_{4\ (aq)}^{2-} + 2\,HR_{(org)} \leftrightharpoons R_2Me_{(org)} + 2\,H^{+}_{(aq)} + SO_{4\ (aq)}^{2-} \quad (9)$$

$$\text{Strippingphase; } R_2Me_{(org)} + 2\,H^{+}_{(aq)} + SO_{4\ (aq)}^{2-} \leftrightharpoons Me^{2+}_{(aq)} + SO_{4\ (aq)}^{2-} + 2\,HR_{(org)} \quad (10)$$

$$Me = Ni, Co, Zn, Cu, Mn\ldots; HR = \text{organic phase}$$

Impurity Solvent Extraction

The filtered PLS solution is pumped to impurity SX where impurities like iron, aluminum, calcium, zinc and manganese are removed from the Co/Ni solution using the extractant Di-(2-ethylhexyl)phosphoric acid (D2EHPA) ~ 20% vol in diluent. The impurity SX has three extraction stages with pH control using soda water (Na_2CO_3) ensuring selective extraction of impurity elements. Sulfur dioxide is added to PLS to keep the redox potential low preventing Co^{2+} and Fe^{2+} oxidation. Purified raffinate continues through an after settler and dual media filters prior to CoSX.

Loaded organic is scrubbed with sulfuric acid removing any co-extracted cobalt. The organic then passes through two stripping stages which also use sulfuric acid; the stripped solution is then pumped to existing bioleaching process utilizing remaining sulphuric acid and zinc content. High strength sulphuric acid is used to prevent gypsum build up, whilst a lower external O/A ratio is used to dilute Ca concentration in stripping solution.

Stripped organic solution is pre-neutralized with soda water before re-entering solvent extraction. Because ferric iron in PLS is extracted and it is then not easily stripped in scrubbing or stripping stages, the process also contains an organic treatment to control iron concentration in organic solution where iron is stripped in one stage (FeS) at lower pH.

Cobalt Solvent Extraction

The feed solution of impurity SX raffinate (~2.5 g/l Co, ~75 g/l Ni) is pumped through active carbon filters to remove any D2EHPA preventing any contamination of reagents. Extractant used in CoSX is Cyanex 272 ~ 20 vol. % in diluent and the temperature of the feed solution is targeted at 55 °C, to promote cobalt selectivity over nickel. The selective extraction of different metals using CYANEX 272 extractant is presented in Fig. 6. CoSX has four extraction stages using ammonia water to ensure the correct pH profile for Co-extraction. Sulfur dioxide is added to feed to keep the redox potential low preventing Co^{2+} oxidation. Raffinate from CoSX is purified Ni containing solution which continues through active carbon filtration prior to NiSX.

Any co-extracted Mg, Mn and other trace impurities which would affect the quality of the end $CoSO_4$ product are scrubbed from the loaded organic in three stages using Co containing solution from stripping. The organic then passes through three stripping stages which use pure sulfuric acid as a stripping agent in order to make highly purified $CoSO_4$ solution. The external O/A ratio is adjusted accordingly to produce ~100 g/l Co solution which is pumped through organic removal dual media filters and then subsequent ion exchange. Stripped organic solution is pre-neutralized with ammonia water before re-entering solvent extraction.

Ion exchange polishes the $CoSO_4$ solution removing the remaining impurities, especially copper to a target copper to a level of less than 1 mg/L. The ion-exchange resin is Lanxess's Lewatit MDS TP 207, which provides selective recovery of Cu over Co at lower pH of 2 to 3. The resin is regenerated with an elution stage of sulphuric acid which converts the resin functional groups back to acid form and washing with purified water. The purified cobalt sulphate solution continues to $CoSO_4$ crystallization plant. Copper containing regeneration solution is collected to separate vessels for further processing.

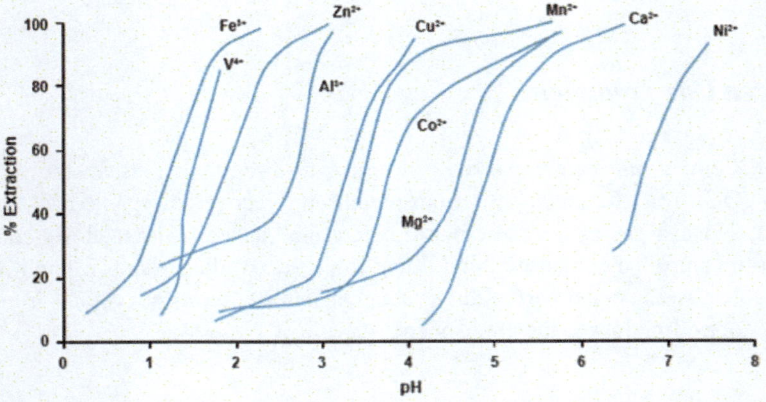

Fig. 6 Selective extraction of metals by CYANEX 272 Extractant from sulfate solution [4]. (Color figure online)

Fig. 7 Installation of modular SX units (left) and the view of multiple connected modules from above (right). (Color figure online)

Nickel Solvent Extraction

The feed solution of CoSX raffinate (~75 g/l Ni) is pumped through active carbon filters to remove any trace amounts of Cyanex preventing any contamination of reagents. Extractant used in NiSX is Versatic acid (V10) ~40 vol. % in diluent. NiSX has four extraction stages using ammonia water to control pH. The depleted raffinate from NiSX is further processed in ammonium sulphate crystallization circuit.

Any co-extracted Mg and other trace impurities which would affect the quality of the end $NiSO_4$ product are scrubbed from the loaded organic in one stage using Ni containing solution from stripping. As before the organic then passes through three stripping stages which use pure sulfuric acid as a stripping agent in order to make highly purified $NiSO_4$ solution. The Ni solution is pumped through organic removal dual media filters before crystallization step and the stripped organic solution is pre-neutralized with ammonia water before re-entering solvent extraction (Fig. 7).

SX Area Gas Treatment

Every SX circuit has its own packed bed scrubbing circuit which treats the volatile gases (VOC) from the mixer settler units, returning recovered organic back to SX circuits. Exhaust gas enters the column below the packed bed and flows counter current to controlled cooled washing liquid to the top of the column where it exits the scrubber. At the upper part of the scrubber, a demister removes droplets from the airflow controlling loss to the atmosphere.

Crystallization of Products

The purified $NiSO_4$ and $CoSO_4$ solutions and the NiSX raffinate from the solvent extraction areas are processed in separate crystallization circuits utilizing third party equipment producing clean battery grade sulphate crystals of Ni, Co as well as ammonium sulfate.

References

1. BASF (2015) Global mining solution, technical information TI/EVH 0152 e
2. International Energy Agency (2019) Global EV outlook 2019
3. Ritcey GA (1984) Solvent extraction, principles and application to process metallurgy. Three volumes
4. Solvay (nd) CYANEX® 272 Extractant
5. Terrafame (nd) Terrafame.com: https://www.terrafame.com/media/terrafame-ltd.-plans-nickel-and-cobalt-chemicals-production-for-battery-applications.pdf
6. The electric vehicle world sales database (2020) Global BEV & PHEV Sales for 2019. EV Volumes: http://www.ev-volumes.com/country/total-world-plug-in-vehicle-volumes/
7. Ymparisto.fi (2018) Terrafame nikkeli- ja kobolttisulfaattien tuotanto ympäristövaikutusten arviointiohjelma. Ymparisto: https://www.ymparisto.fi/fi-FI/Asiointi_luvat_ja_ymparistovaikutusten_arviointi/Ymparistovaikutusten_arviointi/YVAhankkeet/Terrafame_Oyn_nikkeli_ja_kobolttisulfaattien_tuotanto_Sotkamo

Part III
Hydrometallurgy

Alkaline Leaching of Nickel from Electric Arc Furnace Dust Using Ammonia-Ammonium Glutamate as Lixiviant

Erik Prasetyo, Fathan Bahfie, and Anton Sapto Handoko

Abstract Electric arc furnace dust (EAFD) is a potential secondary source of nickel (Ni). Although Ni content (0.5%) is much lower than traditional ore i.e. laterites, EAFD as raw material has advantage in terms of straightforward processing. It was demonstrated that Ni could be directly recovered from EAFD by leaching using glutamate in alkaline condition. In this study, ammonia-ammonium glutamate system as lixiviant would be tested for Ni recovery in terms of kinetic under different leaching parameters, which include stirring speed, oxidant concentration (H_2O_2), ammonia concentration, glutamic acid concentration, and temperature. The results demonstrated that stirring speed had negative effect on Ni rate dissolution. Addition of H_2O_2 substantially increased the rate since the addition imparted oxidative condition, which was favorable for Ni dissolution. Kinetic modelling using several models indicated the dissolution process was affected by interface transfer and diffusion. Calculated activation energy indicated that the effect of temperature on Ni leaching was minimal.

Keywords Nickel · Leaching · Electric arc furnace dust · Glutamic acid · Glutamate · Ammonia

Introduction

EAFD is side product in steel processing and classified as hazardous materials and the amount of the dust produced is significant, which for each ton of steel made, 11–20 kg of dust is generated [1]. The dust content of base metal is substantial e.g. Zn (>40%), Cu (~1%) and Ni (~0.5%). Hence the dust processing is beneficial not only to address environmental problem but also resource depletion problem. Although Ni exists in EAFD in much lower concentration compared to common Ni ore such as laterites, simpler phase composition renders easier and straightforward processing.

E. Prasetyo (✉) · F. Bahfie · A. S. Handoko
Research Unit for Mineral Technology – Indonesian Institute of Sciences, Jl. Ir. Sutami km. 15, Tanjung Bintang, Lampung Selatan 35361, Indonesia
e-mail: erik.prasetyo@lipi.go.id

C. Anderson et al. (eds.), *Ni-Co 2021: The 5th International Symposium on Nickel and Cobalt*, The Minerals, Metals & Materials Series,
https://doi.org/10.1007/978-3-030-65647-8_14

In previous research, Zn and Cu were successfully leached from EAFD using monosodium glutamate in alkaline condition due to strong binding of glutamate to metal ions [2] and selectively enriched undesirable elements such as Fe, Al, Mg and Ca in solid residue [3, 4]. However, the previous research also revealed the disadvantage of using sodium hydroxide as pH modifier, hence in present study sodium hydroxide would be substituted with ammonia to form ammonia-ammonium glutamate leaching system. Aside from ammonia addition, other modifier would be added i.e. H_2O_2 to increase oxidation potential of liquid phase, which is favorable for metal dissolution. Since EAFD also contains Ni, in this study the efficacy of glutamate as complexant for Ni dissolution would be tested in terms of kinetic studies in varied conditions including stirring speed, H_2O_2 concentration, ammonia concentration, glutamate concentration and temperature.

Experimental

Materials and Instrumentation

EAFD sample was obtained from local smelter in Banten Province, Indonesia. 90% of dust passed 270 mesh sieve and used further in leaching studies. Monosodium glutamate (MSG), purchased from PT Ajinomoto Indonesia (99% purity) was precipitated as glutamic acid by acidic precipitation. About 187 gr MSG was dissolved in deionized water and the pH was set to 3 using sulfuric acid. Glutamic acid precipitate was separated from solution using centrifugation, washed with deionized water and dried in oven at 110 °C (about 90% recovery). Chemicals such as ammonia, sulfuric acid, hydrogen peroxide were purchased from Merck, Darmstadt, Germany, all in analytical grade. Deionized water was used throughout the experiment. Elemental analysis was carried out using ICP-OES (Analytik Jena, Plasma Quant 9000 Elite, Germany). XRD analysis to determine the phase present in the raw material was carried out using Panalytical, Expert3 Powder.

Leaching Procedure

Leaching was performed in 2 L glass reactor, equipped with condenser, thermometer, magnetic stirrer and sampling outlet. 250 ml lixiviant was introduced into the reactor and EAFD was added to make liquid solid ratio 20 ml/g. Pregnant leach solution was sampled within certain interval using sampling outlet to determine Ni dissolution rate. Aspirated sample was filtered using syringe filter (Agilent, 25 mm, 0.45 μm pore size), diluted using nitric acid 2% and analyzed using ICP-OES. Ni recovery was calculated using Eq. (1).

$$R(\%) = \frac{C_E \times V}{C_o \times m} \times 100\% \quad (1)$$

C_E Ni concentration in supernatant solution (mg/L).
C_o Ni content in EAF dust (mg/g).
m mass of EAF dust used in leaching (g).
V leaching agent volume (L).

Kinetic Modelling

To describe kinetic leaching process, several kinetic models were tested to fit experimental data including shrinking core model (chemical reaction control) [5] (Eqs. 2–3) and diffuse and transport model as proposed by Dickinson and Heal [6] (Eqs. 4–5).

$$kt = 1 - (1 - R)^{1/3} \quad (2)$$

$$R = [1 - (1 - kt)^3]R_{max} \quad (3)$$

$$kt = (1 - R)^{1/3} - 1 \quad (4)$$

$$R = [1 - (1 + kt)^{-3}]R_{max} \quad (5)$$

In kinetic modelling non-linear equations (Eqs. 3 and 5) are used instead of linear equation, and to evaluate the fitting average relative error (ARE) [7] is applied, which is calculated according to Eq. 6.

$$ARE(\%) = \frac{100}{n-1}\sum_{i=1}^{n}\left(\frac{R_{exp} - R_{mod}}{R_{exp}}\right)^2 \quad (6)$$

k rate constant.
t leaching time.
R Ni recovery at t.
R_{max} Ni maximum recovery at each leaching test.
R_{exp} Ni recovery at t as experimental result.
R_{mod} Ni recovery at t as modelling result.

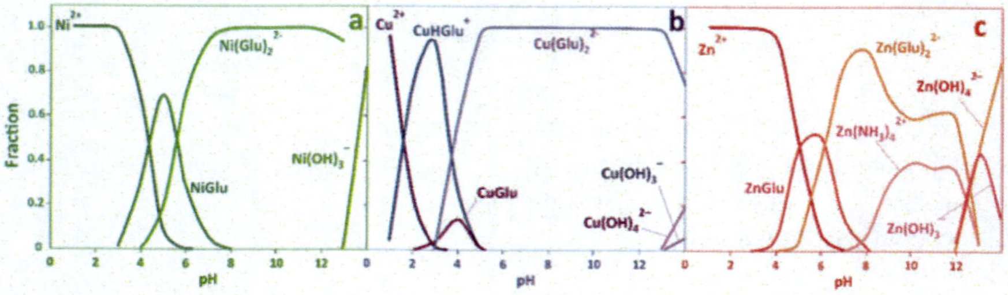

Fig. 1. Species distribution of (**a**) Ni, (**b**) Cu and (**c**) Zn as function of pH (condition cited in the text). (Color figure online)

Results and Discussion

Species Distribution

Leaching of Ni and other elements especially Cu and Zn was dictated by species distribution (Fig. 1). The figure was constructed using Visual MINTEQ 3.1 under condition: Ni 2 mM, Cu 5 mM, Zn 0.3 M, glutamate 1 M and NH3 1 M. It is clear that Ni and Cu speciations were controlled only by glutamate (the Ni and Cu ammine complex were negligible, Fig. 1a and b) at alkaline pH. In the case of Zn, the glutamate complex dominated at neutral and weak alkaline pH and at higher $pH > 8$, Zn tended to exist as ammine complex (Fig. 1c).

Material Characterization

XRD characterization revealed the major phase existed in the EAF dust sample before and after leaching (Figure 2). The raw material consists of zincite (ZnO), CuAl2O4 and metal alloys of Zn-Mn, Fe and Al-Ca-Zn (Figure 2a). After leaching, most of Zn phase was leaching out, shown the decrease of peak intensity in Figure 2b. AAS result analyses to determine quantitative elemental composition are listed in Table 1.

Effect of Initial H_2O_2 Concentration

Hydrogen peroxide was used to increase oxidizing potential of liquid phase to encourage Ni solubilization. In the test, hydrogen peroxide concentration was varied from 0 to 2% v/v with constant variable ammonia concentration 2 M (34 g/L), glutamic acid concentration 1 M, temperature 30 °C and stirring speed 350 rpm. The

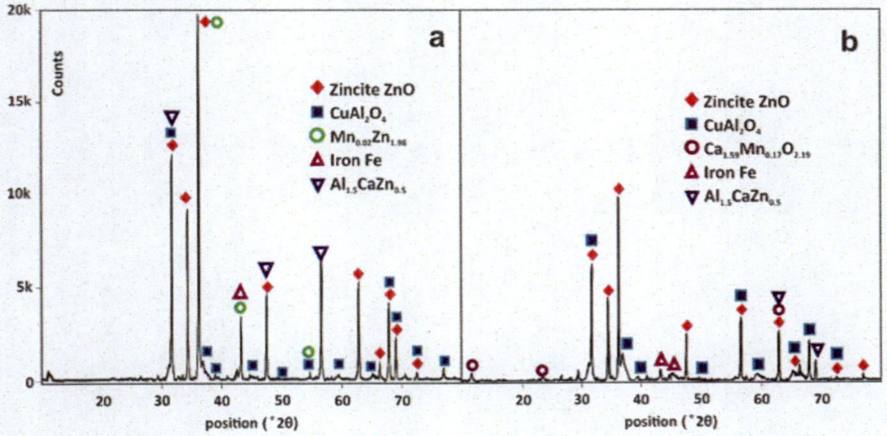

Fig. 2. Phases in raw materials (**a**) and after leaching (**b**) based on XRD scanning. Leaching condition: Glutamic acid 0.5 M, NH_4OH 2 M, H_2O_2 2%, pulp density 50 g/L, leaching time 6 h, room temperature. (Color figure online)

Table 1. Base metal contents n EAF dust sample (after aqua regia digestion and determination using ICP-OES)

Element	Zn	Ni	Cu	Fe
%wt	51.7 ± 0.78	0.54 ± 0.01	0.98 ± 0.01	1.52 ± 0.04

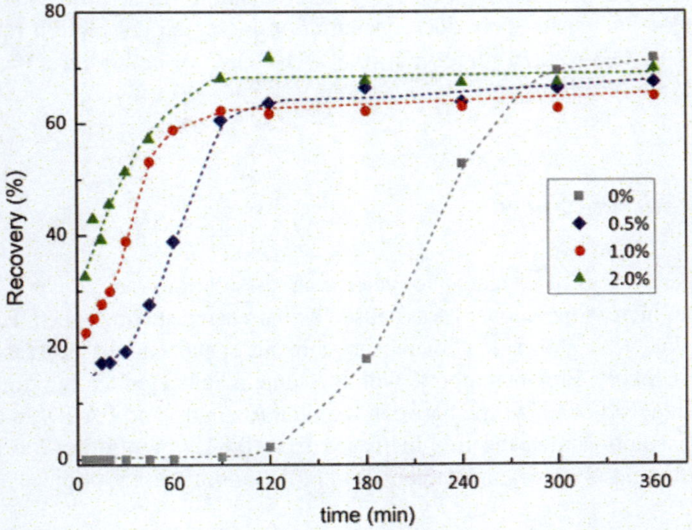

Fig. 3. Recovery of Ni using ammonia-ammonium glutamate lixiviant at different hydrogen peroxide concentration. (Color figure online)

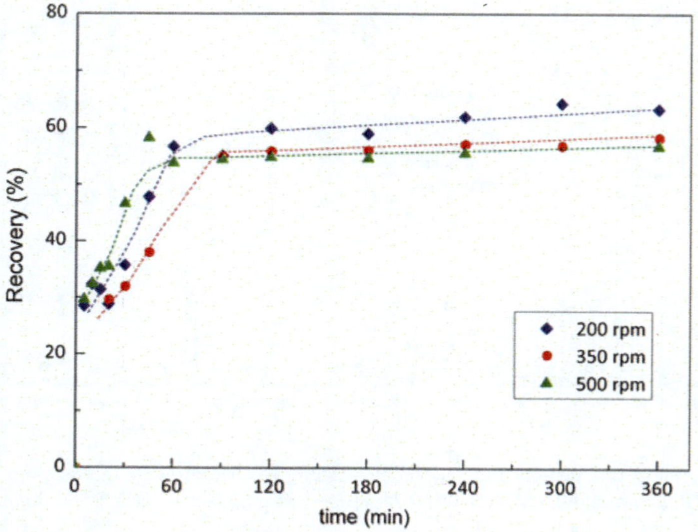

Fig. 4. Recovery of Ni at different stirring speed. Constant variable hydrogen peroxide 0.5%, ammonia concentration 1 M (17 g/L), glutamic acid concentration 0.5 M, temperature 30 °C°C. (Color figure online)

results depicted in Fig. 3 confirm the addition of peroxide in inducing the Ni dissolution. Without the addition of peroxide, Ni solubilization was delayed and occurred at $t > 90$ min. Increasing peroxide concentration increased dissolution rate of Ni from 0 (0% H_2O_2) to 0.16 mmol/min (0.5% H_2O_2), 0.19 mmol/min (1% H_2O_2), 0.28 mmol/min (2% H_2O_2).

Effect of Stirring Speed

Stirring speed is critical factor to determine dissolution rate. Several studies reported that increasing recovery was caused by increasing stirring speed. However, based on Fig. 4, it shows that increasing stirring speed caused decrease in Ni maximum recovery. Previous reports with the same results credited the higher stirring speed reduced contact time between lixiviant (oxidant) and EAFD. In the case of initial dissolution rate, the rate increased from 0.031 mmol/min (200 rpm) to 0.050 mmol/min (350 rpm) and decreased to 0.010 mmol/min (500 rpm).

Effect of Ammonia Concentration

To determine the synergist effect of ammonia, ammonia concentration was varied at 1, 1.25 and 1.5 M, while concentration of hydrogen peroxide, glutamic acid,

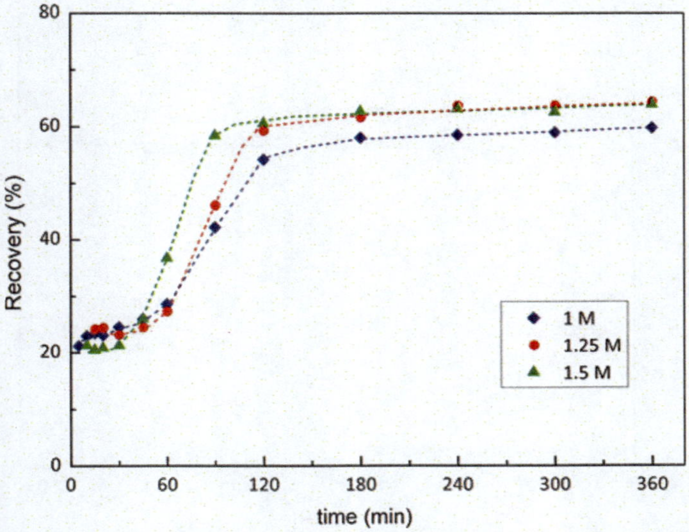

Fig. 5. Recovery of Ni using ammonia-ammonium glutamate lixiviant at different ammonia concentration. (Color figure online)

temperature and stirring speed were set at 0.5%, 1 M, 30 °C and 350 rpm. The results in Fig. 5 confirmed the addition of ammonia positively affected the dissolution rate and maximum recovery of Ni. As modifier, increasing ammonia concentration, increased pH from 8.3 (1 M) to 9.2 (1.25 M) and 9.4 (1.5 M). The effective amount of ammonia added to leaching system dictated by glutamic acid concentration, since the glutamic acid only dissolved in weak acid to alkaline condition (pH > 5).

Effect of Glutamate Concentration

The effect of glutamate as lixiviant was tested at concentration 0.5, 1 and 1.5 M. Figure 6 demonstrated that increasing glutamate suppress Ni dissolution. As explained before the pH of liquid phased was determined by the ratio of ammonia and glutamic acid added. Increasing glutamate (added as glutamic acid) decreased the pH due to neutralization from 9.5 (glutamate 1 M), to 8.3 (1.25 M) and 7.5 (1.5 M). Decreasing pH reduced dissolution rate and maximum recovery of Ni. At lower pH ammonia would be protonated and lost its ability as a synergist to form complex with Ni.

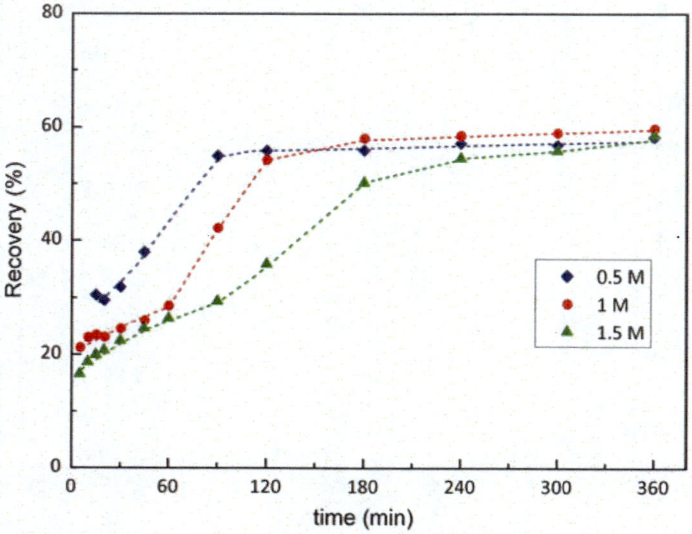

Fig 6. Recovery of Ni at different glutamate concentration. Constant variable hydrogen peroxide 0.5%, ammonia concentration 1 M (17 g/L), temperature 30 °C, stirring speed 350 rpm. (Color figure online)

Effect of Leaching Temperature

The temperature on leaching process was studied at 30, 55 and 65 °C. The concentration of glutamate, ammonia, hydrogen peroxide and stirring speed were speed constant at 0.5 M, 1 M, 0.5% and 350 rpm, respectively, which results were depicted in Fig. 7. The figure demonstrated that increasing temperature improves dissolution rate and maximum recovery of Ni as deduced from increasing slope at the early stage of leaching as temperature increases.

Kinetic Modelling

Table 2 listed the parameter k and R_{max} as modelling results using models. In general shrinking core model fails to model experimental data, while pseudo first order and interface-diffusion fitted better with the data. This indicates Ni dissolution was controlled by diffusion and transfer proses. As explained before Ni dissolution was encourage by oxidative condition, and this condition was partly affected by zinc (matrix) dissolution. Zn dissolution occurred first through diffusion into EAFD grain and increasing Zn in liquid phase increase oxidative potential and initiate Ni dissolution.

Activation energy (E_a) calculation using Arrhenius (Eq. 7) and apparent rate constant (k) obtained from modelling, where A, T, and R are the frequency factor, temperature and gas constant, respectively. Energy activation in this case is

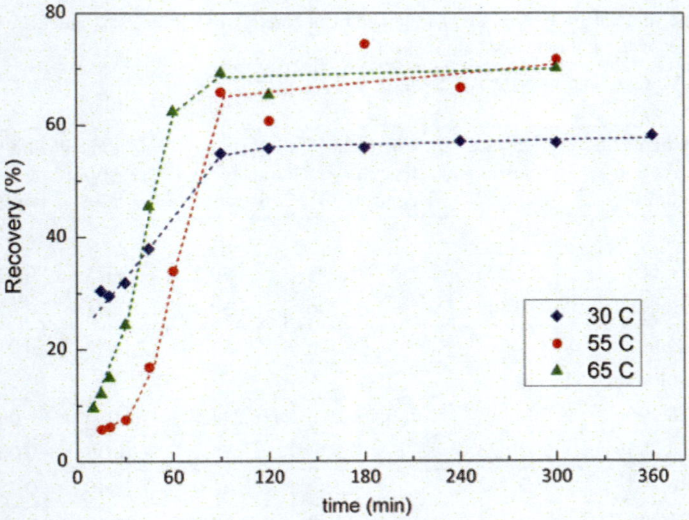

Fig 7. Recovery of Ni using ammonia-ammonium glutamate lixiviant at different leaching temperature. (Color figure online)

24.6 kJ/mol. Low value of activation energy is arbitrarily considered to be controlled by transport process and effect of temperature was minor.

$$k = Ae^{-\frac{E_a}{RT}} \text{ or } lnk = lnA - \frac{E_a}{RT} \tag{7}$$

Conclusion

Efficacy of ammonia-ammonium glutamate as lixiviant in Ni leaching from electric arc furnace dust was demonstrated. The addition of ammonia and hydrogen peroxide as synergist and modifier positively encouraged Ni dissolution. Increasing glutamate concentration depressed Ni recovery since increasing glutamate, added as glutamic acid, decreased solution pH through neutralization of ammonia, and hindered it to form complex with Ni. Recovery Ni decreased as stirring speed increased was probably due to shorter interaction between lixiviant and EAFD during leaching. Modelling of using four kinetic models revealed that pseudo first model and interface transfer and diffusion model fitted better with experimental data. Calculation on energy activation which yield low value (24.6 kJ/mol) indicated the temperature effect on Ni leaching was minimal.

Table 2. The value of ARE, R_{max} and k obtained from kinetic modelling (R_{max} and k are saturation value of Ni recovery and rate constant, respectively)

Variable	Average relative error ARE (%)		R_{max} %		k	
	Chemical reaction control	Interface transfer and diffusion	Chemical reaction control	Interface transfer and diffusion	Chemical reaction control	Interface transfer and diffusion
Stirring speed						
200 rpm	6.601	0.225	75.148	70.219	0.0045	0.0114
350 rpm	6.478	0.348	62.146	58.479	0.0044	0.0112
500 rpm	24.085	1.413	64.356	53.303	0.0046	0.0381
H_2O_2 concentration						
0%	18.196	133.333	186,466.268	418,244.477	0.0000	0.0000
0.5%	0.885	1.623	67.533	78.411	0.0039	0.0038
1%	16.384	2.142	72.332	62.692	0.0045	0.0168
2%	21.520	1.827	79.617	62.225	0.0046	0.0392
Ammonia concentration						
1	3.013	2.044	58.175	62.093	0.0042	0.0054
1.25	1.327	1.519	64.306	74.548	0.0035	0.0035
1.5	0.829	1.318	64.562	73.021	0.0040	0.0042
Glutamate concentration						
0.5	6.478	0.348	62.146	58.479	0.0044	0.0112
1	3.013	2.044	58.175	62.093	0.0042	0.0054
1.5	1.311	0.857	55.455	64.341	0.0030	0.0031
Temperature						
30	6.478	0.348	62.146	58.479	0.0044	0.0112
55	1.415	1.563	71.169	85.598	0.0040	0.0038
65	2.007	4.028	79.107	94.437	0.0040	0.0036

References

1. Suetens T, Klaasen B, Van Acker K, Blanpain B (2014) Comparison of electric arc furnace dust treatment technologies using exergy efficiency. J Clean Prod https://doi.org/10.1016/j.jclepro.2013.09.053
2. Prasetyo E (2012) Simple method of copper analysis using monosodium glutamate and its application in ore analysis. Mineralogia. https://doi.org/10.2478/v10002-012-0002-6
3. Prasetyo E, Anderson C, Nurjaman F, Al Muttaqii M, Handoko AS, Bahfie F, Mufakhir FR (2020) Monosodium glutamate as selective lixiviant for alkaline leaching of zinc and copper from electric arc furnace dust. Metals (Basel). https://doi.org/10.3390/met10050644
4. Prasetyo E, Bahfie F, Muttaqii MA, Handoko AS, Nurjaman FZ (2020) Extraction from electric arc furnace dust using amino acid leaching; AIP conference proceeding: Jakarta; p 4. https://doi.org/10.1063/5.0002159
5. Levenspiel O (1999) Chemical reaction engineering. Ind Eng Chem Res https://doi.org/10.1021/ie990488g

6. Dickinson CF, Heal GR (1999) Solid-liquid diffusion controlled rate equations. Thermochim Acta https://doi.org/10.1016/s0040-6031(99)00256-7
7. Prasetyo E, Toyoda K (2016) Sol-Gel synthesis of a humic acid-silica gel composite material as low-cost adsorbent for thorium and uranium removal. J Radioanal Nucl Chem https://doi.org/10.1007/s10967-016-4861-y

Chemical Leaching of Inactive Gold Mine Tailings as a Secondary Source of Cobalt and Nickel—A Preliminary Case Study

Marouen Jouini, Mathilde Perrin, and Lucie Coudert

Abstract Tailings from inactive gold mines, that are not yet successfully restored (generation of As- and Co-contaminated neutral mine drainage), represent a promising secondary source of strategic metals including Co and Ni. Three different mine tailings (sites A, B and C) from Cobalt Mining Camp were collected and characterized. Preliminary chemical leaching tests were conducted with inorganic acids (HCl, H_2SO_4 and HNO_3) to solubilize Co and Ni at different concentrations (0.01–0.5 N). The influence of the number of the leaching steps on the recovery of Co and Ni was also evaluated. Promising concentrations of Co (0.7%) and Ni (0.3%) were reported in tailings from site A, while lower concentrations were measured in tailings from sites B and C (0.02–0.1%), requiring pre-concentration steps (not tested in this preliminary study) before leaching to reduce operating costs. More than 85% of both Co and Ni were solubilized from tailings from site A after only 30 min using H_2SO_4 (0.25 N) at room temperature. Lower efficiencies (36–62%) were observed for tailings from sites B and C, which can be partially explained by the higher amounts of acid-consuming minerals present in the gangue. Additional experiments are required to better understand the mechanisms involved in Co and Ni solubilization and to optimize operating conditions in terms of Co and Ni recovery.

Keywords Cobalt · Nickel · Arsenic · Mine tailings · Hydrometallurgy · Mine tailings repurposing

M. Jouini
Research Institute of Mines and Environment (RIME), Université du Québec en Abitibi-Temiscamingue (UQAT), 445 Boulevard de l'Université, J9X 5E4 Rouyn-Noranda, QC, Canada
e-mail: Marouen.Jouini@uqat.ca

M. Perrin · L. Coudert (✉)
RIME-UQAT, 445 Boulevard de l'Université, J9X 5E4 Rouyn-Noranda, QC, Canada
e-mail: lucie.coudert@uqat.ca

M. Perrin
e-mail: mathilde.perrin19@gmail.com

C. Anderson et al. (eds.), *Ni-Co 2021: The 5th International Symposium on Nickel and Cobalt*, The Minerals, Metals & Materials Series,
https://doi.org/10.1007/978-3-030-65647-8_15

Introduction

Owing to our transition to a low-carbon economy, our society is becoming increasingly dependent on E-tech or high-technology metals (HTM) including Co and to a lesser extent Ni. Indeed, these elements are key constituents of rechargeable batteries used in phones, electronic goods and electric vehicles [1]. Based on its high economic importance for the sustainable development of our society, Co has been classified as "critical" or "strategic" metal by several countries including Quebec, USA and European Union [2, 3]. Increasing of HTM demands and fluctuations in their market prices, caused by foreign production monopolies, encouraged several countries around the world to identify their own primary (e.g. low-grade ores) or secondary (e.g. mine or industrial waste) sources of HTM [4–6].

Given that Co is mainly extracted as a by-product of base (e.g. Cu, Ni) or precious (e.g. Ag) metal production, metallurgical processes, which are not optimized for Co recovery, entail significant Co losses in tailings and slags after ore processing and refining [7]. Therefore, solid waste from past mining or metallurgical activities is getting more and more attention as promising secondary sources of Co and to a lesser extent of Ni for diversifying sources of supply. Indeed, mine tailings can contain some amounts of Co and are already excavated and finely ground [4, 8, 9]. From both economic and environmental perspectives, the repurposing of non-profitable and potentially harmful mine tailings to recover HTM could be beneficial to give them a second life (i.e. in term of circular economy) and to limit their environmental footprints (i.e. in term of sustainability).

Several studies focused on the development of (bio-)hydrometallurgical processes to efficiently recover Co and/or Ni from mine tailings or metallurgical residues [10–14]. Bioleaching processes, which involved the use of microorganisms to convert insoluble metal sulfides to soluble metal sulfates, showed great potential to recover Co and/or Ni from sulfidic mine tailings due to their relative simplicity of operation, and low capital and operating costs [10, 15–18]. However, these processes required long retention time (few days to several weeks) and their efficiencies can be quite variable, especially when applied to tailings containing low sulfur contents [10, 19]. Previous studies showed that hydrometallurgical processes, which involved the use of inorganic acids and/or oxidizing agents, efficiently solubilized Co (55-99%) from mine tailings or metallurgical waste containing variable sulfide contents [13, 20]. Physico-chemical enrichment processes including gravity or magnetic separation as well as flotation can be used prior to chemical leaching to preconcentrate metals of interest and therefore, to reduce operating costs related to chemicals consumption [20, 21].

Historical exploitation of silver (Ag) in the Cobalt Mining Camp, located in the northern Ontario (Canada), started in the beginning of the 1900s until the mid-1930s and then, restarted in the 1950s until 1989 [22]. Over eight decades of intermittent exploitation of Ag, 18 mills operated within this area to recover 14,000 tonnes of Ag by flotation followed by amalgamation or cyanidation, discharging several millions of tons of waste rock and tailings near processing mill sites [22–24].

Inadequate management of As-rich tailings and/or dams failures led to a widespread contamination of As in this area [22]. The main objective of the present study was to evaluate the performances of hydrometallurgical processes to reduce the environmental impacts of mine tailings originating from inactive mines while recovering some HTM (i.e. Co and Ni), which are still present in aged mine tailings from the Cobalt Mining Camp (Ontario, Canada).

Materials and Methods

Mine Tailings Sampling and Characterization

Mine tailings (approx. 20 kg) were collected from different inactive/abandoned mine sites (Site A, B and C) located in the Cobalt Mining Camp, Ontario, Canada. The selection of sample locations was based on previous physico-chemical characterization campaigns to ensure that sampled mine tailings were representative of the tailing ponds. The samples were oven-dried at 60 °C, homogenized and stored in high-density polyethylene containers until characterization and repurposing experiments.

The particle size distribution (PSD) was determined using a laser particle size analyzer (Malvern Mastersize S 2000). The total carbon and sulfur contents were determined by combustion in an induction furnace coupled to an infrared analyzer (ELTRA-CS-2000) at 1360 °C [25]. The acid generating potential of mine tailings was evaluated using the modified acid-base-accounting method [26]. The chemical composition of mine tailings was determined: (i) on the entire sample, and (ii) on different fractions (<20 μm, 20–53 μm and >53 μm obtained by wet-sieving) to better understand the distribution of metal(loid)s. The concentrations of metal(loid)s were determined using an inductively coupled plasma-atomic emission spectrometry (ICP-AES, Perkin Elmer Optima 3100 RL) after acidic digestion (HCl, HNO_3, $HClO_4$ and HF). The fractionation of Co, Ni and As in mine tailings were evaluated using a sequential extraction procedure [27]. The crystalline phases present in mine tailings were determined by the X-Ray diffraction (XRD–Bruket AXS Advance D8) using CuKα radiation in the range from 5 to 70° with steps of 0.02° in 2θ (detection limit 1%). The DiffracPlus EVA and TOPAS software were used to identify and quantify the abundance of mineral species. Mineralogical characterization was further investigated using a scanning electron microscope (SEM) equipped with a energy-dispersive X-ray spectroscopy probe (EDS) (HITACHI S-3500 N, voltage of 20 kV, amperage 140 μA, pressure 25 kPa, detection limit 1%, and a work distance of 15 mm).

Preliminary Leaching Tests—Influence of Operating Conditions on Cobalt Recovery

Preliminary leaching tests were conduced at room temperature in 500 mL shaker flasks by mixing 20 g of mine tailings with 200 mL of leaching solution (solid/liquid fixed at 10%, w/v) at 150 rpm using an orbital shaker. The leaching solutions were prepared by diluting the adequate amounts of analytical grade inorganic acid (H_2SO_4, HCl or HNO_3) with deionized water. Samples were collected after different retention time and the residual mine tailings were separated from the leachates by filtration through glass fiber filter using a vacuum pump. The concentrations of Co and other elements of interest (e.g. Ni, As) leached were measured by ICP-AES. Three series of preliminary leaching tests were conduced to evaluate the influence of the nature and the concentration of the leaching agent and the number of leaching steps (1 vs. 2 steps). The most promising leaching conditions defined during the previous assays were tested on the different mine tailings collected to evaluate the influence of the mineralogy and metal(loid)s fractionation on the solubilization of Co, Ni and As.

The first series of experiments were performed in the presence of different inorganic acids (H_2SO_4, HCl, HNO_3) with a concentration fixed at 0.5 N to evaluate the influence of the nature of inorganic acid on the solubilization of Co. The second series of experiments was carried out in the presence of the most promising inorganic acid (H_2SO_4) at different concentrations ranging from 0.01 to 0.5 N. In the last series of experiments, the number of leaching steps was evaluated on the solubilization of Co, Ni and As present in the mine tailings from the different sites (A, B and C) collected using a leaching solution of H_2SO_4 at 0.1 N.

Results and Discussion

Physico-Chemical Properties

The main properties of the sampled tailings from Cobalt Camp Mining are presented in Table 1. The tailings from site A, B and C consisted mostly of 72.5%, 88.6% and 81.0% of fine particles (<75 μm) respectively. In consistency with the PSD (not shown), D90 showed that the proportion of the coarser fraction was higher in tailings from site A (~132 μm) compared to sites B and C. Furthermore, results showed that the PSD was semi-spread ($5 < C_U < 20$) for all the samples. The LECO analysis showed low C content in all the tailings (6–9 times lower from site A relative to sites B and C), as an indication of the presence of low carbonates and therefore, low resistance toward acidic solution during leaching experiments. Furthermore, S content did not exceed 0.1 wt% in all samples, with the highest amounts measured in tailings from site B. Overall, low S contents indicated a low presence of sulfate- and/or sulfide-bearing minerals. Considering NNP and NPR values, tailings from sites B

Table 1 Main physico-chemical characteristics of tailings from sites A, B and C

Mine site	A	B	C
Physical properties			
D10 (μm)	3.8	2.5	2.7
D60 (μm)	53.9	33.1	35.0
D90 (μm)	132	79.5	112
Coefficient of uniformity: $C_u = D60/D10$	14.2	13.2	12.9
Chemical composition			
Al (%)	9.4	7.3	7.5
As (%)	> 1	0.4	0.3
Ca (%)	0.7	2.3	2.5
Fe (%)	7.7	5.7	5.4
Mg (%)	5.4	2.8	2.7
Na (%)	3.2	3.6	3.3
Co (%)	0.70	0.12	0.08
Ni (%)	0.26	0.09	0.02
S (%)	0.06	0.09	0.04
C (%)	0.1	0.6	0.9
Cd (mg/kg)	0.44	0.77	0.76
Cr (mg/kg)	207	137	102
Cu (mg/kg)	803	415	202
Pb (mg/kg)	45.9	272	318
Zn (mg/kg)	183	242	177
Static test			
Neutralizing potential—NP (kg $CaCO_3$/t)	8.8 ± 1.9	49.6 ± 7.3	71.8 ± 6.9
Acidic potential—AP (kg $CaCO_3$/t)	1.9 ± 0.3	2.6 ± 1.0	1.0 ± 0.5
Net neutralization potential—NNP (kg $CaCO_3$/t)	6.9 ± 2.0	47.0 ± 6.8	70.7 ± 6.6
Carbonates neutralizing potential/AP ratio (NPR)	4.8 ± 1.7	22.7 ± 9.0	82.1 ± 27.5

and C were considered non-potentially acid-generating (NNP > 20 kg $CaCO_3$/t and NRP > 2.5), while tailings from site A were classified as uncertain (−20 kg $CaCO_3$/t < NNP < 20 kg $CaCO_3$/t and NRP < 2.5).

The chemical analysis showed that tailings were mainly composed of Al (7.3–9.4%), Fe (5.4–7.7%), Mg (2.7–5.4%) and Na (3.2–3.6%) (Table 1). High concentrations of As (0.3–1%) were recorded in all the samples, which is consistent with previous studies highlighting the presence of As-rich tailings in this area [22, 24]. Concentrations of Co (0.07–0.7%) and Ni (0.02–0.26%) were measured in the different tailings, as an indication of the presence of primary and secondary Ni-and Co-bearing minerals [22, 24]. Considering the concentrations of Co and Ni in the different samples and the unit price of these metals (estimated at 41 \$/kg and 17 \$/kg, respectively), these tailings could be considered as a promising secondary source for Co and Ni, especially for those from site A [28]. Indeed, Co and Ni concentrations were ~6–9 and >11 times higher in tailings from site A than those from sites B and C. Hence, there is an important and promising residual potential of Co and Ni that could be recovered in tailings from site A and to a lesser extent from sites B and C. Low concentrations of undesirable elements such as Cd, Cr and Pb were also found.

To better understand the distribution of elements of interest (i.e. metal(loid)s and S), samples were sieved and separated into three fractions: <20 μm, 20–53 μm and >53 μm and their chemical composition was determined (results not shown). Overall, metal(loid)s concentrations seemed to vary by particle size and a consistent pattern was observed for Co, Ni and As in all the samples. The main difference in the distribution of elements between the three fractions is explained by the different occurrence of minerals between each fraction. In tailings from sites A and B, As, Co and Ni were mainly concentrated (47–63%) in the fine fraction (<20 μm) but non-negligible concentrations were also found in the coarser fractions, indicating that the entire sample should be repurposed. For tailings from site C, As, Co and Ni were concentrated to 65–74% in the fine fraction (<20 μm). These results indicated that As, Co and Ni can be easily preconcentrated, therefore reducing the mass of the sample that will undergo chemical leaching (i.e. 55% of the total mass).

Mineralogical Properties

Mineralogical characterization by XRD identified the presence of primary (i.e. non-oxidized) and secondary (i.e. oxidized) minerals (Table 2). These minerals are subdivided into five main groups including: silicates (quartz, albite, muscovite, chlorite, orthoclase, and hornblende), sulfides (arsenopyrite, cobaltite, chalcopyrite, gersdorffite, and pentlandite), arsenides (safflorite and skutterudite), arsenates (Fe-Co-AsO_4 and Ni-Co-AsO_4), and carbonates (dolomite and calcite). Overall, As-, Co- and Ni-bearing minerals identified in the samples are composed of a mixture of primary minerals, that require oxidation to easily extract elements of interest, and secondary phases where these elements are already soluble or could require acid/base extraction. As expected, As-, Co- and Ni-bearing minerals were present in higher amounts in tailings from site A than from sites B and C. Furthermore, carbonates were not detected in tailings from site A, which is in agreement with chemical composition, while dolomite and calcite were identified in tailings from sites B and C (0.1% C; Table 1).

Results of SEM-EDS confirm the presence of arsenates in tailings from site A, already identified by XRD, where a grain appears to be a mineral in annabergite-erythrite series associated with feldspaths (Fig. S1a). Annabergite and erythrite, greenish and pinkish secondary compounds, are typically resulting from the oxidation of Ni–Co–Fe–As minerals [29]. The precipitation of erythrite has been demonstrated to control the mobility of Co and As, and probably Ni, which could replace Co in its

Table 2 Mineral identification by XRD

Minerals		Chemical formula	Composition (wt%)		
			Site A	Site B	Site C
Arsenides	Safflorite	$(Co,Fe)As_2$	–	<1	<1
	Skutterudite	$(Co,Ni)As_{3-x}$	<1	<1	<1
Arsenates	Annabergite	$Ni_3(AsO_4)_2 \cdot 8(H_2O)$	1.0	<1	–
	Erythrite	$Co_3(AsO_4)_2 \cdot 8(H_2O)$	1.6	<1	–
Carbonates	Calcite	$CaCO_3$	–	3.2	1.0
	Dolomite	$CaMg(CO_3)_2$	–	1.7	2.0
Sulfides	Arsenopyrite	FeAsS	<1	<1	<1
	Chalcopyrite	$CuFeS_2$	–	<1	1.0
	Cobaltite	CoAsS	<1	<1	<1
	Gersdorffite	NiAsS	<1	–	–
	Pentlandite	$(Fe,Ni)_9S_8$	–	–	1.0
Silicates	Albite	$NaAlSi_3O_8$	43.8	43.0	43.0
	Chlorite	$(Mg,Fe)_3(Si,Al)_4O_{10}(OH)_2(Mg,Fe)_3(OH)_6$	34.1	9.0	13.0
	Hornblende	$Ca_2Mg_4Al_{0.75}Fe^{3+}_{0.25}(Si_7AlO_{22})(OH)_2$	2.7	3.2	3.0
	Muscovite	$KAl_2(Si_3Al)O_{10}(OH,F)_2$	10.7	4.3	10.9
	Orthoclase	$KAlSi_3O_8$	–	1.3	4.7
	Quartz	SiO_2	5.3	17.5	18.9

structure [30]. In addition, a chalcopyrite grain surrounded by arseniosiderite was also detected (Fig. S1b). Generally, arseniosiderite occurs as a secondary mineral formed by the oxidation of earlier As-bearing minerals, typically arsenopyrite and in association with erythrite. An association of safflorite and clinosafflorite was observed in tailings from site C (Fig. S2a). Safflorite is a Co-arsenate mineral associated with some Fe and Ni, while the clinosafflorite is a near dimorph, i.e. same chemistry but different structure, with the safflorite. Moreover, cobaltiferous arsenopyrite was also detected in tailings from site C (Fig. S2b). The Co occurring within Fe-sulfides such pyrite, arsenopyrite is also reported in the literature [31, 32].

Potential Mobility of Co, Ni, As and Fe

As expected from previous mineralogical investigation, 30–40% of Co and Ni were found in the residual fraction (i.e. silicates and sulfides bound fraction), and more than 50% of these elements were extracted from all residues in the reducible fraction (Fig. 1). Hence, fairly aggressive (i.e. strongly acidic) conditions are deemed necessary to solubilize Co and Ni from their associated minerals. Moreover, As was lowly released in the first fraction (<2% of the total concentration, in all residues), that

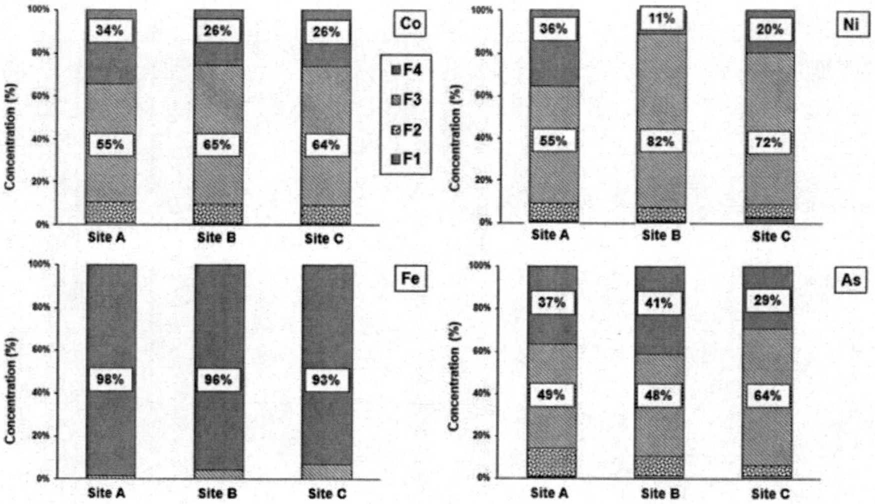

Fig. 1 Concentrations of Co, As, Ni and Fe leached after each step of the sequential extraction for sites A, B and C

corresponds to the release of the exchangeable cations and soluble phases. However, non-negligible amount of As (3–10%) was released in the oxidizable fraction (e.g. sulfides). Besides, most of the As was extracted in the reducible fraction (e.g. >70% in the tailings from site C), indicating a moderate lability of this contaminant. In parallel, for the tailings from site A, ~50% of As was found in the residual fraction. Furthermore, most of the Fe was extracted in the fourth or the last step (>90% of the total concentration, in all residues) of the sequential extraction. As a result, Fe can be considered as non-labile element. Results could probably be explained by the occurrence of Fe- and As-bearing minerals within silicates (as previously mentioned), which require strong acids for their dissolution. In addition, a small amount of Fe was found related to the reducible fraction, confirming the low solubility of Fe from all the residues. Overall, results indicated that Co, As and Ni were mainly associated with the same minerals. Likewise, since Fe was poorly solubilized (i.e. mostly found in the residual fraction), a selective extraction of As, Co and Ni could be conducted by limiting Fe leaching. However, for an efficient separation of Co and As, the presence of low amounts of Fe is sometimes required to remove As from leachates (e.g. by co-precipitation or sorption with Fe-(oxy)hydroxides).

Leaching Test

Effect of the Nature of the Acid on Co Leaching from Site A Tailings

Inorganic acids such as H_2SO_4, HNO_3 and HCl are commonly used as leaching agents in hydrometallurgical processes to improve the solubilization of metal(loid)s (e.g. As, Co, Ni, Mn, Zn) from tailings or residues [13, 33–35]. To identify the most appropriate leaching solution, a series of leaching trial was performed on tailings from site A (Fig. 2). More than 90% of Co was solubilized after only 30 min at room temperature, regardless of the type of the acid. Almost no major differences (~5%) were observed in the efficiency of Co leaching among the acids tested; H_2SO_4 seemed to be a little bit less effective than HNO_3 and HCl. It is worth noting that similar trends were observed for Ni, Fe and As. Lower solubilization of Co, Ni, Fe and As when using H_2SO_4 can be explained by the fact that few amounts of Co and Ni were associated to sulfide minerals (Table 2), which require the addition of an oxidant [13]. Given that H_2SO_4 causes, in general, fewer problems while managing effluent discharges (i.e. no Cl^- and NO_3^- to be treated) and is less expensive among the tested acids, this latter was selected as the best leaching agent. In addition, solubilization of Co and Ni in the presence of H_2SO_4 is mainly due to the formation of soluble complexes (e.g. $CoSO_4$ and $NiSO_4$) [36]. Furthermore, almost no variation in the solubilization of Co and Ni was observed between 2 and 4 h, which can be explained by a rapid dissolution of Co- and Ni-bearing minerals and/or a saturation of the leaching solution. Therefore, for the following tests, the duration of the experiment will be fixed at 2 h.

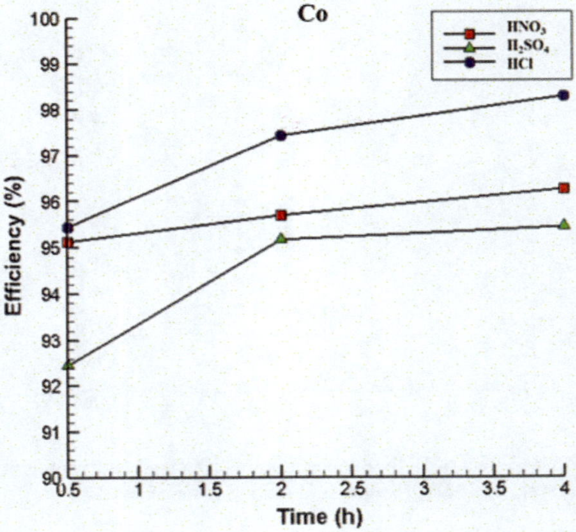

Fig. 2 Effect of H_2SO_4, HNO_3 and HCl on Co leaching from tailings from site A ([inorganic acid] = 0.5 N, T = 25 °C, solid/liquid ratio (S/L ratio) = 10% (w/v)). (Color figure online)

Effect of H_2SO_4 Concentration on Co and Ni Leaching from Site A Tailings

Usually, in hydrometallurgical processes, increasing the concentration of the leaching agent favors the dissolution rate of minerals [13, 34]. Similar results were observed in the present study, as the concentration of H_2SO_4 increased, the leaching of Co and Ni was improved (Fig. 3). Indeed, increasing the concentration of H_2SO_4 above 0.1 N allowed to recover more than 70% of Co and Ni while more than 90% of Co and Ni were solubilized in less than 30 min at concentration higher than 0.25 N (Fig. 3). Increasing H_2SO_4 concentration above 0.25 N showed no improvement in the extraction of Co and Ni. Overall, in addition to elements of interest, it is important to note that other unwanted elements (e.g. As, Pb, Cd) were efficiently leached at higher H_2SO_4 concentrations (not presented).

In the present study, the leaching efficiency is strongly related to the pH of the solution; the lower is the leaching medium's pH, the higher is the transfer of metals to the liquid phase [37]. According to our results, at low concentrations of H_2SO_4 (0.01 N), the final pH was around 4, which can partially explain the lower amounts of Co (~20%) and Ni (~15%) solubilized compared to the higher H_2SO_4 concentrations ($[H_2SO_4] > 0.1$ N) with final pH lower than 2.2. However, it is worthwhile to mention that the more concentrated the acid is, the more expensive the leaching process is. Hence, given the amounts of Co and Ni leached (>70%) at lower concentrations (e.g. 0.1 N) and considering the economic point of view of increasing the H_2SO_4 concentration above 0.1 N, an additional experiment was conduced to evaluate the efficiency of a second leaching step performed under lower acidic conditions (0.1 N) compared to a single step conduced at higher H_2SO_4 concentrations (0.25 N).

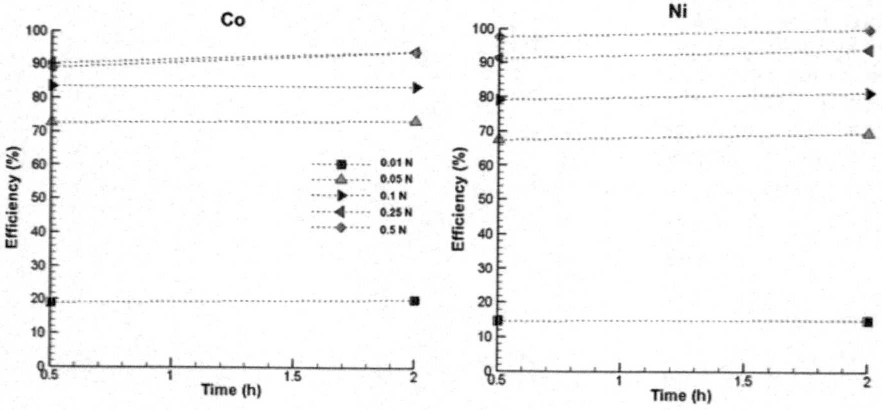

Fig. 3 Effect of H_2SO_4 concentration on Co and Ni leaching from site A tailings (inorganic acid: H_2SO_4, T = 25 °C, solid/liquid ratio (S/L ratio) = 10% (w/v)). (Color figure online)

Effect of a Two-Stage Leaching Process on the Solubilization of Metals

The results, presented in Fig. 4, showed that the second leaching step conduced at low H_2SO_4 concentration (0.1 N) led to an improvement on the amounts of solubilized Co and Ni, with leaching efficiencies similar to those obtained after one leaching step performed at higher concentration ($[H_2SO_4] = 0.25$ N). Therefore, a recirculation of the leaching solution in a reprocessing loop could be considered to decrease acid needs and costs as well as to enhance metals' solubilization. Despite the promising results obtained at low H_2SO_4 concentration of tailings from site A (i.e. 88% for Co, 91% for As and 86% for Ni), relatively low amounts of Co and Ni (<65%) were leached from tailings from sites B and C (Fig. 4). These lower performances can be explained by the higher carbon content measured in tailings from sites B and C (0.6–0.9%) relative to tailings from site A (0.1%). Indeed, higher carbon content indicated the presence of higher amounts of carbonate minerals (confirmed by XRD analysis), that buffer the leaching solution and therefore maintained the pH near neutral values until the depletion of these neutralizing minerals. As a result, leaching of Co, As and Ni from tailings from site C was from 3 to 17 fold higher during the second leaching step in comparison with the first leaching step. Indeed, carbonates, that were abundantly present, were probably consumed during the first leaching step, while during the second leaching step, the pH was enough low to allow the dissolution of As-, Co- and Ni-bearing minerals. These results highlighted the need to preconcentrate As-, Co- and Ni-bearing minerals using mineralurgical processes or flotation to separate them from gangue minerals that contain higher amounts of carbonates, which increased the amount of H_2SO_4 needed to efficiently solubilize Co and Ni.

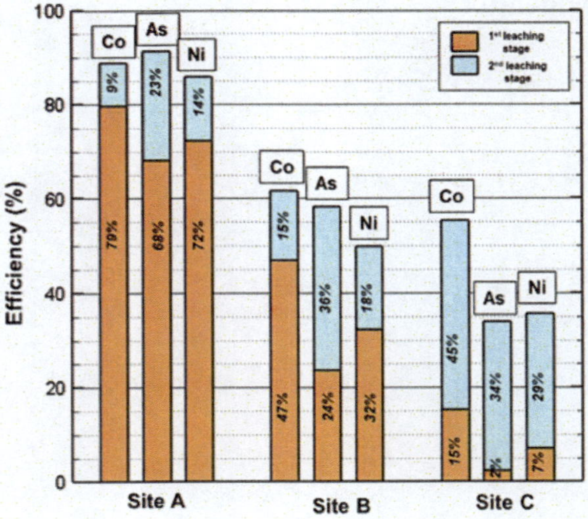

Fig. 4 Effect of the number of leaching steps on metals' solubilization from tailings from sites A, B and C ($[H_2SO_4] = 0.1$ N, T = 25 °C, t = 2 h, S/L ratio = 10% (w/v)). (Color figure online)

Conclusion

In this work, a leaching process was applied to solubilize Co and Ni from low-grade tailings from inactive Ag mines in the Cobalt Mining Camp. The effect of the nature and the concentration of the leaching agent (H_2SO_4, HCl and HNO_3) as well as the leaching time and the number of leaching steps on the solubilization of Co, Ni and As were investigated. The results showed that the old tailings initially contained variable concentrations of Co (0.07–0.7%), Ni (0.02–0.3%), S (0.04–0.09%) and C (0.1–0.9%), while large amounts of problematic metalloids were also measured (e.g. 0.3–1% of As). The mineralogical characterization of the tailings showed the presence of primary (e.g. safflorite, skutterudite, arsenopyrite, cobaltite, gersdorffite) and secondary (e.g. erythrite, annabergite) As-, Co- and Ni-bearing minerals. Due to the low amounts of primary As-, Co and Ni-sulfide minerals, the use of chemical leaching seemed to be a promising approach to recover both Co and Ni from these tailings without requiring roasting as pre-treatment. Similar Co, Ni and As leaching efficiencies were observed in the presence of HCl, HNO_3 or H_2SO_4, indicating that the influence of the nature of the leaching agent was quite negligible. The highest Co (93%), Ni (99%) and As (100%) solubilization was achieved after 2 h in the presence of H_2SO_4 at 0.25 N at room temperature for tailings from site A. Lower efficiencies were observed for tailings from sites B and C, which can be partially explained by the higher amounts of acid-consuming minerals present in the gangue (i.e. carbonates). Additional experiments are required to better understand the mechanisms involved in Co and Ni solubilization and to optimize operating conditions in terms of Co and Ni recoveries and operating costs. The performances of mineralurgical processes or flotation to pre-concentrate Co- and Ni-bearing minerals and to separate them from the acid-consuming minerals present in the gangue should also be evaluated.

Appendix

See Figs. S1 and S2.

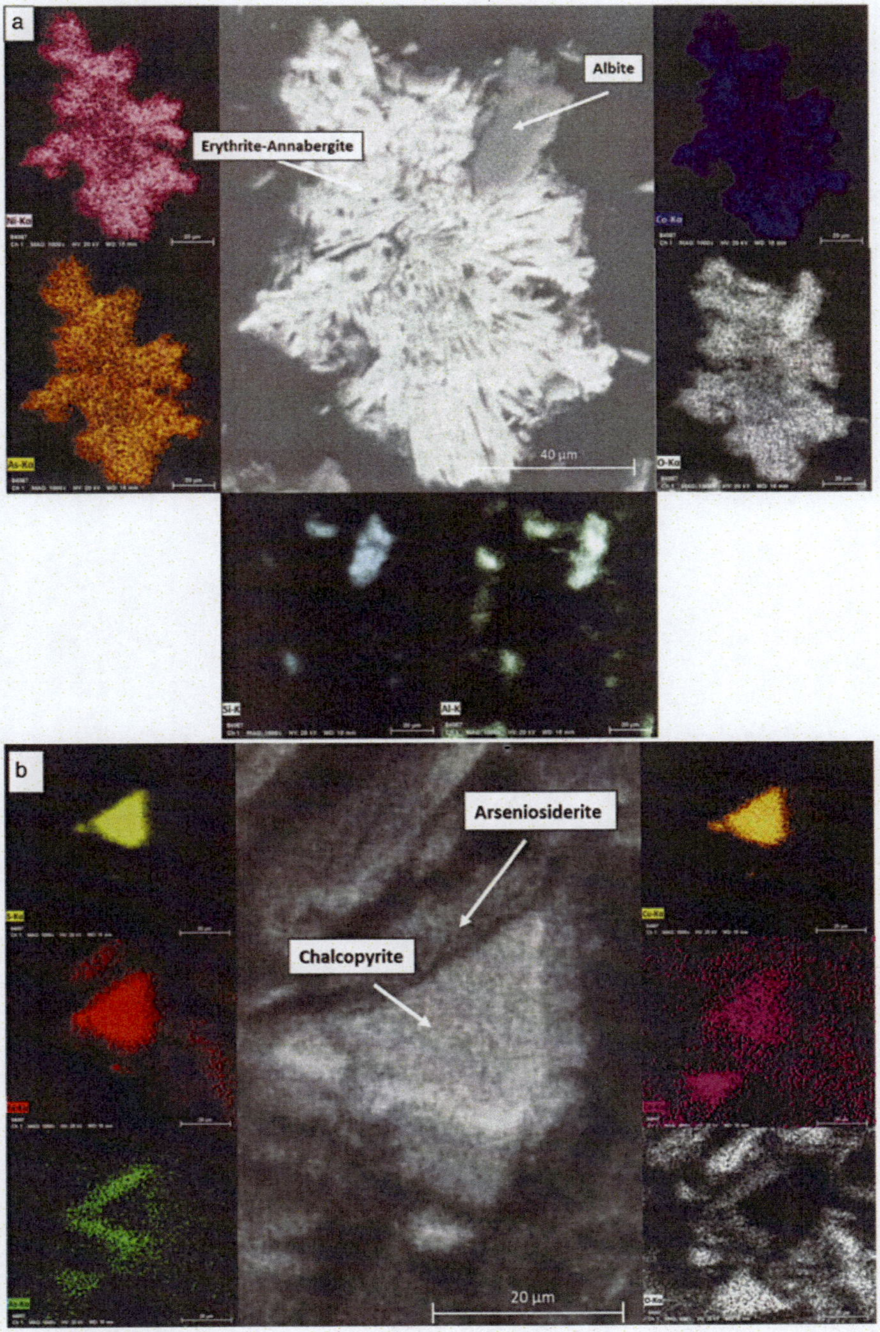

Fig. S1 SEM-EDS secondary electron and X-mapping images of **a** annabergite-erythrite and **b** chalcopyrite and arseniosiderite in tailings from site A. (Color figure online)

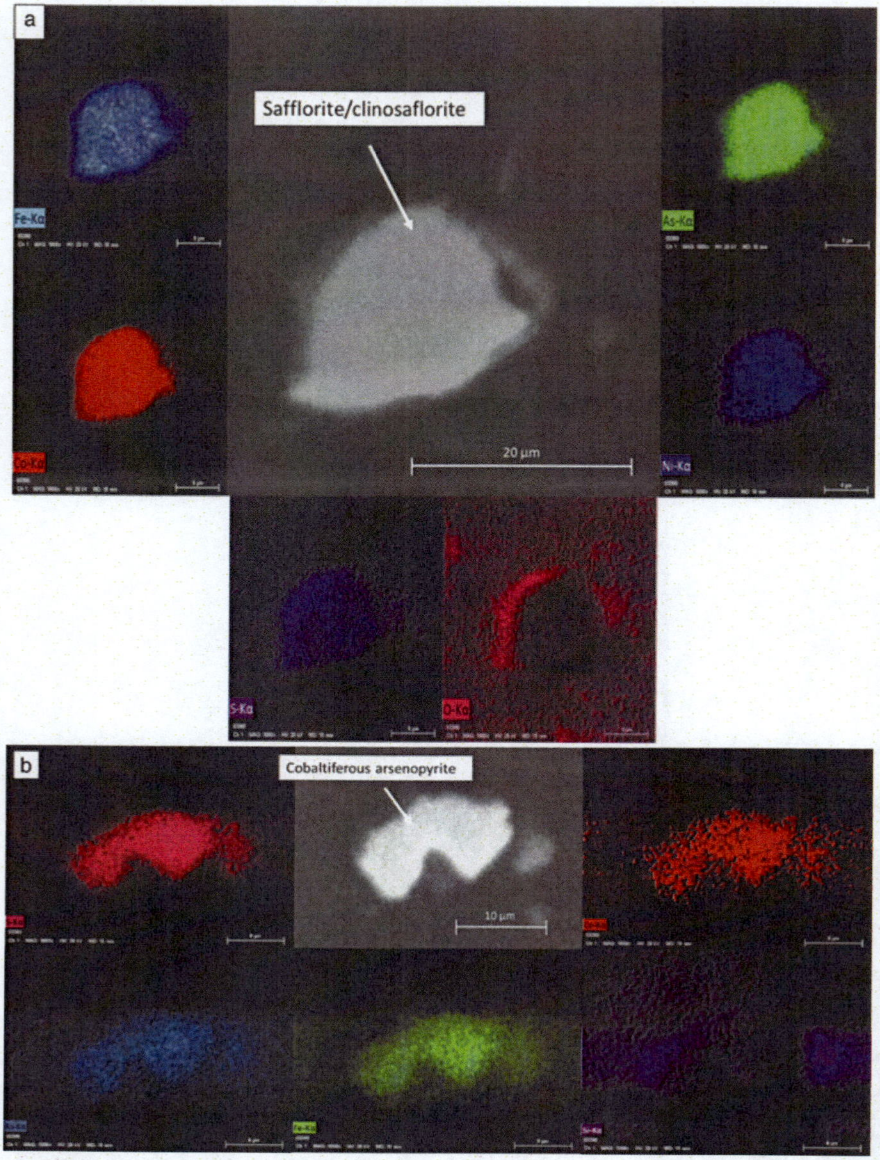

Fig. S2 SEM-EDS secondary electron and X-mapping images of **a** safflorite and clinosafflorite and **b** cobaltiferous arsenopyrite in tailings from site C. (Color figure online)

References

1. Pazik PM, Chielewski T, Glass HL, Kowalczuk PB (2016) World production and possible recovery of from the Kupferschiefer stratiform copper ore. In: E3S web of conferences–mineral engineering conference MEC2016, vol 8, pp 01063, 1–9. https://doi.org/10.1051/e3sconf/20160801063
2. Commission Européenne (2017) Communication de la commission au parlement européen, au conseil, au comité économique et social européen et au comité des régions relative à la liste 2017 des matières premières critiques pour l'UE, p 8
3. MERN (2019) Réflexion sur la place du Québec dans la mise en valeur des minéraux critiques et stratégiques, p 22
4. Falagán C, Grail BM, Johnson DB (2017) New approaches for extracting and recovering metals from mine tailings. Miner Eng 106:71–78. https://doi.org/10.1016/j.mineng.2016.10.008
5. Landenberger A, Arvanitidis N, Jonsson E, Arvidsson R, Casanovas S, Lauri L (2016). Identification and quantification of secondary CRM resources in Europe. SCRREEN, p 129
6. Petavratzi E, Gunn G, Kresse C (2019) Cobalt–British geological survey commodity review, p 72
7. Crundwell F, Moats M, Ramachandran V, Robinson T, Davenport WG (2011) Extractive metallurgy of nickel, cobalt and platinum group metals. Elsevier, p 622
8. Bellenfant G, Guezennec AG, Bodéman F, d'Hugues P, Cassard D (2013) Re-processing of mining waste: Combining environmental management and metal recovery? In: Proceedings of the eighth international seminar on mine closure. Australian Centre for Geomechanics, Cornwall, pp 571–582. https://doi.org/10.36487/ACG_rep/1352_48_Bellenfant
9. Shaw RA, Petavratzi E, Bloodworth AJ (2013) Resource recovery from mine waste–Chapter 2. In: Hester RE, Harrison RM (ed.) Waste as a resource, Royal Society of Chemistry, pp 44–65
10. Ahmadi A, Khezri M, Abdollahzadeh AA, Askari M (2015) Bioleaching of copper, nickel and cobalt from the low grade sulfidic tailing of Golgohar Iron Mine. Iran Hydromet 154:1–8. https://doi.org/10.1016/j.hydromet.2015.03.006
11. Chen G, Yang H, Li H, Tong L (2016) Recovery of cobalt as cobalt oxalate from cobalt tailings using moderately thermophilic bioleaching technology and selective sequential extraction. Miner 6(3):1–11. https://doi.org/10.3390/min6030067
12. Mondal S, Kumar BP, Singh DK, Chakravartty JK (2015) Parametric optimization for leaching of cobalt from Sukinda ore of lateritic origin–a Taguchi approach. Sep Purif Technol 156(2):827–834. https://doi.org/10.1016/j.seppur.2015.11.007
13. Xie Y, Xu Y, Yan L, Yang R (2005) Recovery of nickel, copper and cobalt from low-grade Ni-Cu sulfide tailings. Hydrometallurgy 80:54–58. https://doi.org/10.1016/j.hydromet.2005.07.005
14. Wang Y, Zhou C (2002) Hydrometallurgical process for recovery of cobalt from zinc plant residue. Hydrometallurgy 63:225–234. https://doi.org/10.1016/S0304-386X(01)00213-4
15. Blengini GA, Mathieux F, Mancini L, Viegas HM (2019) Recovery of critical and other raw materials from mining waste and landfills. JRC Science for policy report, European Union, Luxembourg, p 130
16. Cs Gahan, Srichandan H, Kim DJ, Akcil A (2012) Biohydrometallurgy and biomineral processing technology: a review on its past, present and future. Res J Rec Sci 1(10):85–99
17. Mäkinen J, Salo M, Khoshkhoo M, Sundkvist JE, Kinnunen P (2020) Bioleaching of cobalt from sulfide mining tailings: a mini-pilot study. Hydrometallurgy 196:1054181–6. https://doi.org/10.1016/j.hydromet.2020.105418
18. Zhen S, Yan Z, Zhang Y, Wang J, Campbell M, Qin W (2009) Column bioleaching of a low grade nickel-bearing ore containing high magnesium as olivine, chlorite and antigorite. Hydrometallurgy 96:337–341. https://doi.org/10.1016/j.hydromet.2008.11.007
19. Coto O, Galizia F, Hernández I, Marrero J, Donati E (2008) Cobalt and nickel recoveries from laterite tailings by organic and inorganic bio-acids. Hydrometallurgy 94:18–22. https://doi.org/10.1016/j.hydromet.2008.05.017

20. Ozer M (2019) Cobalt and copper recovery from the ancient flotation tailings by selective sulfation roast-leaching process. J Min Metall Sect B-Metall 55(3):315–324. https://doi.org/10.2298/JMMB190304043O
21. Shengo ML, Kime MB, Mambwe MP, Nyembo TK (2019) A review of the beneficiation of copper-cobalt-bearing minerals in the Democratic Republic of Congo. J Sustain Mining 18:226–246. https://doi.org/10.1016/j.jsm.2019.08.001
22. Dumaresq CG (1993) The occurrence of arsenic and heavy metal contamination from natural and anthropogenic sources in the Cobalt area of Ontario. MSc thesis, Carleton University, Ottawa, Canada, p 355
23. Kwong YTJ, Beauchemin S, Hossain MF, Gould WD (2007) Transformation and mobilization of arsenic in the historic Cobalt mining camp, Ontario, Canada. J Geochem Explor 92:133–150. https://doi.org/10.1016/j.gexplo.2006.08.002
24. Percival JB, Kwong YJT, Dumaresq CG, Michel FA (2007) Distribution of As, Ni and Co in tailings and surface waters in the Cobalt area, Ontario. Mining and the Environment IV Conference, Sudbury, Ontario, Canada, October 19–27, p 10
25. CEAEQ (2013) Détermination du carbone et du soufre: méthode de combustion et dosage par spectrophotométrie infrarouge, MA 310-CS 1.0 Rév 3, Ministère du Développement Durable, de l'Environnement, de la Faune et des Parcs du Québec, Québec, Canada, p 8
26. Lawrence RW, Scheske M (1997) A method to calculate the neutralization potential of mining wastes. Environ Geol 32:100–106
27. Ure A, Davidson C (2002) Chemical speciation in soils and related materials by selective chemical extraction. Chem Spec Environ 265–300
28. Investmine (2020) Copper prices and copper price charts. http://www.infomine.com/investment/metal-prices/copper/. Accessed 25 Feb 2020
29. Drahota P, Filippi M (2009) Secondary arsenic minerals in the environment: a review. Environ Int 35(8):1243–1255. https://doi.org/10.1016/j.envint.2009.07.004
30. Jambor JL, Dutrizac JE (1995) Solid solutions in the annabergite–erythrite–hörnesite synthetic system. Can Mineral 33:1063–1071
31. Battaglia F, Morin D, Ollivier P (1994) Dissolution of cobaltiferous pyrite by Thiobacillus ferrooxidans and Thiobacillus thiooxidans: factors influencing bacterial leaching efficiency. J Biotechnol 32(1):11–16. https://doi.org/10.1016/0168-1656(94)90115-5
32. D'Hugues P, Cezac P, Cabral T, Battaglia F, Truong-Meyer XM, Morin D (1997) Bioleaching of a cobaltiferous pyrite: a continuous laboratory-scale study at high solids concentration. Miner Eng 10:507–527. https://doi.org/10.1016/S0892-6875(97)00029-0
33. Tanong K, Coudert L, Mercier G, Blais JF (2016) Recovery of metals from a mixture of various spent batteries by a hydrometallurgical process. J Environ Manage 181:95–107. https://doi.org/10.1016/j.jenvman.2016.05.084
34. Altinkaya P, Liipo J, Kolehmainen E, Haapalainen M, Leikola M, Lundström M (2019) Leaching of trace amounts of metals from flotation tailings in cupric chloride solutions. Mining Metall Explor 36:335–342. https://doi.org/10.1007/s42461-018-0015-9
35. Yaylali B, Yazici EY, Celep O, Deveci H (2016) Extraction of cobalt from a flotation tailngs in different mineral acids under oxidative conditions. In: Proceedings of the 15th international mineral processing symposium, Istanbul, Turkey, October 19–21, p 13
36. Innocenzi V, Veglio F (2012) Recovery of rare earths and base metals from spent nickel-metal hydride batteries by sequential sulphuric acid leaching and selective precipitations. J Power Sources 211:184–191. https://doi.org/10.1016/j.jpowsour.2012.03.064
37. Jouini M, Rakotonimaro TV, Neculita CM, Genty T, Benzaazoua M (2019) Stability of metal-rich residues from laboratory multi-step treatment system for ferriferous acid mine drainage. Environ Sci Pollut Res 26:35588–35601. https://doi.org/10.1007/s11356-019-04608-1

The Adsorption of Heavy Metals from Aqueous Solutions Using Silica Microparticles Synthesized from Sodium Silicate

E. U. Ikhuoria, I. H. Ifijen, P. G. Obiekea, M. Maliki, and A. C. Ehigie

Abstract Silica microparticles synthesized from sodium silicate were characterized and used as an adsorbent for lead and nickel ions (Pb^{2+} and Ni^{2+}) from their separate solutions. Batch adsorption experimental results showed that an increase in the percentage removals of the Pb^{2+} and Ni^{2+} ions from their solutions was favoured by increased contact time, adsorbent dosage, and heavy metal ion concentrations. Optimum adsorptions of 59.40%/64.66%, 61.4%/75.60% and 61.4%/49.58% for Pb^{2+}/Ni^{2+} ions were obtained at contact time of 40 min, an adsorbent dosage of 2.5 g and heavy metal ion concentration of 50/40 mg/l respectively. The adsorption data were subjected to different kinetic models and pseudo-second-order reaction model had the best fit for both metal ions (R^2, 0.985, 0.996). To verify the adsorption performance, the adsorption data were also fitted into the Langmuir and Freundlich isotherms. The Langmuir isotherm had the best fit for both metal ions (R^2, 0.918, 0.872). The maximum metal ion capacity calculated from Langmuir isotherm was 22.3 and 25.5 mg g^{-1} for lead and nickel ions, respectively. This study showed that silica microparticles can bind a substantial amount of lead and nickel ions from aqueous solutions.

I. H. Ifijen
Product Development Laboratory, Rubber Research Institute of Nigeria, P. M. B. 1049, Benin City, Nigeria
e-mail: larylans4u@yahoo.com

E. U. Ikhuoria (✉) · P. G. Obiekea · A. C. Ehigie
Department of Chemistry, University of Benin, Benin City, Nigeria
e-mail: esyikhuoria@yahoo.com

P. G. Obiekea
e-mail: georcious1997@gmail.com

A. C. Ehigie
e-mail: ehigie@gmail.com

M. Maliki
Department of Chemistry, Edo University Iyamho, Iyamho, Edo State, Nigeria
e-mail: muniratu.maliki@edouniversity.edu.ng

C. Anderson et al. (eds.), *Ni-Co 2021: The 5th International Symposium on Nickel and Cobalt*, The Minerals, Metals & Materials Series,
https://doi.org/10.1007/978-3-030-65647-8_16

Keywords Silica microparticles · Adsorption · Pseudo-first-order · Lead ion · Nickel ion

Introduction

The non-biodegradability and the capacity to cause serious illnesses upon accumulation in living organisms have made heavy metal ions highly toxic for aquatic ecosystems even at low concentrations [1]. A major environmental concern is their removal from wastewaters before disposal. Currently, the techniques used for the removal of heavy metal ions from wastewater include membrane processes, chemical and electrochemical techniques, biological treatments, and adsorption procedures [2]. Among these techniques, adsorption has been established as the most economic process in treating contaminated water, due to its high efficiency. Several types of materials have previously been used in the sorption of heavy metals from contaminated water, such as algal species [3], zeolite [4], etc.

Metal oxides are useful materials that can be used in the adsorption of heavy metals [5]. Metals are useful as adsorbents and in various aspects of Physics, Materials Science and Chemistry. Piezoelectric devices, coatings for the passivation of surfaces against corrosion, sensors and as catalysts, etc. [6]. The metallic oxide that has been manipulated into the micro or nano-size range can exhibit mesmerizing properties because of their very small sizes and high density of corner or edge surface sites. The most generally used oxides in the field of catalysis, photocatalysis, solar-cells, gas-sensors and magneto-electronics are the transition metal oxides (TMO) [7]. In this study, our interest is on silica (SiO_2) because of the amazing physical and chemical characteristics of silicon. Spherical-shaped and non-agglomerated silica microparticles were synthesized using sodium silicate solution (SSS) as a precursor in acid medium. The as-synthesized silica particles were used in the adsorption of lead and nickel ions from their solutions

Experimental

Materials

Liquid sodium silicate, concentrated hydrogen chloride (30%), de-ionized water, lead (II) nitrate, nickel (11) nitrate, iodine, sodium thiosulphate, silver nitrate, starch, potassium iodide and potassium iodate were purchased from Sigma Aldrich and used without further purification. Distilled water was used throughout the experiment.

Preparation of Silica Microparticles

The synthesis of silica microparticles was carried out as described by Khdary et al. [8] via sol-gel method. In a typical experiment, Conc. Hydrochloric acid (HCl) was introduced into a beaker containing 500 ml of sodium silicate solution (Na_2SiO_2) at room temperature (29 °C) with constant stirring until a pH of 8 was attained. This resulted in the formation of silica gel. The obtained Gel was allowed to stand for 24 h for complete separation and the water layer is then filtered from the gel. The distilled water used to wash the as-synthesized silica microparticles was tested with dilute $AgNO_3$ solution at different intervals to ensure that the synthesized silica microparticles are completely free of chloride ion. The washed silica microparticles were oven-dried at 90 °C for 18 h.

Characterization of Silica Microparticles

The functional groups of the synthesized silica microparticles were investigated using Fourier transform infrared spectroscopy (FTIR). The morphology of the synthesized silica particles was examined using a scanning electron microscope.

Removal of Lead and Nickel Ions from Their Aqueous Solutions

Equilibrium sorption leads and nickel ions on the synthesized silica microparticles were examined at room temperature using various initial concentrations of the metal ions. Stock solutions of lead (II) nitrate and nickel (II) nitrate were prepared for the adsorption studies. In a typical experiment, 0.5 g of silica microparticles were mixed with 10 ml of known initial metal ion concentration. Thereafter, the mixture was stirred for 10 min, filtered and the residual concentration of the metal ion in the filtrate was determined using an atomic absorption spectrophotometer. The concentrations of the metal ions were varied from 10.0, 20.0, 30.0, 40.0 to 50.0 mg/L, while other parameters remained constant. The adsorption study was also examined as a function of time, using predetermined time intervals (10, 20, 30, 40 and 50 min) to investigate the time required for the heavy metal adsorption to attain equilibrium. The weight of the silica microparticles used in the adsorption studies was also varied from 0.5 to 2.5 g to examine the effect of adsorbent dosage on metal ions adsorption while other conditions remained constant. The amount of metal ions removed from the solution by the silica microparticles were taken as the difference between the initial and residual metal ion concentrations. The silica gel microparticles (adsorbent) were disposed at the end of this study.

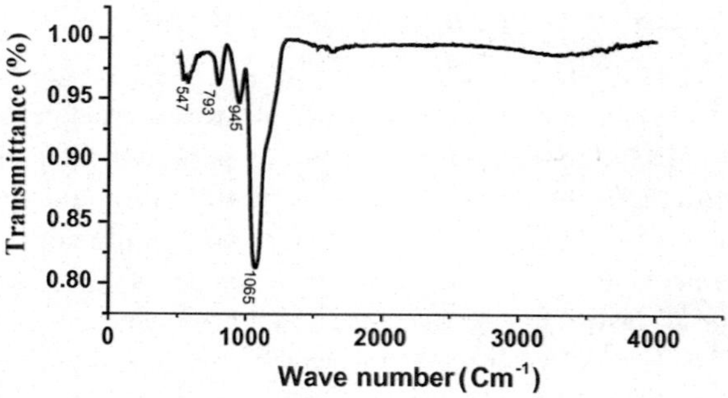

Fig. 1 FTIR spectra of as-synthesized silica microparticles. (Color figure online)

Results and Discussion

Functional Groups of the as-Synthesized Silica Microparticles

Figure 1 showed the functional groups of the synthesized silica microparticles which were examined by FTIR in the range of 3500-1000 cm^{-1}. The absorption peaks at 945 cm^{-1} and 793 cm^{-1} are attributed to the asymmetric vibration of Si–OH and symmetric vibration of Si–O respectively. The peak emanating from ~547 cm^{-1} can be attributed to the rocking motion of oxygen atoms bridging silicon atoms in siloxane bonds (Si-O-Si). The peak at 1065 cm^{-1} is characterized by anti-symmetric motion of silicon atoms in siloxane bonds.

Morphology of the Prepared Silica Particles

The scanning electron micrograph of the as-synthesized silica particles is shown in Fig. 2. The prepared silica particles are spherical. The average particle diameter of the silica particles is estimated to be about 330 nm.

Effect of Contact Time

The effect of contact time on the adsorption of heavy metals from separate solutions of lead and nickel ions using synthesized silica microparticles as adsorbent was examined while leaving other adsorption conditions constant. Figure 3 shows the percentage of heavy metal ions (Pb^{2+} and Ni^{2+}) removed from their solutions at different contact times (10, 20, 30, 40 and 50 min). As observed in Fig. 4, both metal ions (Pb^{2+} and Ni^{2+}) were easily adsorbed onto the as-synthesized silica microparticles and equilibrium was attained at 40 min. At equilibrium, any further increase in the contact time did not affect the amount of heavy metal adsorbed. The short

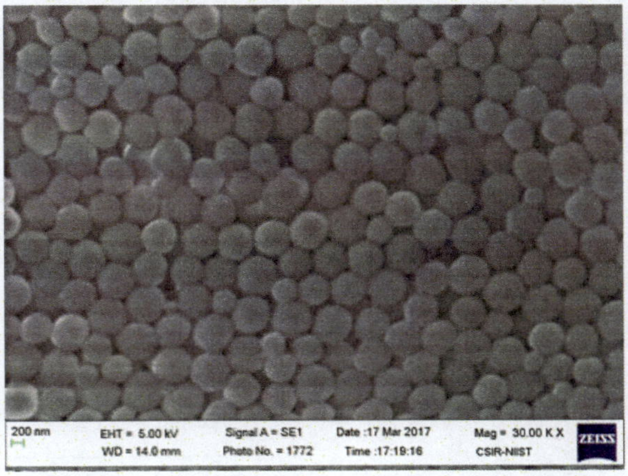

Fig. 2 Scanning electron micrograph of silica microparticles. (Color figure online)

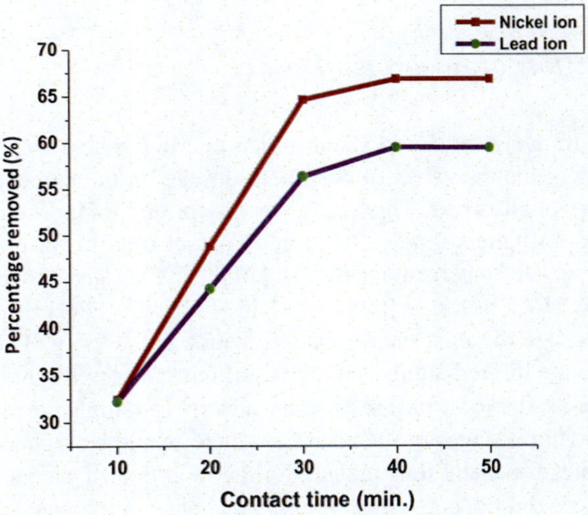

Fig. 3 Effect of contact time on adsorption of Pb^{2+} and Ni^{2+} ion from aqueous solutions. (Color figure online)

contact time may be attributed to the obtained small diameter of the synthesized silica particles, which promotes the dissemination of the metal ions from the bulk effluent sample onto the energetic sites of the silica surface. The results showed an increase in the amount of Pb^{2+} and Ni^{2+}adsorbed as the contact time increased from 10 to 40 min [9]. There was an increase in adsorption from 32.23–59.49% for lead ion and 32.44–67. 5% for nickel ion as the contact times increased from 10 to 40 min.

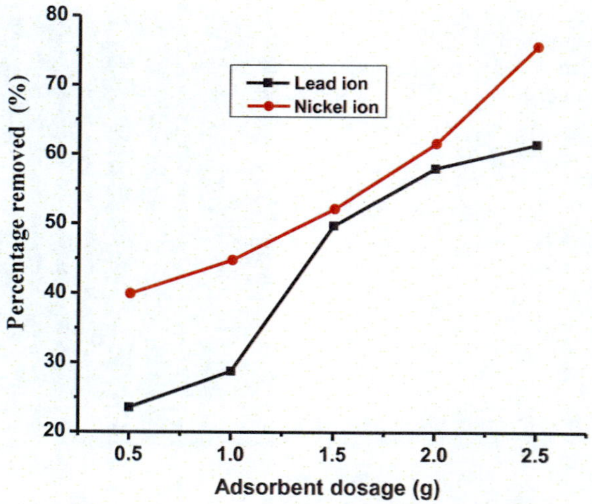

Fig. 4 Effect of adsorbent dosage on adsorption efficiency of silica microparticles from aqueous solution. (Color figure online)

Effect of Adsorbent Dosage

The weights of adsorbent (silica microparticles) were varied from 0.5 to 2.5 g to examine the effect of adsorbent dosage on the percentage of metal ions (Pb^{2+} and Ni^{2+}) adsorbed. This study was performed under the conditions of constant pH 5, room temperature (29 °C) and contact time of 40 min. The results are shown in Fig. 4. The percentage removal of the metal ions was observed to increase from 23.6 to 61.4% for lead (II) ion and 40–75.6% for Nickel (II) ion with an increase in the dosage of silica microparticles from 0.5 and 2.5 g. This increase may be attributed to the limited number of the adsorbing species present for a relatively larger number of available surface sites on the adsorbent at higher dosages. It is a known fact that at higher dosages of the adsorbent there would be higher availability of exchangeable sites for metal ions uptake. Further addition of the adsorbent beyond the optimum dose did not cause any significant change in the metal ions uptake. This is probably due to overlapping of adsorption sites as a consequence of overcrowding of adsorbent particles [10].

Effect of Heavy Metal Ion Concentration

The metal ion concentrations present in the separate solutions of lead and nickel were varied from 10 to 50 mg/L to examine the effect of initial metal ion concentrations on the adsorption process, while other experimental conditions remain constant. Figure 5 showed the effect of increasing metal ion concentrations on the level of removal. The result showed that the adsorption efficiency increased from 22.03 to 32.37% for lead ion from 63.34 to 72.22% for nickel ion as the concentration of

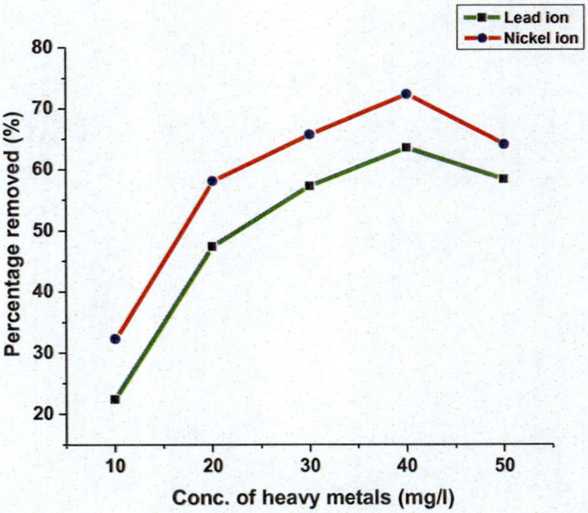

Fig. 5 Effect of concentration of Pb^{2+} and Ni^{2+} ion on the adsorption capacity of silica microparticles. (Color figure online)

metal ions increased from 10 to 40 mg/L. The observed increase in the heavy metal removed could be due to the increase in the collision between the metal ions and adsorbent particles and also, the increased specific area of adsorbent [11]. However, beyond the concentration of 40 mg/L, the adsorption experienced a decrease from 72.22% to 63.95% for nickel ion and 63.34% to 58.37% for lead ion. This is an indication that an additional increase in the lead ion concentration did not bring about any outstanding development in the heavy metal adsorption. This result is in agreement with Ahile et al. [12].

Adsorption Kinetics Model

The adsorption data obtained from the adsorption study of lead and nickel ions from separate aqueous solutions at various time intervals were subjected to pseudo-first and pseudo-second-order kinetic models. The pseudo-first-order and pseudo-second-order equations are expressed in Eqs. 1 and 2 (Dinu et al. 2010);

$$\frac{Ce}{qe} = \frac{1}{\mathrm{k_L q_m}} + \frac{C_e}{\mathrm{q_m}} \tag{1}$$

where q_e is the sorption capacities at equilibrium and time t (mg/g), q_m is the maximum metal ion binding capacity and k_1 is the rate constant of pseudo-first-order sorption ($Lmin^{-1}$). The pseudo-second-order chemisorption kinetic rate equation (Dinu et al. 2010) is expressed as:

$$\log \mathrm{q_e} = \log \mathrm{k_F} + \frac{1}{\mathrm{n}} \log_{\mathrm{Ce}} \tag{2}$$

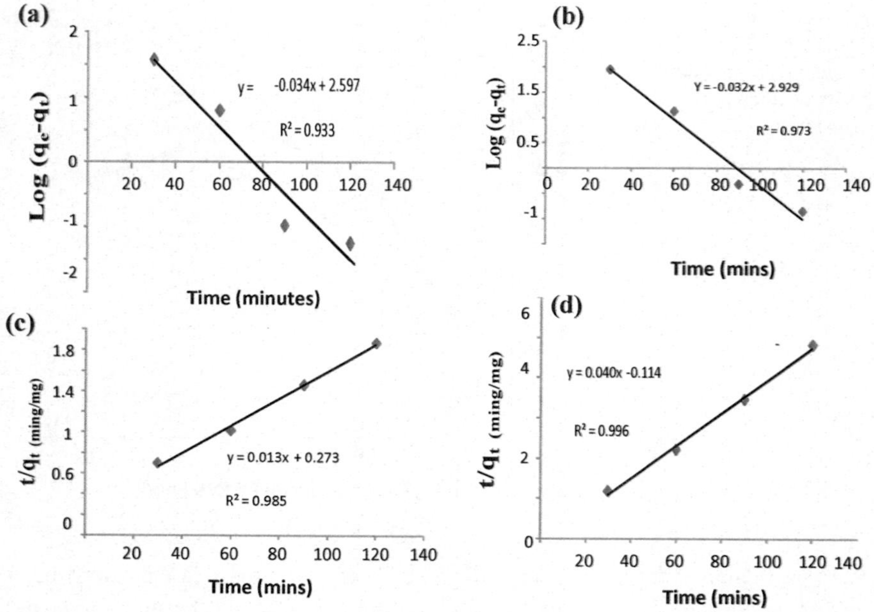

Fig. 6 Adsorption of heavy metals on silica microparticles for (**a**, **b**) pseudo-first-order model and (**c**, **d**) pseudo-second-order model for lead and nickel heavy metal ion. (Color figure online)

where, K_f is the rate constant of the pseudo-second-order sorption (g.mg^{-1}, min^{-1}). The kinetic study showed that the adsorption process of lead and nickel ions fitted best to the pseudo-second-order reaction model (Fig. 6 c and d for Pb^{2+} and Ni^{2+} respectively) compared to the pseudo-first-order model (Fig. 6a and b for Pb^{2+} and Ni^{2+}respectively). Correlation coefficients (R^2) of 0.973 and 0.996 were obtained from the pseudo-second-order model for the nickel and lead ions respectively (Fig. 6c and d). This is an indication that the adsorption process followed a pathway that is similar to chemisorptions. In chemical adsorption, the heavy metal ions stick to the adsorbent surface by forming a chemical (usually covalent) bond and tend to find sites that maximize their coordination number with the surface [13].

Langmuir and Freundlich Isotherm Study

The adsorption data were also fitted into Langmuir and Freundlich isotherm models to verify the adsorption performance. The linear forms of Langmuir and Freundlich equations are shown in Eqs. 3 and 4 [14]:

$$\frac{C_e}{q_e} = \frac{1}{K_L q_m} + \frac{C_e}{q_m} \tag{3}$$

C_e(mg/L) signifies equilibrium adsorbate concentration, q_e(mg/g) represents absorption capacity, q_m is the maximum value of q_e with increasing C_e, K_L is the constant related to the heat of adsorption.

$$\log q_e = \log K_F + \frac{1}{n} \log Ce \quad (4)$$

where, qe is the equilibrium adsorption capacity (mg/g), Ce is the equilibrium concentration of lead and nickel ion in solution (mg L^{-1}); 1/n and $_F$ are constants of Freundlich isotherm which signify the adsorption capability and intensity of adsorption, respectively.

The Freundlich equation gives a detailed interpretation of adsorption data over a specific concentration range; it is normally appropriate for surfaces that are highly heterogeneous and an adsorption isotherm with the absence of plateau, signifying multi-layer adsorption [15, 16]. Significant adsorption that occurs at a low concentration is indicated by values of 1/n less than unity but an elevated amount of adsorbed with concentration becomes insignificant at increased concentration and vice versa [17].

The Langmuir isotherm had the best fit for both Ni^{2+} and Pb^{2+} ions with R^2 values of 0.918 and 0.827 respectively. This inference is drawn from the fact that the correlation coefficients (R^2, 0.918, 0.827) obtained from the Langmuir plots (Fig. 7a and b) for both ions are closer to unity than the 0.721and 0.418 obtained from their Freundlich isotherm plots (Fig. 7c and d). This showed that the maximum coverage of the nickel and lead ions onto the surface of the silica microparticles may have taken place on a monomolecular layer on the adsorbent surface.

It also followed that the metal ions may have been taken up independently on a single binding site in such a way that the uptake of the first metal did not affect the uptake of the next [18].

The Langmuir isotherm equation allows the maximum metal ion binding capacity and binding constants to be predicted from the slope and intercept of linear plots respectively. From the plots in Fig. 7, the maximum metal ion binding capacity of silica microparticles of 22.3 and 25.5 mg/g were obtained for Pb (II) and Ni(II) ions respectively. These results show that silica microparticles can bind a substantial amount of lead and nickel ions from aqueous solution.

This study is economically viable because the studied adsorbent was synthesized using cheap chemicals. It would have practical applicability to industrial/environmental processes that involves the removal of heavy metals from wastewaters before disposal into rivers or land sites. It can also be used to remove heavy metals from portable water for drinking purposes.

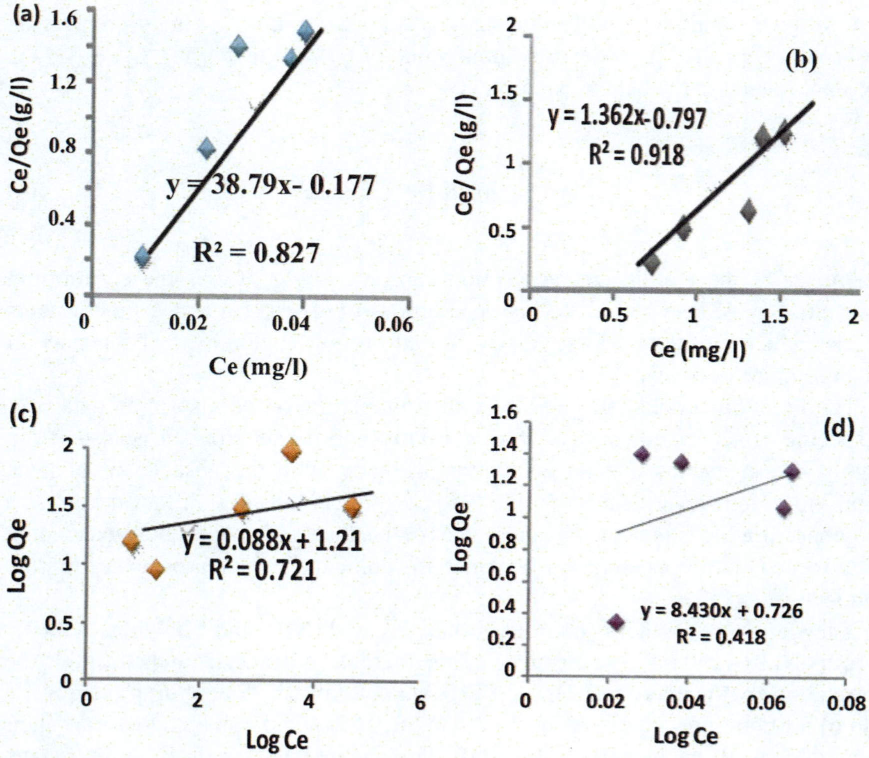

Fig. 7 **a**, **b** Langmuir isotherm plots for nickel and lead ions adsorption; **c**, **d** Freundlich isotherm plots for nickel and lead ions respectively. (Color figure online)

Conclusion

The synthesis and characterizations of spherical shaped silica microparticles for the adsorption of lead and nickel ions from aqueous solution were successfully carried out. The result showed that silica microparticles are a good adsorbent for the removal of Pb^{2+} and Ni^{2+} ion. The adsorbent dosage, heavy metal ion concentration and contact time had considerable effects on the adsorption of the studied metal ions as the metal ion uptake increased with an increase in these variables. Adsorption of Pb^{2+} and Ni^{2+} ion by silica microparticles was found to be extremely rapid as most of the exchange occurred within a matter of minutes. This rapid adsorption reflects the facts that the adsorption is a surface phenomenon and that the surfaces were readily accessible to the ions in solution. This study showed that silica microparticles can be effectively utilized in the treatment of wastewater heavily laden with lead and nickel ions.

Competing Interests The authors declare that they have no competing interests.

References

1. Tchounwou PB, Yedjou CG, Patlolla AK, Sutton DJ (2012) Heavy metal toxicity and the environment. Exp Suppl 101:133–164
2. Barakat MA (2011) New trends in removing heavy metals from industrial wastewater. Arabian J Chem 4:361–377
3. Igiri BE, Okoduwa SIR, Idoko GO, Akabuogu EP, Adeyi AO, Ejiogu IK (2018) Toxicity and bioremediation of heavy metals contaminated ecosystem from tannery wastewater: a review. J Toxicol 2018:16
4. Li J, Zhang C, Lin J, Yin J, Xu J, Chen Y (2018) Evaluating the bioavailability of heavy metals in natural-zeolite-amended aquatic sediments using thin-film diffusive gradients. Aquacultur Fish 3:122–128
5. Singh R, Gautam N, Mishra A, Gupta R (2011) Heavy metals and living systems: an overview. Indian J Pharmacol 43:246–253
6. Gleiter H (1995) Nanostructured materials: state of the art and perspectives. Nanostruct Mater 6:3–14
7. Xu C, Ravi Anusuyadevi P, Aymonier C, Luque R, Marre S (2019) Nanostructured materials for photocatalysis. Chem Soc Rev **48**:3868–3902
8. Khdary NH, Gassim AEH, Howard AG, Sakthivel TS, Seal S (2018) Synthesis and modification of mercapto-submicron scavenger for real-time extraction and preconcentration of As(iii). Anal Methods 10:245–255
9. A.S. Thouria Benzaoui, Djaafar Djabali, Adsorption of copper (II) ions from aqueous solution using bottom ash of expired drugs incineration, Adsorption Science & Technology, 36 (2017) 114–129
10. Dąbrowski A (2001) Adsorption—from theory to practice. Adv Coll Interface Sci 93:135–224
11. Dargahi A, Gholestanifar H, Darvishi P, Karami A, Hasan SH, Poormohammadi A, Behzadnia A (2016) An investigation and comparison of removing heavy metals (Lead and Chromium) from aqueous solutions using magnesium oxide nanoparticles. Pol J Environ Stud 25:557–562
12. Ahile UJ, Ama SO, Mchihi MM, Oteikwu MO, Nyom PN, Utange PI (2017) Removal Of Pb^{2+} from aqueous solution using rice husk as an adsorbent. FUW Trends Sci Technol J 2:886–889
13. Abdus-Salam N, Adekola SK (2018) Adsorption studies of zinc(II) on magnetite, baobab (Adansonia digitata) and magnetite–baobab composite. Appl Water Sci 8:222
14. Nouh ESA (2020) Manganese oxide-coated wool as adsorbent for U(VI) removal from aqueous waste solutions. Int J Environ Anal Chem 1–14
15. Freundlich HMF (1906) Uber die adsorption in Losungen. Zeitschrift fur Physikalische Chemie 57:385–470
16. Lagergren S (1898) About the theory of so-called adsorption of soluble substances. Kungliga Svenska Vetenskapsakademiens Handlingar 24:1–39
17. Ho YS, McKay G (1998) Sorption of dye from aqueous solution by peat. Chem Eng J 70:115–124
18. Okieimen FE, Okundia EU, Ogbeifun DE (1991) Sorption of cadmium and lead ions on modified groundnut (Arachis hypogea) husks. J Chem Technol Biotechnol 51:97–103

Microbial Leaching for Recovery of Nickel and Cobalt from Lateritic Ore: A Review

Lala Behari Sukla, Archana Pattanaik, D. P. Krishna Samal, and Debabrata Pradhan

Abstract The use of heterotrophic fungi (strains of *Aspergillus* and *Penicillium*) and bacteria (strains of *Bacillus* and *Pseudomona*) for metal recovery have been extensively studied. However, in case of nickel recovery from lateritic ores, strains of *Aspergillus* and *Penicillium* are the most preferred microorganisms. *Acidithiobacillus ferrooxidans* a chemolithotrophic bacterium has been reported to solubilize nickel from lateritic ore by microbial reductive method. Recent study reported about the use of DIRB (Dissimilatory Iron Reducing Bacteria) in bio-reduction of lateritic chromite overburden (COB) and enhancement in nickel and cobalt recovery. Subsequent leaching of DIRB pretreated ore by H_2SO_4 results in enhanced recovery of nickel and cobalt. Further work is required to better understand the bioleaching process and identification of more efficient microbial strains.

Keywords Laterite ores · Chromite overburden · Sukinda valley · Bioleaching · Heterotrophic microbes · Chemolithotrophic microbes · DIRB

Introduction

Population growth with increase in demand of metal resulted in diminishing high-grade ores. Thus mineral biotechnology provides greater opportunity for the development of economically viable processes for fulfilling those demands. Low and lean grade ores are present on large amount but recovery of these low-grade metal using conventional techniques is very expensive and causes harm to environment. So implementation of different biological processes involving various group microorganisms are regarded as one of the most promising and revolutionary solutions.

Nickel is a major alloy element, so considerable attention is being paid for its recovery. The country's need for nickel and cobalt is growing due to its annual demand in domestic markets. Due to development in steel, battery and other manufacturing

L. B. Sukla · A. Pattanaik · D. P. K. Samal · D. Pradhan (✉)
Biofuels and Bioprocessing Research Center, Institute of Technical Education and Research, Siksha 'O' Anusandhan (Deemed to be University), Bhubaneswar 751030, Odisha, India
e-mail: dpradhanmetal@gmail.com

C. Anderson et al. (eds.), *Ni-Co 2021: The 5th International Symposium on Nickel and Cobalt*, The Minerals, Metals & Materials Series,
https://doi.org/10.1007/978-3-030-65647-8_17

industries, the consumption of nickel has been increasing rapidly. About 70% of nickel goes into making steel. The remaining goes into production of cast iron, automobiles, storage batteries, wires, electronics, ships and aircrafts. Cobalt is one of the first row transition metals present as a minor component in different copper and nickel minerals.

Two preliminary types of nickel deposited ores have been found: laterites/oxides and sulphides. Laterites ores are predominantly low-grade nickel ores found in tropical climates and it represents approximately 85% of known global reserves. Sukinda mines of Odisha, India are known as the major nickel laterite deposits. Laterite ores contain several kinds of metal elements including nickel, cobalt, iron, silicon, aluminium and chromium. 40% of world's annual nickel production is from low-grade laterite ores [1, 2]. Sulphides are the high-grade nickel ores and typically range from 1 to 4%. Common nickel sulphide is pentadlite ($(Fe, Ni)_9S_8$), which accounts for about 75% of the nickel production [3]. Since the high-grade sulphide ore reserves are being depleted due to the rapid growing industrialization, processing of the low-grade nickel ores is considered as sustainable alternative for the nickel production. The ores like cobaltite, erythrite, glaucodot, and skutterudite are the major minerals of cobalt. But most of the cobalt minerals are obtained by reducing cobalt compounds that occur as by-product of nickel not by active mining of cobalt ores.

Way back the nickel and cobalt containing sulphide ores have been processed using pyrometallurgical techniques by conventional roasting and smelting. Further advances resulted in application of hydrometallurgical techniques for significant nickel purification. Two most common hydrometallurgical processes are high pressure acid leaching (HPAL) and reduction roast-ammoniacal leach (Caron process). Since 1950s under high temperature and acid pressure leaching of nickeliferrous laterite ore has been practiced [4–7]. These processes are energy intensive and not economical [8]. However in reduction roast-ammoniacal leach technique is used for processing of high iron limonitic ores or a mixture of limonite and saprolite. Drying and roasting in a reducing atmosphere followed by low pressure ammonia leaching is the major procedure involved in these techniques [9, 10]. Hence the application of bioleaching to recover low-grade mineral provides an alternative ecofriendly technology compared to hydrometallurgy.

Bioleaching is being used for increasing metal recovery from low-grade and refractory ores using microorganisms. Two groups of microorganisms used in bioleaching are: chemolithotrophic and heterotrophic microorganisms. In commercial operations, acidophilic, iron- or sulphur-oxidizing chemolithotrophic microorganisms have been studied most intensively. *Acidithiobacillus ferrooxidans, Acidithiobacillus caldus* and *Leptospirillum ferrooxidans* are iron- and/or sulphur-oxidising bacteria used in bioleaching of sulphide ores containing nickel. Heterotrophic microorganisms including both bacteria and fungus species have been studied to recover nickel from low-grade nickel laterites [8, 11–13]. *Aspergillus* and *Penicillium* are the two effective heterotrophic strains studied widely in bioleaching. This heterotrophic microorganism secretes different types of organic acids during oxidation process.

Statistics of Nickel Laterite Resources in India

As per UNFC total 189 million tonnes of nickel resources have been estimated in India till 2013. About 175 millions of nickel deposits are reported in Odisha (92%). The remaining are distributed in Jharkhand (9million tonnes), Nagaland (5million tonnes), Karnataka (0.23 million tonnes). Although the nickel deposit mines are quite a few in India but overtime estimates have shown occurrence of over 65 million tonnes of laterite ores in the ore belt of sukinda, Odisha with 0.15–1.2% nickel content [14, 15]. The mineral deposits of Jaduguda mines located in the east Singhbhum district of Jharkhand, India are composed of both nickel and copper with a trace association of uranium. Nickel percentage in these ores varies from 0.3 to 0.8%. In Pokhpur, Tueusang district of Nagaland 1.5% conc. of nickel shows in the goethite matrix within the tubular body of magnetite. Other reported mines of nickels in India are in Sreekandapuram, Kannur district Kerala, Hassan and Rampur district of Karnataka, Khammam chimalpahad ultramafic complex Telengana and Dabachiwadi of Maharashtra.

The underground mining deposits below 65 m in depth in chromite mines in the Sukinda Valley in Odisha are the major reserve of India's lateritic deposit. The only significant deposit of lateritic nickel ore in India, which is in the ultra-basic belt of Sukinda, is yet to be commercially exploited. The overburden (in the present case, it is the overburden of chromite mines) is nothing but the soil and dust removed in order to access an ore or mineral body. The huge amount of overburden (nearly 8–10 times that of the ore) that is generated during chromite mining and dumped nearby has found a very little use so far. The mineralogical studies indicated that there was no separate nickel bearing mineral phase in the lateritic nickel ore. Goethite is the main iron bearing phase or host, which contains most of the nickel in the raw lateritic ore.

In the mid nineteenth century Canada is known as the world's leading nickel producer and Sudbury Basin of Ontario contains one of the largest nickel deposits. Nickel is relatively un-reactive in nature; it reacts only with strong acids. So in 1922–1981 it is used for coin producing.

Nickel/Cobalt Mineralogy

The mineralogical studies indicated that there was no separate nickel bearing mineral phase in the lateritic nickel ore. Goethite is the main iron bearing phase or host, which contains most of the nickel in the raw lateritic ore. Nickel laterites are becoming an increasingly important source of nickel and cobalt. 70% of the world's on shore nickel resources are laterite ores [16, 17]. Lateritic deposits are often located very close to the surface typically less than 50 m in depth so there recoveries are easy by open pit techniques. According to mineralogy nickel laterites are classified as oxide deposit, clay silicate deposit and hydrous silicate deposit [9, 18]. Nickel sulphide ores are millerite

(NiS), nickeline (NiS), violarite ((Ni, Fe)$_3$S$_4$), bravolite ((Ni, Fe)S$_2$) and Pentlandite ((Fe, Ni)$_9$S$_8$) [3]. Laterite ores occur as saporite, smeotite and limonite. Nickel closely embedded with iron mineral matrixes are called goethite, FeO(OH) [19]. Cobalt is usually found associated with ores of copper and nickel. Common cobalt bearing ores are erythrite ($Co_3(AsO_4)_2.8H_2O$), skutterudite (Co,Ni)As_3, cobaltite (CoAsS), linnaeite($CoCo_2S_4$), etc. Extractions of cobalt from these ores are not economic hence most of the cobalt is recovered from the other mineral associated ores. A few studies have been reported that 90% of nickel and 34% of cobalt is extracted from lateritic nickel ore [20].

Microbes Involved in Nickel and Cobalt Recovery

Bioleaching involves the utilization of microorganisms and their metabolic products to dissolve metals from low-grade ores. This mechanism directly depends upon some of the factors that are the size of the microbial population, its metal tolerance capacity and adaptation towards the mining environment. Temperature, pH, oxygen supply, stirring rate and nutritional composition of the medium have a direct influence on the leaching efficiency of microbes. According to the deviation of temperature and pH the distribution of microorganisms also varied. So generally used microorganisms which are engaged in metal solubilisation are mostly thermophilic and acidophilic in nature. Both autotrophic and heterotrophic microorganisms possess the potential to recover nickel from its ores.

The acidophilic chemolithotrophs present in the mining area produce different inorganic and organic acids eventually facilitate the bioleaching [21]. They fix the carbon dioxide available in the atmosphere and utilise it as an energy source. These microbes grow autotrophically in mining areas and use reduced organic sulphur as an electron donor and oxygen as electron acceptor [22, 23]. Chemolithotrophic microorganisms oxidize the sulfide minerals and produce sulphuric acid which can dissolve different metals present in the nickel laterites. Acidophilic sulphur oxidizing chemolithotrophs are mostly thermophillic in nature. *Acidithiobacillus caldus, Leptospirillum ferrooxidans* and *Acidithiobacillus ferrooxidans* bacteria are engaged in nickel solubilisation. *Comamonas testosteroni, Alicyclobacillus ferrooxydans* and *Pantoea septic* have abilities to recover nickel from limonite and saprolite ores using stirred tank bioleaching process [24]. *Acidithiobacillus ferrooxidans* a chemolithotrophic bacterium has been reported to solubilize nickel from lateritic ore by microbial reductive method. This bacterium reduces ferric iron (Fe^{3+}) in the goethite to ferrous iron (Fe^{2+}) in anoxic condition with elemental sulfur as electron donor, thereby producing sulfuric acid which generates acidity in the medium [25–27] and is responsible for dissolution of nickel.

Most non-sulfidic ores contain no energy source for the growth of chemolithoautotrophs, so heterotrophic leaching is considered as a better substitute. More than 30 years ago heterotrophic microorganisms have been used for bioleaching of Ni-laterites [28]. Heterotrophic microorganisms have an absolute dependence on

Table 1 Microorganisms produced organic acids

Organism name	Type of organic acid produced	References
Bacillus megaterium (bacteria)	Citric acid	[34]
Pseudomonas putida (bacteria)	Citric acid, gluconic acid	[34]
Paenibacillus polymyxa (bacteria)	Oxalic acid, acetic acid	[35]
A. niger(fungi)	Oxalic acid, citric acid, gluconic acid	[32, 36, 37]
P. simplicissimum(fungi)	Citric acid, oxalic acid, gluconic acid	[38, 39]
Fusarium sp(fungi)	Oxalic acid, malic acid, pyruvic acid, oxaloacetic acid	[40]

autotrophs. They obtain their carbon and other energy source of nutrition solely from autotrophs. *Aspergillus* and *Penicillium* genera are generally found to be most effective fungal species used in biomining [29–31]. *Aspergillus* and *Penicillium* have ability to produce hydroxycarboxylic acids, especially citric acid [11] (Table 1). The acids produced have dual activities that are lowering the pH of media and increase in efficient microbial load. Different organic acids are secreted by several heterotrophic fungus and bacteria engages in bioleaching process such as citric, lactic, gluconic, formic, oxalic (Table 1). 90% of nickel recovery had been investigated by *Aspergillus* sp. and *Penicillium* sp [20]. Agatzini and Tzeferis [32] reported that 60% and 50% of Ni and Co can be extracted by *Aspergillus* and *Penicillium* sp from nonsulfidic nickel ores. Bacteria from the genera *Bacillus* and *Pseudomonas* can also leach nickel from silicate ores [33]. Up to 36,54, and 0.76% of Ni, Co, and Fe recovered from low-grade laterite ores [13].

However, under commercial and large scale application, use of heterotrophic organisms shows higher possible contamination. Generation of some undesirable organisms is there because of the vast amount of carbon sources.

Dissimilatory iron reducing bacteria (DIRB) are microbes which have ability to reduce Fe (III) coupled to the oxidation of organic compounds. They can be isolated from different environmental habitats and they have the capacity to reduce several metal ions. Commonly obtained DIRB strains belong to *Shewanella*, *Geobacter*, *Geovibrio*, etc. [41].

Numerous studies had shown that Chemolithotrophic microorganisms along with heterotrophic microorganisms are likely to exist together under similar conditions as a bacterial consortium and mixture of sulphuric acid and citric acid enhanced the dissolution of nickel laterite ore. In these cases heterotrophic microorganisms can also feed on organic matter generated by chemolithotrophic microorganisms to produce organic acids. Mixotrophic bacterial consortia consisting of *Comamonas testosteroni, Alicyclobacillus ferrooxydans and Pantoea septic* resulted in bioleaching yields of 2.5% Ni extraction [28, 42].

Microbial Leaching Mechanism

The dissolution of different metals by the acidophilic microorganisms follows two mechanisms, i.e. direct and indirect leaching mechanisms. In the direct leaching mechanism there is an attachment between the microbial cells and the surface of sulphide minerals [43]. Then the oxidation of mineral surface takes place by numerous enzymatic catalysis steps. In the indirect leaching mechanism there is no specific contact between the microorganisms and the surface of sulfide minerals. Microorganisms have the capacity to produce required leaching reagent lixiviant which is a chemically sulphide mineral. Metal leaching by heterotrophic microorganisms generally involves an indirect process through microbial production of organic acids, amino acids and other metabolites.

$MS + 2O_2 \rightarrow MSO_4$ (Direct mechanism)

$MS + Fe_2\,(SO_4)_3 \rightarrow MSO_4 + 2FeSO_4 + S^o$ (Indirect mechanism)

(where M is a divalent metal ion)

Leaching effects of bacteria and fungi on low-grade mineral ores are mainly founded on principles: acidolysis, redoxolysis, complexolysis and bioaccumulation [44, 45]. Reductive dissolution of oxidized Nickel laterites ores were also noticed [46]. Extraction of cobalt from pyrite concentrate has also been reported in some of the cases [47–49].

Microorganisms during metabolism convert the carbon source into variety of organic acids like citric, oxalic, gluconic acids [13, 50, 51],which lead to lowering the pH [52]. The production of these acids partially dissolved the goethite matrix and released the nickel into the solution. Therefore the load of soluble metals in the solution is increased and this process helps in better metal dissolution by chelating into soluble organometallic complexes [1, 53]. Several studies suggest that citric acid and oxalic acid are the two major fungal metabolites to have a major role in nickel bioleaching.

Carbon sources (glucose/sucrose) + Fungal/bacterial species $\rightarrow$ organic acids (oxalic, citric, etc.)

Ni/Co + citric/oxalic acids $\rightarrow$ Ni/Co-oxalate/citrate.

The leaching rate of nickel is directly proportional to the dissolution of iron matrix by the organic acids secreted by the *A.niger* [54].

Recent study reported about the use of DIRB (Dissimilatory Iron Reducing Bacteria) in bio-reduction of lateritic chromite overburden (COB) and enhancement in nickel and cobalt recovery. DIRB have the ability to utilize Fe (III) as terminal electron acceptor during bio-reduction of lateritic mineral. In this process DIRB consortium reduces the goethite phase to hematite and magnetite with the exposure of nickel oxide. Subsequent leaching of DIRB pretreated ore by H_2SO_4 results in enhanced recovery of nickel and cobalt. However, bioleaching of nickel and cobalt from lateritic ore is in laboratory scale only. It was observed that in case of pretreated

COB the extraction rate is much more compare to raw COB. Around 6–7 million tonnes of lateritic chromite overburden (COB) is generating annually in chromite mining at Sukinda valley. Nickel content in this COB is 0.5–1.0% [55].

Effect of Pretreatment on Low-Grade Ni-Laterite

Pretreatment of ores is a major factor in determining the rate and extent of any leaching. Recent years microwave and ultrasound pretreatment are improved better extraction of low-grade ores [56]. Pretreatments of ores create crack on surface area of the ores that have led to ore strength reduction and weaken of coarse particle. In comparison with the untreated ores the pretreated ores have showed improved metal extraction in reduced contact time [57, 58]. For instance the nickel dissolution rate increased from 8.1 to 26.4% upon the application of pretreatment of a nickel laterite sample [28]. With increasing microwave power and preheating time Ni-laterite ore, Ni extraction has been also increased about 80% [56]. Some experimental studies also described that effect of ultrasound pretreatment on nickel leaching using *A.niger* 95% nickel could be leached along with 12.5% of iron from laterictic ore [59]. Behera et al. [54] evaluated that 18% of nickel was recovered from pretreated Indian chromite (FeCr2O4) in shake flasks using the *Aspergillus niger* strain. Up to 15.3% of nickel recovery is shown in pretreated DIRB as compared to untreated ones.

Limitations and Future Aspects

One of the major drawbacks in bioleaching mechanism is it is a time consuming process. Application of genetic engineering technique upon potential strains and practice of some enzymatic process may solve this. The fungal heterotrophic strain used in lateritic mineral processing shows optimum growth in neutral pH. But there is an increased chance of contamination by other microbes in that pH range which is a major drawback associated with this process. Other added concerns to this process are maintenance of sterile environment, need for enormous organic carbons and production of undesirable excessive fungal biomass. One of the technical problems of carrying out these bioleaching processes is the difficulty in separating some of the fungal species and their mycelia from the ore substrate. The organic wastes generated from different agriculture and food industries can be used as the growth substrates for the purpose of fungal leaching, which will definitely decrease the process cost upon implementation of the technique in the industrial scale. Further work and observation is needed to know better about bioleaching mechanisms and pretreatment processes by which recovery of Ni/Co becomes higher. These are including adequation of a physical pretreatment, adoption of continuous leaching. Identify other potential strains that can be used in mixotrophic cultures for improved extraction rate.

Conclusion

Interest in low-grade Ni-laterite ores has increased in recent years as high-grade Ni-sulfide deposits are being quickly depleted. However, processing of Ni-laterites has proven technically difficult and costly, and the development of alternative low-cost biotechnologies for Ni solubilization has been encouraged. Nickel reduction from lateritic ore through microbe-mineral interaction is easier due to the production of organic acids and chelating compounds by the microbes. Microbial nickel cobalt recovery process by using DIRB and chemolithotrophic bacteria can provide a developed commercial approach towards low-grade ores extraction of nickel and cobalt. The heterotrophic fungal strains used in mineral processing show high-quality metal recovery but some challenges are there. Over the period of time practicing of these methods tremendously increased due to its cost effectiveness and environmentally safe perspective. Application of processing of pretreatment of laterite ores in biological leaching process should been implemented industrially.

Acknowledgements The authors are grateful to Prof. (Dr.) Manojranjan Nayak, President, Siksha 'O' Anusandhan University, for providing infrastructure and encouragement throughout.

References

1. Mohapatra S, Bohidar S, Pradhan N, Kar RN, Sukla LB (2007) Microbial extraction of nickel from Sukinda chromite overburden by *Acidithiobacillus ferrooxidans* and *Aspergillus* strains. Hydrometallurgy 85:1–8
2. Simate GS, Ndlovu S, Walubita LF (2010) The fungal and chemolithotrophic leaching of nickel laterites–challenges and opportunities. Hydrometallurgy 103:150–157
3. Giaveno A, Donati E (2001) Bioleaching of heazelwoodite by *Thiobacillus* spp. Process Biochem 36:955–962
4. Li J, Li X, Hu Q, Wang Z, Zhou Y, Zheng J, Liu W, Li L (2009) Effect of pre roasting on leaching of laterite. Hydrometallurgy 99:84–88
5. Luo W, Feng Q, Ou L, Zhang G, Chen Y (2010) Kinetics of saprolitic laterite leaching by sulphuric acid at atmospheric pressure. Miner Eng 23:458–462
6. McDonald RG, Whittington BI (2008) Atmospheric acid leaching of nickel laterites review. Part II: Chloride and bio-technologies. Hydrometallurgy 91:56–69
7. Girgin I, Obut A, Ucyildiz A (2011) Dissolution behaviour of a Turkish lateritic nickel ore. Miner Eng 24:603–609
8. Tang JA, Valix M (2006) Leaching of low grade limonite and nontronite ore by fungi metabolic acids. Miner Eng 19:1274–1279
9. Elias M (2002) Nickel laterite deposits-geological overview, resources and exploitation. In: Cooke CR, Pontgratz J (eds) Giant ore deposits: characteristics, genesis and exploitation 2002. CODES Special Publication, Centre for Ore Deposit Research, University of Tasmania, pp 205–220
10. Gleeson SA, Butt CRM, Elias M(2003) Nickel laterites: review. Society of Economic Geologists (SEG) Newsletter
11. Tzeferis PG (1994) Leaching of a low grade hematitic laterite ore using fungi and biologically produced acid metabolites. Int J Miner Process 42:267–283

12. Valix M, Usai F, Malik R (2001) Fungal bioleaching of low grade laterite ores. Miner Eng 14(2):197–203
13. Valix M, Tang JY, Cheung WH (2001) The effects of mineralogy on the biological leaching of nickel laterite ores. Miner Eng 14(12):1629–1635
14. Sukla LB, Das RP (1987) Kinetic of nickel dissolution from roasted laterites. Trans Indian Inst Metals 40(4):351
15. Kanungo SB, Jena PK (1988) Reduction leaching of manganese nodules of Indian Ocean origin in dilute hydrochloric acid. Hydrometallurgy 21(1):41
16. Dalvi AD, Bacon WG, Osborne RC (2004) The past and the future of nickel laterites. In: PDAC 2004 international convention-trade show and investors exchange
17. Watling HR (2007) The bioleaching of nickel-copper sulphides. Hydrometallurgy 91:70–80
18. Brand NW, Butt CRM, Elias M (1998) Nickel laterites: classification and features. J Aust Geol Geophys 17(4):81–88
19. Swamy YV, Kar BB, Mohanty JK (2003) Physico-chemical characterization and sulphatization roasting of low-grade nickeliferous laterites. Hydrometallurgy 69:89–98
20. Sukla LB, Panchanadikar V (1993) Bioleaching of lateritic nickel ore using a heterotrophic microorganism. Hydrometallurgy 32:373–379
21. Brierley JA, Brierley CL (2001) Present and future commercial applications of biohydrometallurgy. Hydrometallurgy 59(5):233–239
22. Hanford GS, Vargas T (2001) Chemical and electrochemical basis of bioleaching processes. Hydrometallurgy 59:135–145
23. Schippers A, Sand W (1999) Bacterial leaching of metal sulphides proceeds by two indirect mechanism via thiosulphate or via polysulphides and sulphur. Appl Environ Microbiol 65:319–321
24. Chaerun, SK, Alting, SA, Mubarok, MZ, Sanwani, E(2016) Bacterial bioleaching of low grade nickel limonite and saprolite ores by mixotrophic bacteria. In: E3S web of conferences EDP sciences
25. Johnson DB (2009) Extremophiles:acid environments. In: Schaechter M (ed) Encyclopaedia of microbiology 2009. Elsevier, pp 107–126
26. Rawlings DE, Coram NJ, Gardner MN, Deane SM (1999) *Thiobacillus caldus* and *Leptospirillum ferrooxidans* are widely distributed in continuous flow biooxidation tanks used to treat a variety of metal containing ores and concentrates, In: Amils R, Ballester A (eds) Biohydrometallurgy and the environment toward the mining of the 21st century Part A1999. Elsevier Press, Amsterdam, pp 777–786
27. Vian M, Creo C, Dalmastri C, Gionni A, Palazzolo P, Levi G (1986) *Thiobacillus ferrooxidans* selection in continuous culture. In: Lawrence RW, Branion RMR, Ebner HG (eds) Fundamental and applied biohydrometallurgy 1986. Elsevier, Amsterdam, pp 395–406
28. Giese EC, Carpen HL, Bertolino LC, Schneider CL (2019) Characterization and bioleaching of nickel laterite ore using *Bacillus subtilis* strain. Biotechnol Prog e2860. https://doi.org/10.1002/btpr.2860
29. Rastegar SO, Mousavi SM, Rezaei M, Shojaosadati SA (2014) Statistical evaluation and optimization of effective parameters in bioleaching of metals from molybdenite concentrate using Acidianus brierleyi. J Ind Eng Chem 20(5):3096–3101
30. Rasoulnia P, Mousavi SM, Rastegar SO, Azargoshasb H (2016) Fungal leaching of valuable metals from a power plant residual ash using *Penicillium simplicissimum*: evaluation of thermal pretreatment and different bioleaching methods. J Waste Manag 52:309–317
31. Vakilchap F, Mousavi SM, Shojaosadati SA (2016) Role of Aspergillus niger in recovery enhancement of valuable metals from produced red mud in Bayer process. Bioresour Technol 218:991–998
32. Agatzini S, Tzeferis P (1997) Bioleaching of nickel- cobalt oxide ores. Aus IMM Proc 1:9–15
33. Castro IM, Fietto JLR, Vieira RX, Tropia MJM, Campos LMM, Paniago EB, Brandao RL (2000) Bioleaching of zinc and nickel from silicate using *Aspergillus niger* culture. Hydrometallurgy 57:39–49

34. KrebsW Brombacher C, Bosshard PP, Bachofen R, Brandl H (1997) Microbial recovery of metals from solids. FEMS Microbiol Rev 20:605–617
35. Deo N, Natarajan KA (1997) Interaction of *Bacillus polymyxa* with some oxide minerals with reference to mineral beneficiation and environmental control. Miner Eng 10:1339–1354
36. Bosshard PP, Bachofen R, Brandl H (1996) Metal leaching of fly ash from municipalwater incineration by *Aspergillus niger*. Environ Sci Technol 30:3066–3070
37. Singh OV, Sharma A, Singh RP (2001) Optimization of fermentation conditions for gluconic acid production by a mutant of *A. niger*. Ind J Expt Biol 39:1136–1143
38. BurgstallerW Strasser H, Wobking H, Schinner F (1992) Solubilization of zinc oxide from filter dust with *Penicillium simplicissimum*: bioreactor leaching and stoichiometry. Environ Sci Technol 26(2):340–346
39. Burgstaller W, Zanella A, Schinner F (1994) Buffer-stimulated citrate efflux in *Penicillium simplicissimum*. Arch Microbiol 161:75–81
40. Bosecker, K(1989) Bioleaching of valuable metals from silicate ores and silicate waste products. In: Salley J, McGready RGL, Wichlacz PL (eds) Biohydrometallurgy 1989, CANMET, Ottawa, pp 15–24
41. Behera SK, Mulaba-Bafubiand AF (2015) Advances in microbial leaching processes for nickel extraction from lateritic minerals-a review. Korean J Chem Eng 32:1447–1454
42. Chaerun SK, Minwal WP, Mubarok MZ (2017) Indirect bioleaching of low-grade nickel limonite and saprolite ores using fungal metabolic organic acids generated by *Aspergillus niger*. Hydrometallurgy 174:29–37
43. Ahmadi A, Schaffie M, Petersen J, Schippers A, Ranjbar M (2011) Conventional and electrochemical bioleaching of chalcopyrite concentrates by moderately thermophilic bacteria at high pulp density. Hydrometallurgy 106(1):84–92
44. Brandl H (2001) Microbial leaching of metals. In: Rehm HJ, Reed G (eds) Biotechnology 2001. Wiley-VCH, Weinheim, 10, pp 191–224
45. Wu HY, Ting YP (2006) Metal extraction from municipal solid waste (MSW) incinerator fly ash—Chemical leaching and fungal bioleaching. Enzyme Microb Technol 38(6):839–847
46. Johnson DB (2013) Development and application of biotechnologies in the metal miningindustry. Environ Sci Pollut Res 20(11):7768–7776
47. Bosecker K (1997) Bioleaching: metal solubilization by microorganisms. FEMS Microbiol Rev 20(3–4):591–604
48. Watling HR, Shiers DW, Collinson DM (2015) Extremophiles in mineral sulphide heaps: some bacterial responses to variable temperature, acidity and solution composition. Microorganisms 3(3):364–390
49. Petersen J (2016) Heap leaching as a key technology for recovery of values from low-grade ores-a brief overview. Hydrometallurgy 165(1):206–212
50. Bosecker K (1986) Leaching of lateritic nickel ores with heterotrophic microorganisms. In: Lawrence RW, Branion RMR, Ebner HG (eds) Fundamental and applied biohydrometallurgy 1986. Elsevier Press, Amsterdam, pp 367−382
51. Franz A, Burgstaller W, Schinner F (1991) Leaching with *P. simplicissimum*: Influence of metals and buffers on proton extrusion and citric acid production. Appl Environ Microbiol 57:769–774
52. Marafi M, Stanislaus A, Furimsky E (2010) Handbook of spent hydroprocessing catalysts: regeneration, rejuvenation, reclamation. Environment and Safety, Elsevier
53. Burgstaller W, Schinner F (1993) Leaching of metals with fungi, mini review. J Biotechnol 27:91–116
54. Behera SK, Panda PP, Singh S, Pradhan N, Sukla LB, Mishra BK (2011) Study on reaction mechanism of bioleaching of nickel and cobalt from lateritic chromite overburdens. Int. Biodeter Biodegr 65:1035–1042
55. Swain PK, Chaudhury GR, Sukla LB (2007) Dissolution kinetics of chromate overburden by using mineral acids. Korean J Chem Eng 24(6):932–935
56. Agacayak T, Koseler M (2015) Effect of microwave heating on the leaching of lateritic nickel ore in perchloric acid. J Chem Soc Pak 37(2):230–235

57. Yang K, Li S, Zhang L et al (2017) Microwave roasting and leaching of an oxide-sulfide zinc ore. Hydrometallurgy 166:243–251
58. Zhai XJ, Wu Q, Fu Y, Ma LZ, Fan CL, Li NJ (2010) Leaching of nickel laterite ore assisted by microwave technique. Trans Nonferrous Metals Soc China 20(1):77–81
59. Sukla LB, Swamy KM, Narayana KL, Kar RN, Panchanadikar VV (1995) Bioleaching of Sukinda laterite using ultrasonics. Hydrometallurgy 37:387–391

Sulfuric Acid Leaching for Low Nickel Matte Under Atmospheric Pressure

Wanhai Xiao, Fenglong Sun, Xuheng Liu, and Zhongwei Zhao

Abstract This research focuses on leaching low nickel matte through sulfuric acid under atmospheric pressure, in which the Ni/Fe/Co enter into solution, noble elements and Cu_2S remain in the residue and the sulfur in Ni/Fe/Co sulfide is converted to H_2S. The subsequent process can be divided into three parts: (1) Fe in the solution can be removed by forming hematite or goethite and Ni/Co can be separated by organophosphorus extractants; (2) the noble elements and Cu_2S can be further treated in copper smelting process; and (3) H_2S can be transformed into elemental sulfur by Claus process. Compared with pyrometallurgy for copper-nickel sulfide ores, this method has advantages, such as low equipment requirement of atmospheric leaching, high recovery for noble elements and cobalt, convenient storage of sulfur than sulfuric acid and producing nickel sulfate that can be directly used in ternary materials. The whole process achieves the comprehensive utilization of low nickel matte with technical and economic benefits.

Keywords Low nickel matte · Acid leaching · Cobalt · Noble metal

Introduction

In recent years, the demand for nickel, cobalt and platinum group element (PGE) has greatly increased with the development of new energy batteries materials and catalyst for automobile exhaust purification [1–3]. And these elements mainly exist in nickel sulfide mineral. Most of the nickel sulfide minerals are currently treated by pyrometallurgy, which mainly includes matte smelting for low nickel matte and matte converting for high nickel matte [4]. After that, the Ni sulfide and Cu sulfide were separated by milling and floatation, and Ni metal was obtained by electrolysis

W. Xiao · F. Sun · X. Liu · Z. Zhao (✉)
School of Metallurgy and Environment, Central South University, Changsha 410083, Hunan, China
e-mail: zhaozw@csu.edu.cn

Key Laboratory of Hunan Province for Metallurgy and Material Processing of Rare Metals, Changsha 410083, Hunan, China

C. Anderson et al. (eds.), *Ni-Co 2021: The 5th International Symposium on Nickel and Cobalt*, The Minerals, Metals & Materials Series,
https://doi.org/10.1007/978-3-030-65647-8_18

after casting the Ni sulfide as anode. In the converting process, about half cobalt is lost to slag. Moreover, PGE is scattered because they enter into Ni sulfide, Cu sulfide, and some alloy separately in the grinding and floatation process. Besides that, as pyrometallurgy of copper, large amount of SO_2 is released in this process. And even if SO_2 is converted to H_2SO_4, the equipment for storage and low price of H_2SO_4 will eventually bring huge pressure to the nickel smelters.

To solve these problems, researchers turn their attention to hydrometallurgy. According to the leaching reactant, the hydrometallurgy methods can be divided into sulfide mineral leaching, low nickel leaching, and high nickel leaching; and depending on different conditions, it can also be grouped into atmospheric pressure acid leaching and oxygen pressure leaching [5–8]. Remarkably, encapsulation of sulfur in oxygen pressure leaching process needs high temperature and pressure, which raises new issues about leaching conditions and equipment.

According to the above analysis, the existing problems are mainly caused by converting and grinding and flotation process. Therefore, if the low nickel matte is directly treated by other methods than converting, loss of cobalt and dispersion of PGE can be avoided. In this paper, the difference between acid leaching and oxidation leaching are compared, and the characteristics of atmospheric H_2SO_4 leaching process are described.

Experiment

Materials

In this study, low nickel matte was obtained from Jinchuan Group Ltd of China. The samples were ground into powder and sieved to desired size fractions. All reagents used in this work were analytical grade and can be used without further purification.

Characterization

The phase structure of the low nickel matte and leach residue was determined by X-ray diffractometer (XRD, Rigaku D/max-2500). The surface morphologies of the samples were characterized by scanning electron microscope (SEM, JSM-5600, JEOL, Tokyo, Japan) and mineral liberation analyzer (MLA, FEI MLA 650, USA), respectively.

Results and Discussion

Leaching Principle

A thermodynamic description of the Me-S-H_2O system was developed. The Gibbs free energies of formation of the elements and inorganic compounds in low nickel matte are shown in Table 1.

According to the literature, the thermodynamic calculation of metal sulfides was as follows [10]:

$$\text{(1) } Me^{2+} + S + 2e^- = MeS, \tag{1}$$

with the equilibrium potential according to Eq. 2.

$$\varphi = (-\Delta_r G_m^\Theta + 2.303RT \lg a_{Me^{2+}})/2F \quad (T = 298\text{ K}, a_{Me^{2+}} = 1, \varphi_{298} = -\Delta_r G_m^\Theta/2F) \tag{2}$$

where φ is equilibrium potential (V), $\Delta_r G_m^\Theta$ is the Gibbs free energy of formation ($kJ \cdot mol^{-1}$), R is molar gas constant (J/(mol·K)), T is the temperature (K), $a_{Me^{2+}}$ is the activity of Me^{2+}, F is Faraday constant (C/mol).

$$\text{(2) } MeS + 2H^+ = Me^{2+} + H_2S \tag{3}$$

with the equilibrium pH according to Eq. 4.

$$\begin{aligned} pH = &- \Delta_r G_m^\Theta/(2 \times 2.303RT) - \frac{1}{2}\lg a_{Me^{2+}} - \frac{1}{2}\lg(p_{H_2S}/p^\Theta) \\ &(a_{Me^{2+}} = 1,\ p_{H_2S}/p^\Theta = 1,\ pH = -\Delta_r G_m^\Theta/(2 \times 2.303RT)) \end{aligned} \tag{4}$$

where pH is the equilibrium pH value, $\Delta_r G_m^\Theta$ is the Gibbs free energy of formation ($kJ \cdot mol^{-1}$), R is molar gas constant (J/(mol·K)), $a_{Me^{2+}}$ is the activity of Me^{2+}, p_{H_2S} is the partial pressure of H_2S (kPa), p^Θ is the standard pressure (kPa).

The data of Gibbs energies of formation were substituted into Eqs. 2 and 4, the equilibrium potential and pH of metal sulfides are shown in Table 2.

Table 1 The Gibbs energies of formation of the elements and inorganic compounds [9]

Substance	Cu^{2+}	Cu^+	Ni^{2+}	Fe^{2+}	Co^{2+}	CuS
$-\Delta_f G^\Theta/kJ \cdot mol^{-1}$	65.52	50	−45.6	−78.87	−54.4	−53.7
Substance	Cu_2S	NiS	FeS	CoS	H_2S	S
$-\Delta_f G^\Theta/kJ \cdot mol^{-1}$	−86.2	−79.5	−100.4	−96.775	−33.4	0

Table 2 The thermodynamic calculation of metal sulfides

No.	Chemical equations	$\Delta_r G_m^{\ominus}$/kJ·mol^{-1}	φ_{298}/V	pH
1	$Cu^+ + \frac{1}{2}S + e^- = \frac{1}{2}Cu_2S$	−93.1	0.96	–
2	$Cu^{2+} + S + 2e^- = CuS$	−119.22	0.62	–
3	$Co^{2+} + S + 2e^- = CoS$	−42.375	0.22	–
4	$Ni^{2+} + S + 2e^- = NiS$	−33.9	0.18	–
5	$Fe^{2+} + S + 2e^- = FeS$	−21.53	0.11	–
6	$Cu_2S + 2H^+ = 2Cu^+ + H_2S$	152.8	–	−13.39
7	$CuS + 2H^+ = Cu^{2+} + H_2S$	85.82	–	−7.52
8	$CoS + 2H^+ = Co^{2+} + H_2S$	8.975	–	−0.79
9	$NiS + 2H^+ = Ni^{2+} + H_2S$	0.5	–	−0.044
10	$FeS + 2H^+ = Fe^{2+} + H_2S$	−11.87	–	1.04

The data in Table 2 are plotted and the results are shown in Fig. 1. It shows that the low nickel matte can be directly treated by oxygen pressure acid leaching. The leaching reaction can be expressed as follows: $MeS + 2H^+ + 0.5O_2 = Me^{2+} + S + H_2O$, where Me represents iron, cobalt, nickel and copper. In the process, copper, nickel and cobalt will enter the solution in the form of sulfate. Moreover, iron will generate in the form of ferric oxide; sulfur in the low nickel matte enters the residue and is recovered as elemental sulfur. Compared with the traditional process, although it simplifies the process and improves the recovery rate of cobalt, there are still some problems. The noble metals will enter the residue together with

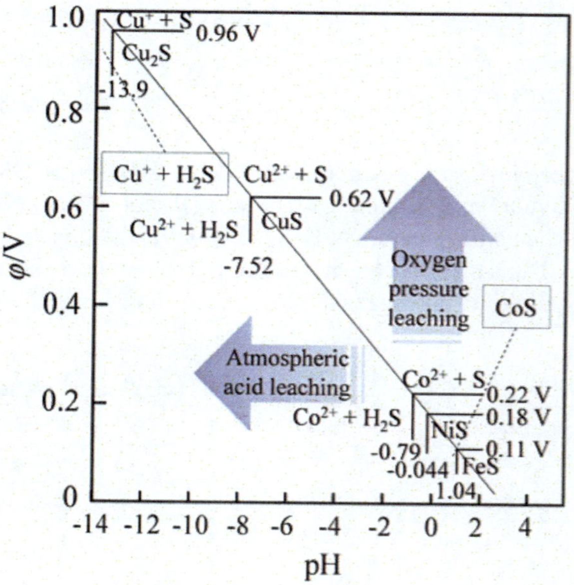

Fig. 1 The potential-pH diagram of Me-S-H₂O. (Color figure online)

ferric oxide and sulfur, which makes it difficult to recover. The elemental sulfur is easy to form on the surface of unreacted minerals and impede the leaching process, thus affecting the leaching efficiencies of copper, nickel and cobalt. Moreover, the subsequent separation of copper and nickel is difficult.

In addition, the curves reveal that the equilibrium pH of FeS, CoS and NiS is close to zero, which indicates that they can be decomposed by simple atmospheric acid leaching reaction without the use of oxidant. However, Cu_2S with -13.39 of equilibrium pH is difficult to be dissolved by acid. The low nickel matte is composed of metal sulfides and a small amount of alloys, therefore separation of copper from iron, cobalt and nickel can be achieved by simple acid leaching. Based on the above analysis, the simple leaching reaction by sulfuric acid can be expressed as follows: $MeS + 2H^+ = Me^{2+} + H_2S(g)$, where Me represents iron, cobalt and nickel. However, the process will produce a large amount of hydrogen sulfide. Generally, the production of H_2S in metallurgical process is relatively small, and it is mostly treated by alkali absorption. In the chemical industry, especially in the exploitation and utilization of oil and natural gas, there is a large amount of hydrogen sulfide, which is mostly used to prepare elemental sulfur by Claus process. It is a relatively mature process and the recovery of sulfur can reach more than 99%.

Atmospheric H_2SO_4 Leaching

Based on the characteristics of metal sulfides, the atmospheric non oxidizing sulfuric acid leaching of low nickel matte was studied. Through the factors of liquid-solid ratio, sulfuric acid concentration, temperature and time, the overall recovery of iron, cobalt and nickel were 99.1%, 99.4 and 99.5%, respectively. The loss of copper was less than 0.2% and hydrogen sulfide generated continuously in the leaching process prevent Cu^+ from entering the leaching solution leading to unmeasurably low concentration of copper. The weight percentage of leach residue was about 23.3%. By analyzing the composition content of raw materials and leach residue, the copper content in the residue is more than 70%, and the residue can be directly sent to the existing copper smelting. The enrichment of multiple noble metals such as gold, silver, platinum, palladium occurred by ca. factor of 4.3, corresponding to the weight percentage of 23.3%. It shows that noble metals are not leached in the decomposition process, and all of them are enriched in the leach residue, which is conducive to the subsequent treatment.

Figure 2 shows the comparison of XRD patterns between low nickel matte and leach residue. It can be seen from the figure that the main phases in the leach residue are chalcocite and chalcopyrite. The peaks of nickel mineral disappear completely, which indicates that nickel in low nickel matte has been well decomposed. Meanwhile, the surface of the leach residue changed obviously in the morphology and the roughness and porosity increased (Fig. 3).

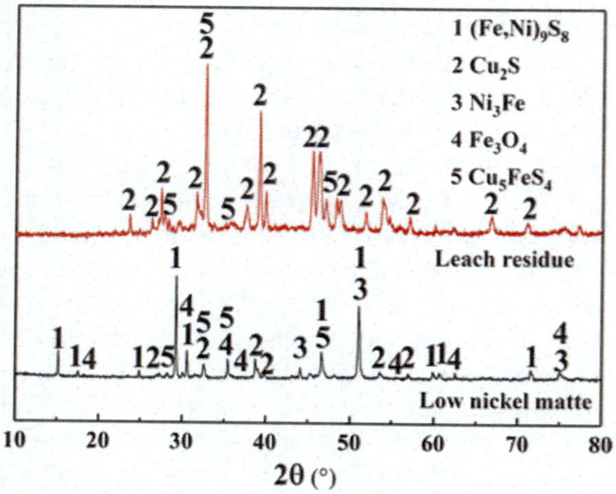

Fig. 2 XRD pattern of low nickel matte and leach residue. (Color figure online)

Fig. 3 SEM images of low nickel matte (**a**) and leach residue (**b**)

Compared with the traditional process, a promising method for extraction of valuable metals from low nickel matte with H_2SO_4 under atmospheric pressure was proposed. The flowsheet of the process is shown in Fig. 4. In the process, sulfur dioxide produced by matte smelting is used to prepare sulfuric acid, and acid leaching reaction is carried out with low nickel matte under atmospheric pressure to obtain leach residue, leaching solution and hydrogen sulfide.

The principle of sulfur preparation by Claus method is to make hydrogen sulfide ignite in air or oxygen. The incomplete combustion process will produce sulfur dioxide, and then the generated SO_2 will react with hydrogen sulfide to form elemental sulfur. Actually, a large amount of SO_2 is generated in the matte smelting process of nickel concentrate and hydrogen sulfide can be directly reacted with sulfur dioxide to produce elemental sulfur by Claus method, which can further simplify the Claus reaction process.

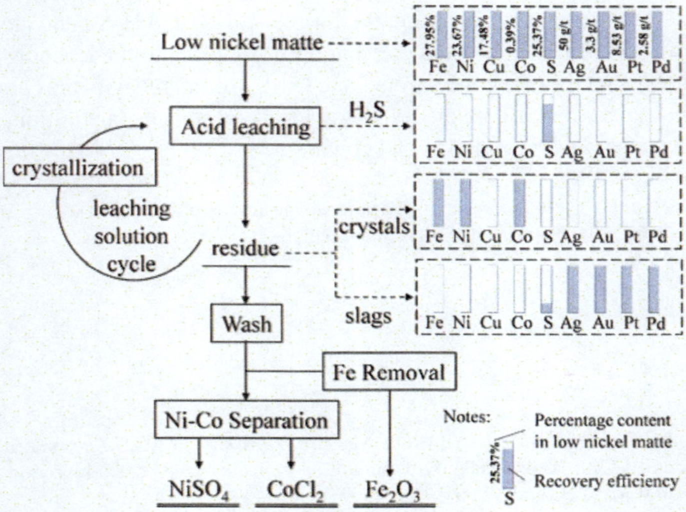

Fig. 4 The flowsheet of atmospheric H_2SO_4 leaching of low nickel matte. (Color figure online)

The leach residue containing Cu_2S and noble metals can be sent to the copper smelting. The noble metals can be recovered from copper anode slime. Moreover, the leaching solution can be used as the pregnant liquor for the next leaching. After several cycles, the sulfates will be crystallized out with the increase of concentration in leaching solution due to their low solubility in high-concentration sulfuric acid. Thus, iron, cobalt and nickel can be separated from sulfuric acid, which is conducive to the subsequent iron removal operation. The leaching residue can be washed with deionized water to obtain cuprous sulfide and solutions containing iron, cobalt and nickel. The solution will be used to remove iron before the separation of nickel and cobalt. In the traditional hydrometallurgical process, iron removal methods include recovery as jarosite, goethite or hematite. Generally, iron is trivalent after oxidation leaching, which needs to be reduced first and then slowly oxidized. However, here iron is divalent in the process, and no reduction is needed. Iron can be removed by hematite or goethite. After the operation of iron removal and cobalt extraction, iron products, cobalt solutions and nickel sulfate can be obtained, respectively.

Conclusion

A promising method for extraction of valuable metals from low nickel matte was proposed. Thermodynamic analysis and laboratory experiments verified the feasibility of using sulfuric acid under atmospheric pressure to separate Fe, Co, Ni and Cu. A leaching efficiency of more than 99% of nickel, cobalt and iron was obtained.

99.8% of the copper was recovered in the form of Cu_2S, and H_2S generated continuously during the leaching process prevented Cu^+ entering to the leaching solution. Also, the noble metals were concentrated into Cu_2S. Compared with the conventional process, the separation of copper and nickel has been solved. In addition, iron was successfully separated from nickel and cobalt.

Acknowledgements The authors gratefully acknowledge the National Key R&D Program of China (2018YFC1901603) for financial support of the research.

References

1. Nshizirungu T, Agarwal A, Jo YT, Rana M, Shin D, Park J-H (2020) Chlorinated polyvinyl chloride (CPVC) assisted leaching of lithium and cobalt from spent lithium-ion battery in subcritical water. J Hazard Mater 393:122367
2. Mpinga CN, Eksteen JJ, Aldrich C, Dyer L (2018) A conceptual hybrid process flowsheet for platinum group metals (PGMs) recovery from a chromite-rich Cu-Ni PGM bearing ore in oxidized mineralization through a single-stage leach and adsorption onto ion exchange resin. Hydrometallurgy 178:88–96
3. Qi S, Wu D, Dong Y, Liao J, Foster CW, O'Dwyer C, Feng Y, Liu C, Ma J (2019) Cobalt-based electrode materials for sodium-ion batteries. Chem Eng J 370:185–207
4. Habashi F (1997) Handbook of extractive metallurgy, vol 2. Wiley-VCH, New York, pp 715–791
5. Park K, Mohapatra D, Hongin K, Xueyi G (2007) Dissolution behavior of a complex Cu–Ni–Co–Fe matte in CuCl2–NaCl–HCl leaching medium. Sep Purif Technol 56:303–310
6. Sheng MW, Ji CQ, Zhu CL, Cai W (2012) Experimental study on oxygen pressure and water leaching of low nickel matte. YunNan Metall 41:32–34
7. Hofirek Z, Kerfoot DGE (1992) The chemistry of the nickel-copper matte leach and its application to process control and optimisation. Hydrometallurgy 29:357–381
8. Hofirek Z, Nofal PJ (1995) Pressure leach capacity expansion using oxygen-enriched air at RBMR (Pty) Ltd. Hydrometallurgy 39:91–116
9. James GS (2005) Lange's handbook of chemistry, 16th edn. McGraw-Hill, New York, pp 1237–1279
10. Li HG (2005) Metallurgical principle. Science Press, Beijing, pp 172–189

Cobalt and Nickel Separation in Hydrometallurgy Using Modified Clinoptilolite with Dialkyl Phosphoric and Ethylenediaminetetraacetic Acid as an Ion Exchanger

M. Banza, H. Rutto, and J. Kabuba

Abstract Natural Clinoptilolites were modified with dialkyl phosphoric acid and EDTA for cobalt (II) and nickel (II) separation in hydrometallurgy effluent. The unmodified and modified materials used were characterized using XRF, SEM and FTIR analysis. The operating parameters such as pH, initial concentration, particle size, contact time, clinoptilolites dosage and temperature were studied. A higher separation efficiency was obtained under the following optimal conditions: pH 6, initial concentration of 600 mg/L, contact time of 360 min, particles size of 1000–1400 μm, dosage of 10 g/250 mL and a temperature of 25 °C. Modified clinoptilolites showed a good separation for Cobalt (II) from Nickel (II) in an aqueous solution mixture. Adsorption isotherms, kinetic and thermodynamic studies were investigated. The Langmuir model was successful in describing the isotherm data for the considered metal ions with higher R^2. The pseudo-second order and the intra-particle model described the kinetic data from R^2 and Marquardt's percent standard deviation (MPSD), which indicates that a chemical reaction is responsible for the rate-limiting step.

Keywords Clinoptilolites · Hydrometallurgy · Dialkyl phosphoric acid

Introduction

The hydrometallurgical industry produces numerous metal waste contaminated streams and then requires effective methods for recovering these metals. The recovery of heavy metals from the hydrometallurgical effluent stream is economical and prevent water pollution. The effective recovery of these metals is possible only if the separation process is selective enough [1, 2].

Cobalt and nickel are applied in different industries due to the great properties they possess. Nowadays the demand for nickel is very high, with more than half of the

M. Banza (✉) · H. Rutto · J. Kabuba
Department of Chemical Engineering, Vaal University of Technology, Private Bag X021, Vanderbijlpark, South Africa
e-mail: banzajeanclaude@gmail.com

C. Anderson et al. (eds.), *Ni-Co 2021: The 5th International Symposium on Nickel and Cobalt*, The Minerals, Metals & Materials Series,
https://doi.org/10.1007/978-3-030-65647-8_19

world's production used in the manufacturing of stainless steel. Cobalt can be used in electroplating, alloys for airplane engines parts and to treat cancer and anemia. Cobalt is basically obtained as the product of the extraction of copper, nickel, lead and zinc from their sulphide and oxide ores/concentrates [3–5]. The high demand for high purity nickel and cobalt products makes the separation process of cobalt from nickel a crucial factor in hydrometallurgy, using classic, economical methods such as precipitation or oxidation [3, 4, 6].

The separation of cobalt from nickel in aqueous solutions has been studied for many years [7–9]. Many accomplishments for selective separation of cobalt from nickel have been reported in fields of solvent extraction [8, 10–12], liquid membrane electrodialysis [2], ion exchange [9, 13–17] and electrowinning [18].

In addition, more attention has been paid to ion exchange with polymeric resins, which shows productive, straight and recyclable purification in comparison to other methods. Ion exchange with polymeric resins has the disadvantage of the loss of organic material through entrainment, very complicated operation and high cost of the filtration process for the selectivity of nickel over cobalt which is a weakness of solvent extraction and liquid membrane electrodialysis [18]. An electrochemical process such as electrowinning consumes high electrical energy and releases toxic compounds which is harmful to the environment [15].

Among these methods, ion exchange is considered to be cost-effective and straightforward if natural zeolites as ion exchangers are used [19]. The main problem in heavy metal separation by unselective ion exchange is the saturation of sorbent by unwanted ions that reduces the sorbent ability for selective ion removal. Currently, natural zeolites, as low-cost ion exchanger, are used in wastewater treatment [20].

Clinoptilolite (natural zeolite) belongs to the heulandite group of minerals with the formula $(Na, K)_6Si_{30}Al_6O_{72}.nH_2O$ and is found in sedimentary rocks of volcanic origin. It is known as high silica heulandite because of its high content of silicon. The first building block of the zeolite framework is the tetrahedron, where the centre is occupied by aluminium or silicon, with four atoms of oxygen, and where each oxygen is divided between two tetrahedral [21].

The negative charge of Clinoptilolites comes from the tetrahedrally coordinated aluminium which is counteracted by divalent cations (Mg^{2+}, Ca^{2+}) and monovalent cations (K^+, Na^+), which move freely in the channels of the lattice framework because they do not occupy fixed positions. These ions act as counter ions and can be substituted by other cations [22]. Studies have identified the high potential of Clinoptilolites which is eco-friendly, cost-effective and practical for a broad range of concentrations of different heavy metals. The availability and low cost of Clinoptilolites have triggered further research for new application [23, 24]. Clinoptilolites can be modified by introduction of the new functional group to improve its selectivity and activity for removal of various substances. Efficient removal of cobalt (II) from aqueous solution was observed with Clinoptilolites nanoparticles modified with glutamic acid. Clinoptilolites nanoparticles have been modified with dimethylglyoxime to remove nickel (II) from aqueous solution [15].

This study investigates the modification of Clinoptilolites with dialkyl phosphoric acid and Ethylenediaminetetraacetic acid for separating cobalt (II) nickel (II) in an aqueous metal mixture.

Experimental

Materials

The natural Clinoptilolites used in this study were supplied by Pratley mining company in South Africa. The synthetic solution was prepared using cobalt chloride hexahydrate and nickel chloride hexahydrate. The pH of the solutions was appropriately adjusted with hydrochloric acid and sodium hydroxide. Modification of Clinoptilolites was done using dialkyl phosphoric and Ethylenediaminetetraacetic acid. All the chemicals used were analytical grade reagents from the Sigma-Aldrich Corporation.

Pre-treatment and Modification of Clinoptilolites

Clinoptilolites were first washed with distilled water to remove fines and impurities, then dried in the oven at 50 °C for 8 h after which it was crushed in an agate mortar and screened trough analytical sieves, to obtain an average particles sizes <100 μm for characterization and (+1000 to +2500 μm) for modification. The Clinoptilolites powder was heated at 70 °C in distilled water on a magnetic stirrer for 8 h to remove water soluble impurities. In order to reach a fixed water content, after centrifuging and drying, the powder was stored in a desiccator for 1 week.

To prepare the modified samples some experiments were done as follows: 100 g (+1000 to +2500 μm) of natural Clinoptilolites was added to 250 mL of 2 M dialkyl phosphoric acid and 250 mL of 2 M Ethylenediaminetetraacetic acid. The mixture was agitated using a thermostatic shaker for 6 h and then dried at 60 °C for 24 h. The modified Clinoptilolites samples were washed with distilled water and dried at room temperature.

Characterization Methods

Cobalt (II) and nickel (II) content in the solutions were determined using flame atomic absorption spectroscopy (AAS, thermos electrical corporation 3300). The elemental composition of the Clinoptilolites was determined using X-Ray fluorescence (XRF, Philips Magix pro). The functional groups present in the Clinoptilolites were determined using Fourier transform infrared spectroscopy (FTIR, Varian 7000). The morphological structure of the Clinoptilolites was studied using scanning electron microscopy (SEM, Philips XL 30FEG model).

Batch Adsorption Procedure

The removal of Cobalt (II) and Nickel (II) from aqueous solutions by modified Clinoptilolites was performed in a Shaking Incubator at 200 rpm for a period of 6 h. 10 g of modified Clinoptilolites were placed in a conical flask with 250 ml solution containing Cobalt (II) and Nickel (II) at the same ratio. Metal concentration varying from 50 to 1000 mg/L, temperature between 25 and 55 °C, particles size from +1000 to +2500 μm and a pH between 3 and 9 were used for the study. Every hour, 10 ml of the suspension was taken and filtered from the mixture and the resultant solutions were analyzed using atomic absorption spectroscopy (AAS). Cobalt (II) and nickel (II) percentage removal were calculated by the following equation:

$$\text{Metal removed }\% = \frac{(C_i - C_{eq})}{C_i} \times 100 \quad (1)$$

where C_i and C_{eq} are the initial and equilibrium concentration of Cobalt (II)/Nickel (II) in (mg/L), respectively.

The mass of adsorbed metal ion of the clinoptilolites (mg/g) was calculated by the following equation:

$$qe = \frac{(C_{i-C_e}) \times V}{M} \quad (2)$$

where V and M are the volume of solution (L) and amount of clinoptilolites (g), respectively [24]. All removal experiments were reproduced three times, and the mean values were used.

Error Analysis

To demonstrate the adequacy of the pseudo-first-order and pseudo-second-order models, Marquardt's percent standard deviation (MPSD) test of statistical analysis was applied

$$MPSD = 100\sqrt{\frac{1}{N-P}\sum_{i=1}^{n}\left(\frac{\left(q_{e(\exp)} - q_{e(pred)}\right)^2}{q_{e(\exp)}}\right)_i} \tag{3}$$

where N is the number of measurements, P stands for the number of parameters in the model $q_{e(pred)}$ and $q_{e(exp)}$ are predicted and experimental uptake rates, respectively.

Results and Discussion

Characterization Studies on Modified and Non-modified Clinoptilolites

XRF Analysis

XRF analysis was done to determine the chemical composition of Clinoptilolites before and after modification. Table 1 shows the results of chemical analysis of the natural Clinoptilolites and modified Clinoptilolites with dialkyl phosphoric and Ethylenediaminetetraacetic acid. Natural Clinoptilolites contain a large amount of SiO_2, Al_2O_3 and K_2O, while the other oxides, such as Fe_2O_3, Na_2O, K_2O, TiO_2 and CaO are present in much lower amounts. It was observed that the chemical composition of the natural Clinoptilolites changed during the acid modification. The acid modification resulted in an increase in the amounts of SiO_2 octahedral, Al_2O_3 tetrahedral and Fe_2O_3. Most of the exchangeable cations such as Mg, K, Ca, and Na in the octahedral structure of Clinoptilolites was decreased and hence, the quantity of the CaO, MgO K_2O, and Na_2O contents in the modified Clinoptilolites decreased.

Table 1 Chemical composition of natural and modified Clinoptilolites

Metal oxide in (%)	Natural clinoptilolites	Modified clinoptilolites
SiO_2	65.67	76.01
Al_2O_3	12.15	12.35
K_2O	8.85	5.20
Fe_2O_3	3.93	4.07
CaO	3.89	1.21
Na_2O	2.42	1.32
MgO	1.98	0.53
Ratio Si/Al	5.37	6.41

FTIR Analysis

To determine the qualitative characteristic of a surface functional group of Clinoptilolites, the FTIR was used in the range as shown in Fig. 1.

In the FTIR spectrum of natural Clinoptilolites in Fig. 1A, the a peak at 607 cm^{-1} is due to the bending vibration of (Si–O–Al) and (Si–O–Si), while the peak at 795 cm^{-1} is due to the (Si–O) stretching vibration. The asymmetric vibration of (Si–O–Si) appeared at 1000 cm^{-1}. The peak at 2840 cm^{-1} is attributed to free hydrogenation water and intermolecular hydrogen bonding (zeolitic water) and the peak at 3458 cm^{-1} are attributed to the bending and stretching vibration of hydroxyl groups (O–H).

FTIR spectrum in Fig. 1B is shows the modified Clinoptilolites peaks at 465 cm^{-1}. Skeleton deformation vibration of the alkyl molecule, peaks at 710 and 890 cm^{-1} are attributed to (N–O) stretching vibration and (C–N–O) bending vibration mode, respectively. The peak at 1000 cm^{-1} is due to (Si–O(Al) and Si–O(Si) stretching vibration this main peak was observed in the natural Clinoptilolites because it is possible for an ion exchanger/adsorbent to maintain some of its properties after modification [31].

Peaks at 1220 and 1345 cm^{-1} are attributed to asymmetrical and symmetrical vibration of CH_3, peaks in the range of 1500 and 3000 cm^{-1} are attributed to (N–O) and (C = O) stretching vibration and peak at 3465 cm^{-1} is attributed to the bending and stretching vibration of (O–H). Between 1500 cm^{-1} and 3000 cm^{-1} the modified Clinoptilolites showed an increase in intensity. This gives evidence of the modification..

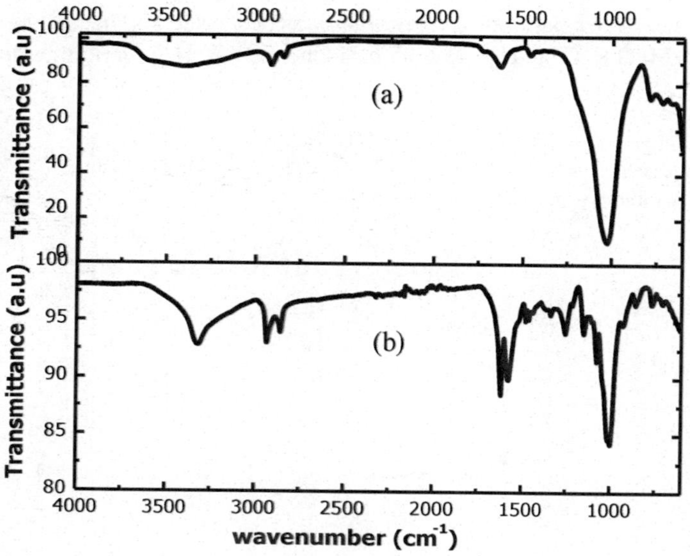

Fig. 1 FTIR spectra of (**A**) natural Clinoptilolites and (**B**) modified Clinoptilolites with dialkyl phosphoric and EDTA

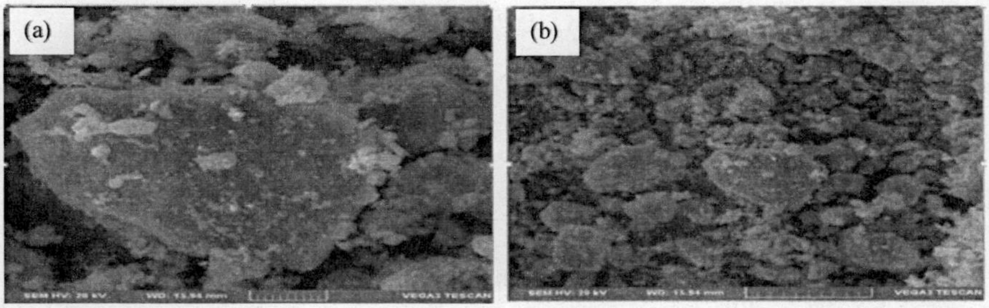

Fig. 2 SEM image of natural clinoptilolites (**a**) and modified clinoptilolites with dialkyl phosphoric and EDTA (**b**)

SEM Analysis

SEM was used to observe the change in the morphological structure of the natural and modified Clinoptilolites, the SEM image of natural and modified Clinoptilolites with dialkyl phosphoric and ethylenediaminetetraacetic acid was presented in Fig. 2a and b, respectively. The SEM image of the natural Clinoptilolites was different from that of the modified Clinoptilolites, the natural Clinoptilolites indicates the presence of bigger globular particles with irregular shapes. After modification, cavities of different shapes and sizes, and larger pores between the particles could be observed. The image reveals a well-developed and small glassy spherical particle with an external surface which is full of cavities this is as a result of modification.

The Effect of Process Variables on the Separation of Cobalt (II) from Nickel (II) in Batch Method Using Modified Clinoptilolites

Effect of Contact Time on the Separation of Cobalt (II) from Nickel (II) in an Aqueous Mixture

The effect of contact time on the separation of Cobalt (II) from Nickel (II) using modified Clinoptilolites with dialkyl phosphoric acid and Ethylenediaminetetraacetic acid is shown in Fig. 3a upon modification of natural clinoptilolites with dialkyl phosphoric acid and Ethylenediaminetetraacetic acid, the percentage removal at 300 min for Cobalt (II) and Nickel (II) by natural clinoptilolites was 96% and 27%, respectively, and therefore this shows good separation of Cobalt (II) from Nickel (II) ions [9].

Effect of PH on the Separation Co (II) from Ni (II) Solution in an Aqueous Mixture

The results obtained from batch experiments are shown in Fig. 3a. The pH solution had a huge effect on the removal of Co (II) and Ni (II) onto modified clinoptilolites with the highest removal efficiency of 96% for Co (II) and 26% for Ni (II) was observed at a pH 6. This shows good separation between cobalt and nickel. Because of the competition of proton ion in strongly acidic medium with metallic cations and also the effect of hydroxyl anions to precipitate metallic cations at a high pH. At lower pH values the removal efficiency was lessened, the low removal at low pH values is due to the increase (protons) in positive charge density on the surface sites, and the repulsion occurs between the metal ions and the groups with a positive charge on the surface.

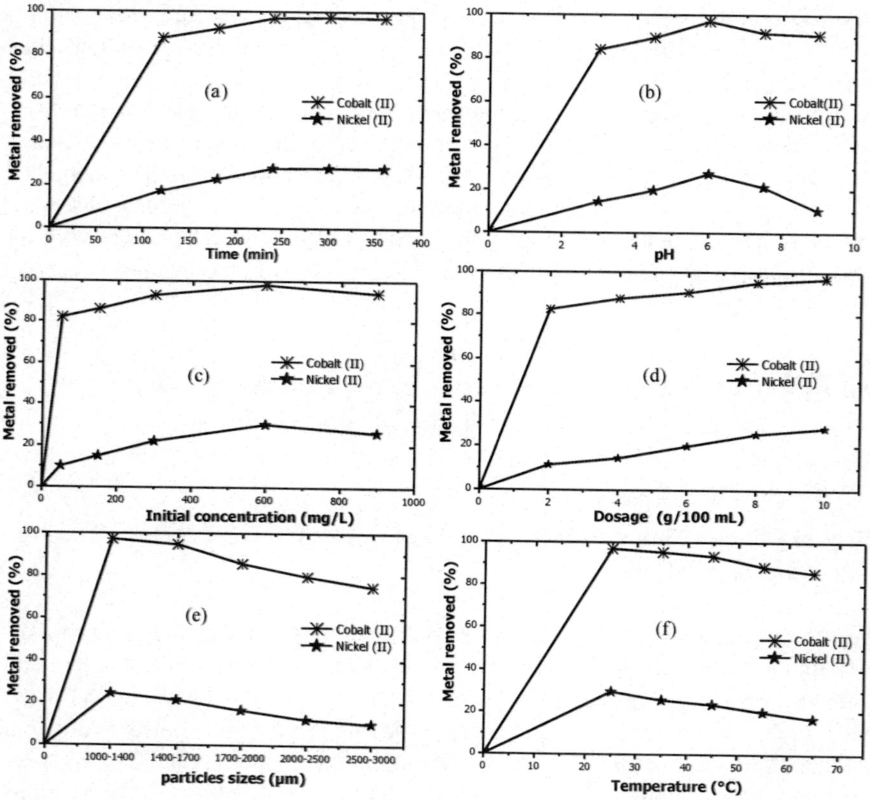

Fig. 3 Effect of contact time (**a**), pH (**b**), initial concentration (**c**), mass of dosage (**d**), particles size (**e**) and temperature (**f**) on the separation of Cobalt (II) from Nickel (II) in an aqueous mixture

Effect of Initial Concentration on the Separation of Cobalt (II) from Nickel (II) in an Aqueous Mixture

As depicted in Fig. 3b the Cobalt (II) and Nickel (II) removal efficiency increases from 77 to 97% and from 20 to 25%, respectively, when the initial metal concentration mixture was increased from 50 to 600 mg/L. This is due to the equilibrium nature of ion exchange and cobalt-modified clinoptilolites complexation processes. With an increase in the concentration of cobalt in the solution phase, according to Le Chatelier's principles, equilibrium goes forward and the removal efficiency of dissolved cobalt increases. Another explanation is as follows, at low initial metal concentration there is low competition for the active sites on the surface of the modified clinoptilolites and this means there will be more chances ion exchange to take place hence a higher removal efficiency is expected at low initial metal concentration.

Effect of Particles Size on the Separation of Cobalt (II) from Nickel (II) in an Aqueous Mixture

As shown in Fig. 3c, the removal percentage of Cobalt (II) and Nickel (II) decreases with increase in particles size. With a particle size of 1000–1400 μm the removal percentage was 96% and 34% for Cobalt (II) and Nickel (II), respectively, then decreased to 85% for Cobalt (II) and 21% for Nickel (II) with a particle size of + 2500 to +3000 μm. The increase in removal percentage is attributed to the increase in surface area of the ion exchanger available for removal. The reduction of particle size of modified clinoptilolites causes greater external areas available for interaction and short diffusion path lengths reducing mass transfer resistance in ion exchange, resulting in more ion exchange sites available for ions.

Effect of Clinoptilolites Dosage on the Separation of Cobalt (II) from Nickel (II) in an Aqueous Mixture

As shown in Fig. 3d, the dosage of modified clinoptilolites dosage was increased from 2 to 10 g, the removal percentage raises from 79 to 98% for Cobalt (II) and 16–27% for Nickel (II), the percentage removal of Cobalt (II) and Nickel (II) increases with an escalation of modified clinoptilolites dosage. This is due to the increase in the active site on the modified clinoptilolites making adsorption more favourable to Cobalt (II) as compared to Nickel (II). The highest removal percentage was obtained at 10 g with 98% of Cobalt (II) and 37% Nickel (II).

Effect of Temperature on the Separation of Cobalt (II) from Nickel (II) in an Aqueous Mixture

As shown in Fig. 3e temperature increased from 25 to 55 °C, the removal percentage of Cobalt (II) decreased from 96% to 85% and Nickel (II) from 37 to 21%. Temperature is an important variable in this process as it can affect the solution/solid interface and the kinetic characteristics of the process. The percentage removal will decrease with an increase in temperature, due to the mobility and competition of Cobalt (II) and Nickel (II) ions at higher temperatures onto modified Clinoptilolites. High temperature causes an increase in energy which results in a collision of molecules and reduces the probability of Cobalt (II) and Nickel (II) being attached to the available active sites

Adsorption Isotherms Studies

Equilibrium is reached when the capacity of the exchanger material is achieved. Generally, the removal capacity of an ion exchanger material or adsorbent can be obtained from isotherms such as the Temkin, Freundlich and Langmuir models. The corresponding isotherms at different temperatures are shown in Fig. 3.

Table 2 Parameter values of Temkin, Freundlich and Langmuir isotherm model

Metal ions	Isotherm model	Parameters	298 K	308 K	318 K	328 K
Cobalt (II)	Temkin	B	21.621	20.952	14.835	10.513
		R^2	0.916	0.920	0.929	0.939
	Freundlich	n	3.861	2.745	2.254	2.218
		K_F	37.715	17.645	11.214	10.473
		R^2	0.925	0.933	0.927	0.938
	Langmuir	K_L	0.125	0.121	0.108	0.107
		qm	121.958	99.001	73.545	61.346
		$\mathbf{R^2}$	**0.995**	**0.995**	**0.998**	**0.992**
Nickel (II)	Temkin	B	11.22	9.872	6.721	5.355
		R^2	0.915	0.925	0.917	0.908
	Freundlich	n	1.851	1.812	1.758	1.706
		K_F	1.995	1.904	1.205	1.188
		R^2	0.899	0.910	0.923	0.921
	Langmuir	K_L	0.021	0.016	0.014	0.011
		qm	47.845	46.363	30.301	24.874
		$\mathbf{R^2}$	**0.989**	**0.992**	**0.997**	**0.991**

In Table 2 the summarised results imply that nickel and cobalt were adsorbed in a monolayer on the surface of the modified clinoptilolites with higher R^2. Agreeable fitting of Langmuir model to the adsorption data of cobalt and nickel on different adsorbents have been reported by different researchers [20–24].

Kinetics Studies

The kinetic studies play an important role because it gives an important insight into the reaction mechanism and also determine whether the behavior of metal ion can be explained by a predictive model. Kinetic data collected were described with pseudo-first-order, kinetic and pseudo-second-order models in order to investigate the time course of Cobalt (II) and Nickel (II) onto modified clinoptilolites. Furthermore, it is important to determine whether the behavior of metal ion uptake can be clarified by a predictive model.

The values of the parameters from kinetic models in this study with the R^2 and MPSD values are presented in Table 3. High R^2 values can be observed from Table 3, but the R^2 values are close to 1 for the pseudo-second-order model. The predicted q_e values from the pseudo-second-order model agreed with the experimental q_e values, while the predicted q_e values from the pseudo-first-order model were not in agreement with the experimental q_e values. The low MPSD values (0.35–0.58) for the pseudo-second-order kinetic model signifies coherence to predicted and experimental data and the high MPSD values (21.39–24.30) from pseudo-first-order signifies poor coherence to predicted and experimental data. This statistical model of analyzing data demonstrates the suitability of the pseudo-second-order model.

Table 3 Rate constants (K_1 and K_2), MPSD and R^2 values obtained from pseudo-first order, pseudo-second order and intra-particle diffusion constants for cobalt (II) and nickel (II)

Metal ions	Pseudo-first order			Pseudo-second order	
		MPSD	Experimental values		MPSD
Cobalt (II)	$q_{e(pred)}$(mg/g) 53.305	24.30	$q_{e(exp)}$(mg/g) 97.95	$q_{e(Pred)}$(mg/g) 97.87	0.35
	$K_1(min^{-1})$ 3.814			$K_2(min^{-1})$ 0.0202	
	R^2 0.851			R^2 0.998	
Nickel (II)	$q_{e(pred)}$(mg/g) 8.814	21.39	$q_{e(exp)}$(mg/g) 19.01	$Q_{e(pred)}$(mg/g) 19.30	0.58
	$K_1(min^{-1})$ 2.612			$K_2(min^{-1})$ 0.012	
	R^2 0.835			R^2 0.997	

Conclusion

Ion exchanger was developed using clinoptilolites modified with dialkyl phosphoric acid and EDTA. Results from XRF, FTIR and SEM analyses showed that chemical modification took place upon addition of dialkylphosphinic acid and EDTA. The batch experiment showed that the removal of Cobalt (II) and Nickel (II) depends on initial concentration, pH, the mass of clinoptilolites and the temperature. The particles size of the modified clinoptilolites also showed a significant role in the removal efficiency.

Cobalt (II) and Nickel (II) removal were favoured by an increase in the pH of the solution up to 6. High percentage removal of Cobalt (II) and Nickel (II) were obtained at an initial concentration of 600 mg/L, mass of adsorbent 10 g, particle sizes of 1000–1400 μm and a temperature of 25 °C. The equilibrium data at various temperatures fitted well with the Langmuir isotherm. The adsorption process was fast, and the kinetic data showed a great fit to the pseudo-second-order kinetic model that indicates that the limiting-rate step for the process involves a chemical reaction. It can be concluded that the modification of clinoptilolites with dialkylphosphinic acid and EDTA is efficient for the separation of cobalt and nickel from hydrometallurgy effluent.

References

1. Lett DS (2004) Chem Sustain Dev 12:81–91
2. Sadyrbaeva TZ (2015) Desalination 365:167–175
3. Surucu A, Eyupoglu V, Tutkun O (2012) Ind Eng Chem 18:629–634
4. Juang R, Wang Y (2003) Ind Crops Prod 37:845–852
5. Lutandula MS, Maloba B (2013) J Environ Chem Eng 1:1085–1090
6. Rodriguez-Iznaga I, Petranovskii V, Rodriguez-Fuentes G (2014) Environ. Chem. Eng. 2:1221–1227
7. Xu Y, Xie Y, Liu J, Yan L, Yang R (2009) Hydrometallurgy 95:28–32
8. Mubarok MZ, Hanif LI (2016) Procedia Chem 19:743–750
9. Borandegi M, Nezamzadeh-Ejhieh A (2015) Colloids Surf Physicochem Eng Asp 479:35–45
10. Sun X, Ji Y, Zhang L, Chen J, Li D (2010) Hazard Mater 182:447–452
11. Wen JJ, Zhang QX, Zhang GQ, Cao ZY (2010) Trans Nonferrous Met Soc China 20:1534–1540
12. Li L, Xu S, Ju Z, Wu F (2009) Hydrometallurgy 100:41–46
13. Hadi P, Barford J, McKay G (2013) Chem Eng 228:140–146
14. McKevitt B, Dreisinger D (2012) Hydrometallurgy 125–126:1–7
15. Nezamzadeh-Ejhieh A, Kabiri-Samani M (2013) Hazard Mater 260:339–349
16. Mendes FD, Martins AH (2004) Selective sorption of nickel and cobalt from sulphate solutions using chelating resins 74:359–371
17. Zainol Z, Nicol MJ (2009) Hydrometallurgy 99:175–180
18. Li B, Liu F, Wang J, Ling C, Li L, Hou P, Li A, Bai Z (2012) Chem Eng 195–196:31–39
19. Wang S, Peng Y (2010) Chem Eng 156:11–24
20. Nosuhi M, Nezamzadeh-Ejhieh A (2017) Electrochim Acta 223:47–62

21. Dziedzicka A, Sulikowski B, Ruggiero-Mikołajczyk M (2016) Catal Today 259:50–58
22. Lin H, Liu QL, Dong YB, He YH, Wang L (2015) Microporous Mesoporous Mater 218:174–179
23. El-Kamash AM (2008) Hazard Mater 151:432–445
24. Nezamzadeh-Ejhieh A, Shahanshahi M (2013) Ind Eng Chem 19:2026–2033

Part IV
Pyrometallurgy

One-Step Extraction of Nickel from Nickel Sulfide Concentrates by Iron Addition

Fanmao Wang, Sam Marcuson, Leili Tafaghodi Khajavi, and Mansoor Barati

Abstract The conventional nickel smelting and refining is the dominant strategy for nickel production from the sulfide nickel ores. The main disadvantages are the significant amount of SO_2 emissions during smelting and complex refining processes. In this paper, the authors investigated a solid-state nickel extraction method that recovered nickel values into ferronickel alloy and simultaneously retained the bulk of sulfur of the nickel sulfide concentrate in the solid iron sulfide, thus mitigating the potential SO_2 emissions. The results showed that after heat treatment at 1073 K (800 °C) for 240 min, the residual nickel concentration in the resulting sulfide was only 0.6 mass pct and that in the ferronickel alloy was averaging 16 mass pct. The experimental data were in good agreement with the thermodynamic evaluations. Furthermore, the materials and energy balance were assessed for the thermal treatment process.

Keywords Nickel sulfide concentrate · Ferronickel · Sulfur dioxide emissions · Thermal treatment

Introduction

As an essential alloying element, more than 65 pct of the total nickel production is used in the stainless steels annually. The nickel production is currently dominated by the sulfide nickel ores and lateritic nickel ores. The former constituents approximately 40 pct of the total ore reserves and account for half of the nickel production worldwide [1, 2]. Smelting followed by refining is the conventional technology to extract nickel from the sulfide ores as metallic nickel. However, during the smelting process, significant amounts of sulfur dioxide are generated [3]. When the off-gas of

F. Wang (✉) · S. Marcuson · M. Barati
Department of Materials Science and Engineering, University of Toronto, 184 College Street, Toronto, ON M5S 3E4, Canada
e-mail: fanmao.wang@mail.utoronto.ca

L. T. Khajavi
Department of Materials Engineering, University of British Columbia, 309-6350 Stores Road, Vancouver, BC V6T 1Z4, Canada

C. Anderson et al. (eds.), *Ni-Co 2021: The 5th International Symposium on Nickel and Cobalt*, The Minerals, Metals & Materials Series,
https://doi.org/10.1007/978-3-030-65647-8_20

the smelting process contains 10–12 vol pct SO_2, then it can be efficiently captured [4]. The process of the produced SO_2 gas is costly and the product, e.g., sulfuric acid, needs transportation to the market, hence increasing the cost. Below the 10 vol pct, the SO_2 fixations is prohibitively expensive and the gas is sometimes vented to the atmosphere directly. Furthermore, the complex following refining processes of Ni increases the overall cost of nickel extraction.

The authors have proposed a virtually SO_2-free and one-step nickel extraction method, by which Ni values in the sulfide concentrate can be extracted into a ferronickel (FeNi) alloy with sulfur remaining in the solid as iron sulfide. Therefore, the evolution of SO_2 gas is significantly eliminated [5, 6]. In this paper, the authors studied the solid-state extraction of nickel from nickel sulfide concentrate at 1073 K (800 °C) by metallic iron addition.

Experimental

A commercial nickel sulfide concentrate containing 18.8 mass pct Ni with characteristic particle size of $d_{80} = 30\ \mu m$ was used as the raw material. The additive metallic Fe was over 99 pct pure and 100 pct smaller than 74 μm. Before the thermal treatment, a mixture of 10 g Ni sulfide concentrate and 12 g of metallic Fe was pelletized into two briquettes with 16 mm D × 12 mm H. Then the briquettes were isothermally treated at 1073 K (800 °C) for 240 min in a horizontal tube furnace under an Ar atmosphere, as shown in Fig. 1. After that, the samples were moved to the cold zone of the furnace tube for quenching, where the temperature was lower than 423 K (150 °C).

The microstructure and composition of the thermal treatment products were examined by scanning electron microscopy (SEM) and the electron probe micro-analyzer (EPMA).

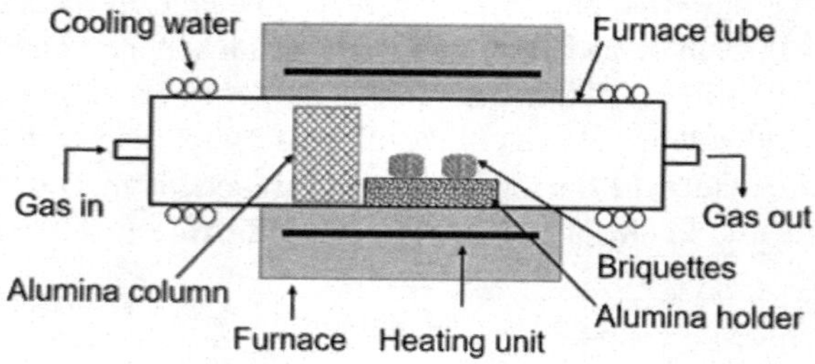

Fig. 1 Thermal treatment setup [5]. (Color figure online)

Results and Discussion

Thermodynamic Evaluation

The Ni sulfide concentrate is essentially a Fe–Ni–S system, and the phase diagram at 1073 K (800 °C) is shown in Fig. 2. As seen, the initial equilibrium of the Ni sulfide concentrate lies in the Fe-Ni sulfide phase domain, suggesting that heating the Ni concentrate alone does not yield ferronickel alloy. However, adding metallic Fe into the system modifies the overall composition into the FeNi-containing phase domain in the direction of the dashed arrow. As a result, the FeNi alloy is precipitated with the residual Ni-depleted iron sulfide.

The equilibrium phases and their composition resulting from the 12 g Fe and 10 g Ni concentrate are presented in Table 1. As seen, theoretically, 59 mass pct of FeNi is in equilibrium with 41 mass pct of iron sulfide at 1073 K (800 °C). The FeNi should contain 15.5 mass pct Ni while that in the iron sulfide is only 0.4 mass pct Ni. Based on the Ni distribution in the FeNi and iron sulfide, the maximum Ni recovery to FeNi was calculated to be 97 pct.

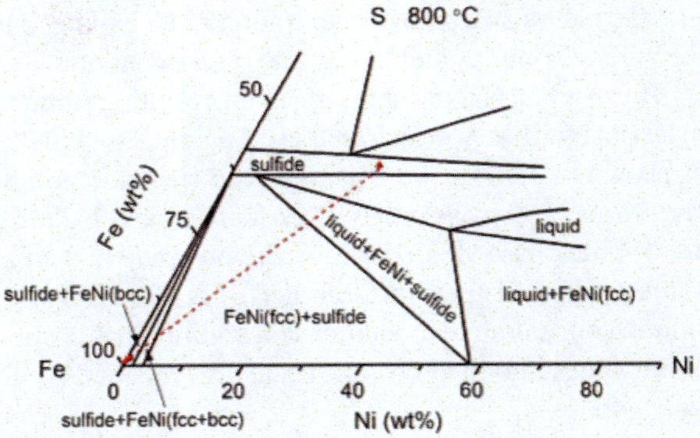

Fig. 2 The Fe-Ni-S phase diagram at 800 °C, calculated by FactSage™ 6.4 [7]. Sulfide: solid Fe-Ni sulfide; FeNi: Fe-Ni alloy; Liquid: metal-rich sulfide. (Color figure online)

Table 1 Equilibrium phases of 12 g Fe and 10 g Ni concentrate at 1073 K

Phase	Mass pct	Composition (mass pct)		
		Ni	Fe	S
FeNi	59	15.5	83.2	4.2E-4
Iron sulfide	41	0.4	62.7	36.5

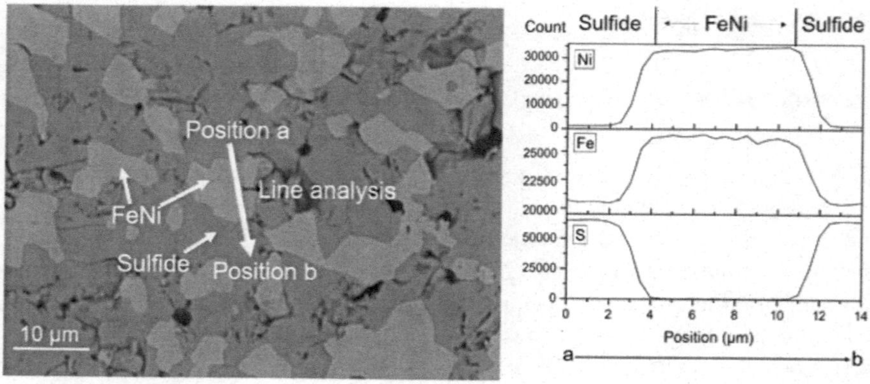

Fig. 3 Ni, Fe, and S concentration profiles in a FeNi particle and the adjacent sulfide phase. (Color figure online)

Microstructure and Composition

Figure 3 depicts the line scan analysis of a FeNi particle and the adjacent sulfide phase. The bright phase is FeNi, and the dark grey phase is the sulfide. It is evident that the FeNi has a high concentration of Ni and Fe, but a low level of S. In contrast, the sulfide contains high S but low Ni concentration.

The residual Ni concentration in the resulting sulfide phase was measured to be 0.6 mass pct by EPMA, which is in good agreement with the thermodynamic evaluation, as shown in Table 1. Furthermore, the Ni grade in the FeNi particles was found to vary from 3–59 mass pct by EPMA. The mass balance of Ni, Fe, and S in the FeNi and sulfide phase yielded an average Ni grade in the FeNi of 16 mass pct, also close to the predicated grade in Table 1. This confirms the one-step thermal treatment of Ni sulfide concentrate by metallic iron addition has successfully extracted the majority of Ni values into the FeNi alloy and also retained the S in the solid and Ni-depleted iron sulfide.

Materials and Energy Balance

A materials and energy balance for the thermal treatment process was carried out as detailed in Table 2. The additive iron for the Ni extraction was provided by Fe_2O_3 and coal. The fixed carbon of the coal was 80 mass pct. The S loss during the treatment was assumed as 3 mass pct in the form of SO_2. The Ni concentrations in the FeNi (16 mass pct) and sulfide (0.6 mass pct) were the results in Sect. 3.2 (Microstructure and composition). The volume ratio of CO to CO_2 in the off-gas was estimated to 1, comparable to that in the blast furnace iron making and direct reduced iron making process. Furthermore, the input and output materials were assumed at 298 K (25 °C)

Table 2 Materials balance and energy balance of the thermal treatment process at 1073 K (800 °C). Pn: pentlandite; Pyrr: pyrrhotite; Py: pyrite; Cpy: chalcopyrite. Qr = MW· ΔH formation, and Qs = MW·(H_T −H_{298})

	Phase	Mass/tpd	Element matrix/tpd									MW (g/mol)	ΔH formation (KJ/mol)	H_T −H_{298} (KJ/mol)	Qr (x10^4 MJ)	Qs (x10^4MJ)
			Ni	Cu	Co	Fe	S	C	O	Si (Al/Mg)	H					
Materials In	Pn	1140	376		8.2	602	598					771	−896.8	0	−132.8	0
	Py + Pyrr	420										88	−105	0	−50.3	0
	Cpy	100		32								183	−195	0	−10.6	0
	Gangue	340						76	192	74		60	−903	0	−511.5	0
	Fe_2O_3	3432				2400			1032			160	−826	0	−1770.8	0
	Coal (Reductant)	674						540	72	63		12	0	0	0	0
	Natural gas	225						169			56.3	16	−75	0	−105.3	81.6
	O_2 (air)	897							898			32	0	0	0	0
	N_2 (air)	2926										28	0	0	0	0
	Total in	10154	376	32	8.2	3002	598	785	2194	137	56.3				−2581.4	81.6
Materials Out	FeNi	2344	366	18	4.7	1923	0.7					57	0	24	0	99.6
	FeS	1698	10	14	3.5	1066	579					88	−102	53	−197.2	102.5
	CO	748						320	427			28	−111	25	−296.3	66.8
	CO_2	1423						389	1035			44	−394	39	−1287.9	125.7
	Gangue	474				13	0.3	76	264	137		60	−902	50	−713.9	39.8

(continued)

Table 2 (continued)

	Phase	Mass/tpd	Element matrix/tpd									MW (g/mol)	ΔH formation (KJ/mol)	$H_T - H_{298}$ (KJ/mol)	Qr (x10^4 MJ)	Qs (x10^4MJ)
			Ni	Cu	Co	Fe	S	C	O	Si (Al/Mg)	H					
	H_2O	506							450		56.3	18	−242	30	−680.6	84.9
	N_2 (air)	2926										28	0	25	0	258.8
	SO_2	35					18		18			64	−297	40	−16.7	2.2
	Total out	10154	376	32	8.2	3002	598	785	2194	137	56.3				−3179	780
In - Out		0	0	0	0	0	0	0	0	0	0					

and 1073 K (800 °C), respectively, and the heat loss was assumed to be 20 pct of the thermal load.

As seen in Table 2, with the addition of natural gas, Q(in)/Q(out) = (Qr in + Qs in)/(Qr out + Qs out) = 1, where Qr is the amount of the required energy for the formation reaction, and Qs is the sensible heat. The energy balance shows that the proposed solid-state nickel extraction at 1073 K (800 °C) is an endothermic process. To produce 2344 tpd FeNi from 2000 tpd Ni sulfide concentrate with Fe_2O_3 (3432 tpd) and coal (674 tpd), approximately 3.2E5 m^3pd natural gas is required to provide the additional heat. Replaced by coal, the required amount of fossil fuel is about 430 tpd.

Conclusions

This paper investigated the feasibility of nickel extraction from nickel sulfide concentrate by metallic iron addition. It was found that at 1073 K (800 °C), the addition of metallic iron recovered nickel into ferronickel alloys with the remaining sulfur in the solid iron sulfide, thus eliminating SO_2 emission. The materials and energy balance of the solid-state extraction at 1073 K (800 °C) indicated the endothermic process.

Acknowledgements The authors wish to acknowledge the financial support from the Natural Science and Engineering Research Council of Canada (NSERC, STPGP 479533-15), Process Research Ortech Inc., and technical support from XPS Consulting & Testwork Services, Glencore and Vale Canada. Fanmao Wang was partially supported by the China Scholarship Council (CSC, No. 201708530245).

References

1. Michele ME (2020) In: Nickel statistic and information. United States Geological Survey. https://pubs.usgs.gov/periodicals/mcs2020/mcs2020-nickel.pdf. Accessed 4 Aug 2020
2. Minerals Engineering Technical Service (2015) In: The rise of nickel laterites and advances in sulphide processing. Midas Engineering Group. https://www.metsengineering.com/wpcontent/uploads/2015/07/METS_Resources_2014_Nickel-Processing-and-The-Rise-of-Laterites_JBM.pdf. Accessed 4 Aug 2020
3. Warner AEM, Díaz CM, Dalvi AD, Mackey PJ, Tarasov AV, Jones RT (2007) JOM world nonferrous smelter survey part IV: nickel: sulfide. JOM 59(4):58–72
4. Crundwell FK, Moats MS, Ramachandran V, Robinson TG, and Davenport WG (2011) Sulfur dioxide capture in sulfuric acid and other products. In: Extractive metallurgy of nickel, cobalt and platinum group metals 2011, Elsevier, pp 247–257
5. Wang F, Liu F, Elliott R, Rezaei S, Khajavi LT, and Barati M (2020) Solid state extraction of nickel from nickel sulfide concentrates. 822:153582
6. Wang F, Marcuson S, Khajavi LT, and Barati M (2020) Optimum treatment time for solid-state extraction of nickel from nickel sulfide concentrates at 1073 K. Metallurgical and Materials Transactions B 51(6):2642–2652
7. Bale CW, Bélisle E, Chartrand P et al (2016) FactSage thermochemical software and databases, 2010-2016. CALPHAD: Comput Coupling Phase Diagrams Thermochem 54:35–53

Refractory Challenges in Nickel and Cobalt Processing Furnaces

Dean Gregurek, Jürgen Schmidl, and Alfred Spanring

Abstract The primary metallurgy of nickel and cobalt, as well as copper, is closely interlinked, and follows processing of sulfidic ore concentrates. Commonly these furnaces are lined with magnesia-chrome-based materials. The selection of the processing route, furnace type and slag system will be dictated by the specific ore type available; this will determine the individual refractory wear mode. This paper evaluates the common refractory wear mechanisms of infiltration, spalling and chemical attack by various slag systems, as well as sulfur attack as observed when processing primary sulfidic ores. All these wear parameters discussed in the paper lead to a severe degeneration of the brick microstructure and a decreased lining life. In addition to post-mortem studies, thermochemical calculations and corrosion testing provide better understanding of the wear mechanisms. Based on such research results, combined with specific process knowledge, appropriate brick lining solutions for nickel and cobalt processing furnaces can be recommend.

Keywords Refractories · Wear mechanisms · Post-mortem studies

Introduction

The metallurgical production of cobalt and nickel and of copper is closely interlinked to the mineralogy of the respective ores: cobalt is found in African Cu-Co ores as well as in Co-rich nickel ores (laterites and sulfides). Also, Co-bearing secondary raw materials contain the mentioned elements in different contents. In pyrometallurgical processing of sulfidic ores, cobalt is generally recovered as a by-product during the production of copper and nickel. Besides various hydrometallurgical production

D. Gregurek (✉)
RHI Magnesita, Technology Center Leoben, Magnesitstrasse 2, 8700 Leoben, Austria
e-mail: dean.gregurek@rhi-ag.com

J. Schmidl · A. Spanring
RHI Magnesita, Kranichberggasse 6, 1120 Vienna, Austria

C. Anderson et al. (eds.), *Ni-Co 2021: The 5th International Symposium on Nickel and Cobalt*, The Minerals, Metals & Materials Series,
https://doi.org/10.1007/978-3-030-65647-8_21

routes, the pyrometallurgical processes involved comprise smelting (i.e., production of a Cu-Ni-Co-Fe matte in electric or flash furnaces), converting and refining (i.e., removal of Fe and S as well as other elements from matte by oxidation). The ongoing trend to use higher oxygen enrichment in process air enables higher intensity processes but also increases the oxidation potential in the vessel. Therefore, during smelting and especially during the oxidative treatment in converting and refining, some amounts of valuable metals are also transferred to the slag in their oxides form. A subsequent slag treatment is then applied for recovering valuable metals from the slag, namely allowing an economic material stream handling and minimizing metal losses. This is often done in a slag reduction furnace. Additionally, copper and nickel are very good collectors for precious metals and platinum group metals (PGMs)—therefore, depending on the content of these metals, an additional refining process is used to recover these valuable metals from the matte or metal phase [1–3].

The present paper describes the refractory wear as observed in copper, nickel and cobalt production furnaces, using samples from different production stages and metallurgical furnace types. The laterite smelting is not considered. The type of furnace operation (i.e., continuous or batch-wise) as well as the slag type used (i.e., mainly fayalite slags, but also some special processes using calcium-ferrite slag) have a significant influence on the dominating refractory wear mechanism.

Analytical Procedure

Every single post mortem study starts with the visual inspection carried out on the brick cut section followed by selection of samples for chemical analyses and mineralogical investigation. The chemical analyses are carried out by using X-ray fluorescence analysis (Bruker S8 TIGER). The mineralogical investigation is performed on polished sections using reflected light microscope, X-ray diffraction (Bruker D8 ADVANCE) and scanning electron microscope (SEM) (JEOL JSM-6460) combined with an energy-dispersive and wavelength-dispersive X-ray analyser. The specimen for mineralogical investigation is embedded into the epoxy resin and polished on the lead blade by using of diamond spray.

Refractory Wear

The observed wear phenomena can generally be subdivided into continuous and discontinuous wear [4, 5].

Continuous wear is characterized by continuous mass loss as a function of time due to chemical, thermal and mechanical load (single loads or interrelated). In most cases, particularly for the magnesia-chromite bricks, the most frequent continuous wear is corrosion of the refractory by dissolution in melts/slags with or without additional hot erosion. Discontinuous wear is mainly characterized by mechanical failure as a consequence of thermal and chemical load occurring discontinuously over time, and is, in most cases, associated with mass loss. Other discontinuous wear phenomena include structural spalling, thermal shock failure and bursting reactions.

The common refractory wear mechanisms of infiltration, chemical attack by various slag systems, as well as sulfur attack and non-oxide infiltration are briefly introduced and discussed in this paper.

Chemical Attack by Fayalite Slag and Forsterite Bursting

The most frequent wear of the refractory lining in the nickel furnaces such as flash smelting furnaces, Electric Furnace, PS converters or slag reduction furnaces for high cobalt slags is corrosion by acidic slag. The chemical composition of the main slag types from Ni-sulphide smelting vessels can roughly be expressed by the systems FeO-MgO-SiO_2(-CaO-Al_2O_3-Cr_2O_3) for smelting and FeO-CaO-SiO_2(-MgO-Al_2O_3-Cr_2O_3) for converting (Fig. 1).

Generally, the corrosion of the refractories by slag attack manifests itself in three ways [5]:

(a) Dissolution reaction occurring at the immediate brick hot face: The dissolution process, at least in the closed system, will continue until the liquid slag has reached saturation. However, in practice, the point of saturation is never reached and dissolution continues until the entire refractory has been consumed.
(b) Dissolution and chemical reaction within the refractory microstructure: Infiltrating slag will dissolve magnesia especially from the fines according to the respective phase equilibrium. This will not directly contribute to corrosive wear which takes place at the immediate refractory hot face. Nevertheless, it will contribute to wear by preparing hot erosion due to a loss of brick bonding.

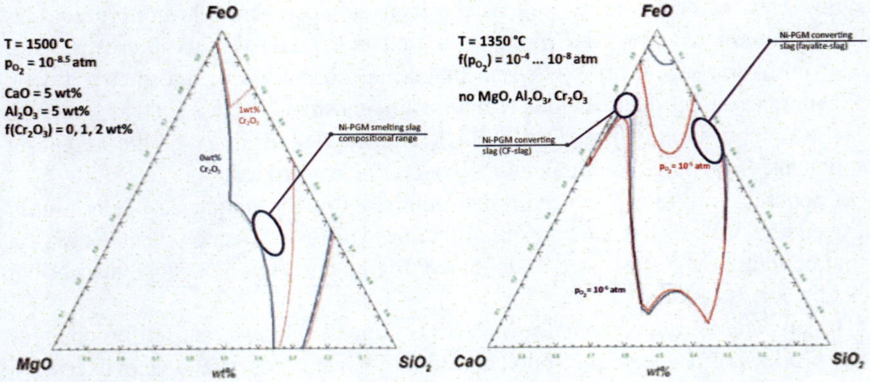

Fig. 1 Typical chemical compositional range of Ni-PGM smelting ("FeO"-MgO-SiO_2) and converting slags ("FeO"-CaO-SiO_2). (Color figure online)

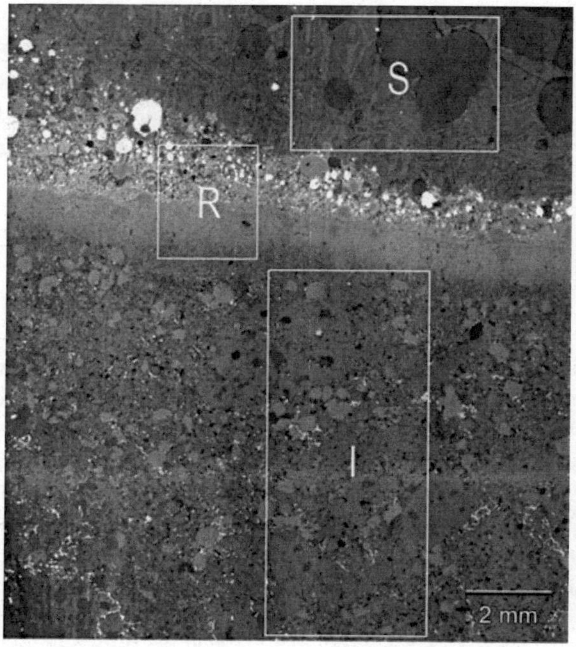

Fig. 2 Photograph of the immediate brick hot face taken with reflected light microscopy showing a microstructural overview of a used magnesia-chromite brick from a flash smelting furnace: slag coating (S), reaction zone (R) and infiltrated and corroded brick microstructure (I)

(c) Kinetics of slag infiltration: Kinetics of slag infiltration causing processes mentioned above depends on several parameters like temperature, viscosity, pore size distribution and wetting angle.

At a microscopic level, several zones can be distinguished, as shown in Fig. 2. Below the slag coating at the immediate brick hot face, a thin reaction zone is usually present followed by an infiltrated and corroded brick microstructure zone.

Within the reaction zone, the magnesia brick component is frequently dissolved leaving relics of chromite as well as primary and secondary chromite precipitations. Below the reaction zone in the infiltrated and corroded brick microstructure, due to corrosion of the brick-inherent magnesia (coarse grains and matrix fines), the main reaction products include (calcium)-magnesium silicates such as forsterite (Mg_2SiO_4) and monticellite ($CaMgSiO_4$). The interstitial phase of the magnesia component, especially the di-calcium silicate, is also corroded.

In general, chromite is not corroded compared to the magnesia but is highly enriched with iron, nickel and copper oxides due to diffusion phenomena. The reason is that periclase (MgO) is more basic than chromite and therefore more susceptible to acidic corrosion.

The atypical and exceptionally high SiO_2 content of slags, caused by changes in the processing or the uncontrolled addition of silica sand, results in an excessive corrosion of the magnesia and subsequent formation of forsterite. In severe cases, the associated volume expansion associated with forsterite formation can lead to the so-called "forsterite bursting" with deterioration of the brick structure.

Chemical Attack by Calcium-Ferrite Slag

Generally, the fayalite (Fe-silicate) slag is the most common slag type in copper, nickel and cobalt processing furnaces. Additionally, Ca-ferrite (Ca–Fe-oxide) slag is used in Ni-matte converting [6]. An example of a spalled-off magnesia-chromite brick from a refining furnace using such Ca-ferrite slag is shown in Fig. 3. The cold end of the refractory brick is missing. Several cracks formed in the infiltrated brick microstructure can be observed macroscopically.

Due to a lower viscosity and lower melting point in comparison to fayalite slag, the Ca-ferrite slag is able to penetrate much deeper into the brick microstructure. Within a deeply infiltrated brick microstructure, severe corrosion of both brick components, i.e., magnesia and chromite, occurs. Additionally, a high supply of iron oxide from the slag resulted in the formation of low melting Mg–Fe-oxide, identified as magnesio-wustite, (Mg,Fe)O, under the given processing conditions (see Fig. 4).

Fig. 3 Cross sectional view of a used magnesia-chromite brick from a refining furnace with its brick hot face spalled-off and showing cracks formed in the infiltrated and corroded brick microstructure, as indicated by the arrows

Fig. 4 Photograph taken by scanning electron microscopy showing the microstructural details of a used magnesia-chromite brick from a refining furnace: corroded magnesia (**1**) with Fe-oxide enrichment and formation of magnesio-wustite (**2**), di-calcium ferrite (**3**) and di-calcium silicate (**4**)

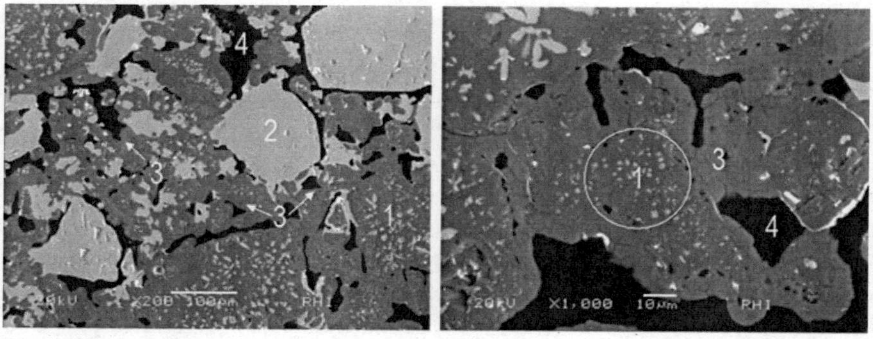

Fig. 5 Photograph taken by scanning electron microscopy showing the microstructural details of a used magnesia-chromite brick from a Peirce-Smith converter: corroded magnesia (**1**), chromite (**2**), (Ca)-Mg-sulfate (**3**) and pore (**4**)

Chemical Attack by Sulfate

Another very common type of chemical attack is corrosion by high sulfur supply when processing sulfidic ores. The penetration of gaseous SO_2 from the oxidation of sulfidic matte creates the prerequisite for SO_3 to react with basic oxides of the magnesia-chromite brick at temperatures below approximately 1100°C, leading to the formation of earth alkaline sulfates in the system $MgSO_4$ and $CaSO_4$ (Fig. 5).

The intensity of the sulfate corrosion depends, on one hand, on the amount of pSO_3, which is a product of the pSO_2 and pO_2, surplus of acidic versus the basic components of the infiltrate such as alkaline compounds, reaction temperature and time, and on the other hand on brick properties such as porosity, bonding strength, type of bonding, brick composition, etc. [7]. For instance, in the case of overheated salt melts, the solubility of CaO and MgO drastically increases and the corrosion can be much more severe.

At higher temperature, typically above 1100°C, $MgSO_4$ dissociates and forms again fine crystalline MgO [8].

Non-oxide Infiltration

In addition to slags, mattes also infiltrate the brick microstructure, as illustrated in Fig. 6. Generally, non-oxide infiltration only densifies the bricks microstructure without any corrosive attack of the brick components. The degree of infiltration depends on the temperature, surface tension, the boundary angle in contact with the refractory oxides, the metal density, the bath height and the size and distribution of the brick pores [8].

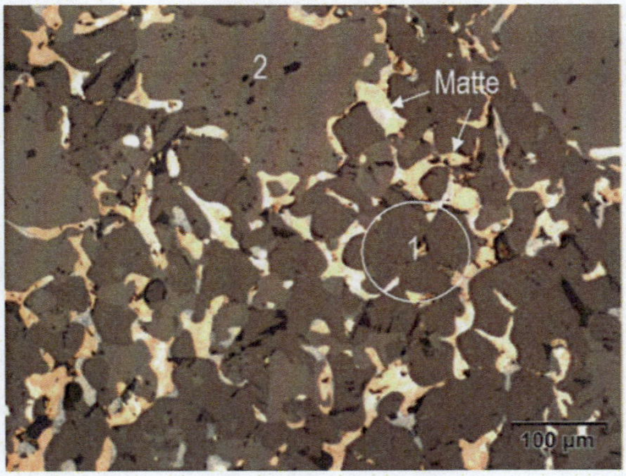

Fig. 6 Photograph taken by reflected light microscopy showing the brick microstructure of a used magnesia-chromite brick from a refining furnace, which is completely infiltrated with nickel-copper-cobalt matte: magnesia (**1**) and chromite (**2**). (Color figure online)

Conclusion

The wear mechanisms observed in post-mortem analysis and discussed in this paper demonstrate that a combination of slag attack, infiltration and corrosion of the bricks inherent components lead to a softening of the brick microstructure and a loss of flexibility and brick strength. This weakened microstructure is then susceptible to continuous wear by hot erosion. Additionally, due to the changes in the thermo-mechanical properties of the refractory, thermal shock leads to crack formation, primarily at the interface between the infiltrated and non-infiltrated brick areas and finally, to discontinuous wear by spalling.

In the case of magnesia-chromite bricks, the massive SiO_2 supply characteristic of fayalite slags resulted in the formation of forsterite with its associated volume expansion—so-called forsterite bursting—causing a subsequent weakening of the brick structure. The acidic fayalite slag mainly corrodes the MgO component of the brick but not the chromite. In comparison to the fayalite slag, contact with Ca-ferrite slags leads to a much higher infiltration depth and the corrosion of the basic magnesia as well as the chromite. Low melting Mg–Fe-oxide (magnesio-wustite) is formed as a main reaction product with the magnesia component.

The high sulfur supply, typically occurring when processing sulfidic materials under oxidizing conditions, leads to corrosion of the brick-inherent magnesia and of the interstitial CaO-containing secondary phase within the magnesia. Chromite generally shows a higher corrosion resistance against both fayalite slag and sulfur attack.

Non-corrosive infiltration of the brick microstructure by metals or sulfidic components (matte) dramatically changes the thermal conductivity of the brick, thus increasing the susceptibility to crack formation and spalling, which is intensified by thermal shocks.

A detailed investigation of wear mechanisms is an important prerequisite for the refractory producer as it provides the basis for both product recommendations and innovative product development. Such post-mortem investigations can clearly highlight which specific stresses affect the refractory products in the various copper, nickel and cobalt processing furnaces. On the basis of the investigation results, combined with its long-term service experience, a refractory producer can recommend the most appropriate choice of furnace brick lining for its clients' specific operational parameters. This is frequently enhanced through active collaborations with the customer.

References

1. Elvers B, Hawkins S, Russey W (1997) Ullmann's Encyclopedia of industrial chemistry 5th edn. John Wiley & Sons
2. Davenport WG, King M, Schleisinger M, Biswas AK (2002) Extractive metallurgy of copper. Kidlington, Oxford OX5 1 GB, UK, Elsevier Science Ltd, pp 1–10
3. Crundwell FK, Moats MS, Ramachandran V, Robinson TG, Davenport WG (2011) Extractive metallurgy of nickel, cobalt and platinum-group metals. Kidlington, Oxford OX5 1 GB, UK, Elsevier Science Ltd pp 21–47
4. Routschka G, Wuthnow H (2012) Handbook of Refractory Materials, 4th edn. Vulkan-Verlag, Essen
5. Harmuth H, Vollmann S (2014) Refractory corrosion by dissolution in slags–challenges and trends of present fundamental research. Iron Steel Rev 58(4):157–170
6. Nelson L, Georgalli G, Hines K, Hundermark R (2018) Converter processing of platinum group materials, AUSIMM, pp 1–26
7. Gregurek D, Reinharter K, Majcenovic C, Wenzl C, Spanring A (2015) Overview of wear phenomena in lead processing furnaces. J Eur Ceramic Soc 35:1683–1698
8. Barthel H (1981) Wear of chrome magnesite bricks in copper smelting furnaces. Interceram 30:250–255

Continuous Improvement of Process Advisor Optimizing Furnace Model

Peter Björklund, Mikko Korpi, David Grimsey, and Miikka Marjakoski

Abstract The Outotec® Process Advisor has successfully been used for automatic control of flash smelting furnace setpoints continuously for 5 years. This is one contribution that has enabled record long furnace campaign life due to the improved matte temperature control, which has prevented matte infiltration into the hearth. Building on this success, the Process Advisor has been enhanced to further utilize the online dynamic heat and material balance model also for other purposes than direct process control. This paper covers some of the latest improvements taken into use after the initial startup. These include, for example, an online maximum feed rate estimator, which continuously updates the maximum feed rate possible for the operator and an online matte level estimator to get a better matte level estimate also between sounding.

Keywords Automation · Digitalization · Industry 4.0 · Information management · Process control · Pyrometallurgy · Smelting · Expert system · Process advisor

Introduction

As part of the Outotec vision for a fully automated smelter, a closed loop control of the flash furnace is required. The Process Advisor has been used for this purpose successfully for 6 years. Improved temperature control enables longer campaign life of the furnace and improved matte grade control improves metal recoveries.

The functional layout of the Process Advisor is shown in Fig. 1.

P. Björklund (✉) · M. Korpi
Outotec (Finland) Oy, Rauhalanpuisto 9, 02231 Espoo, Finland
e-mail: peter.bjorklund@mogroup.com

D. Grimsey
BHP Nickel West Kalgoorlie Smelter (NKS), Smelterman Dr, Feysville, WA 6431, Australia

M. Marjakoski
Boliden Harjavalta Smelter, Teollisuuskatu 1, FI-29200 Harjavalta, Finland

C. Anderson et al. (eds.), *Ni-Co 2021: The 5th International Symposium on Nickel and Cobalt*, The Minerals, Metals & Materials Series,
https://doi.org/10.1007/978-3-030-65647-8_22

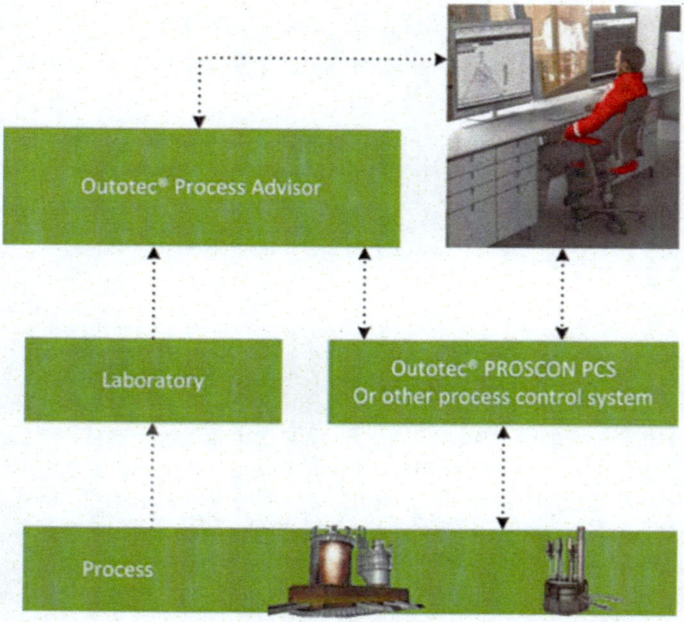

Fig. 1 General layout of the Process Advisor setup [1]. (Color figure online)

The Process Advisor is linked to both the automation system and the laboratory information system. The Advisor reads the actual setpoints from the automation system, in order to be able to simulate the process typically with 5 or 10 min intervals. The feed assays can be laboratory assays or commercial assays of larger batches. The melt assays are laboratory assays. Additionally melt temperatures, levels and amount of tapped melt can be fed back to the Advisor.

Example Models

A typical setup of a flash furnace process model units for matte smelting may look like the image in Fig. 2.

The feedforward part consists of the feed mixture from the blending plant to the point of entry to the furnace. The assay is taken at the discharge of each bin or the other option is that the assay is already known before entering the bins. A more detailed description of the feedforward model is given in the next chapter.

The feedback part consists of tanks for the slag and matte, which model the melt delays and melt mixing and analysis pipe delays to describe the delays of sampling and analysis. The temperature feedback is either through the settler slag temperature

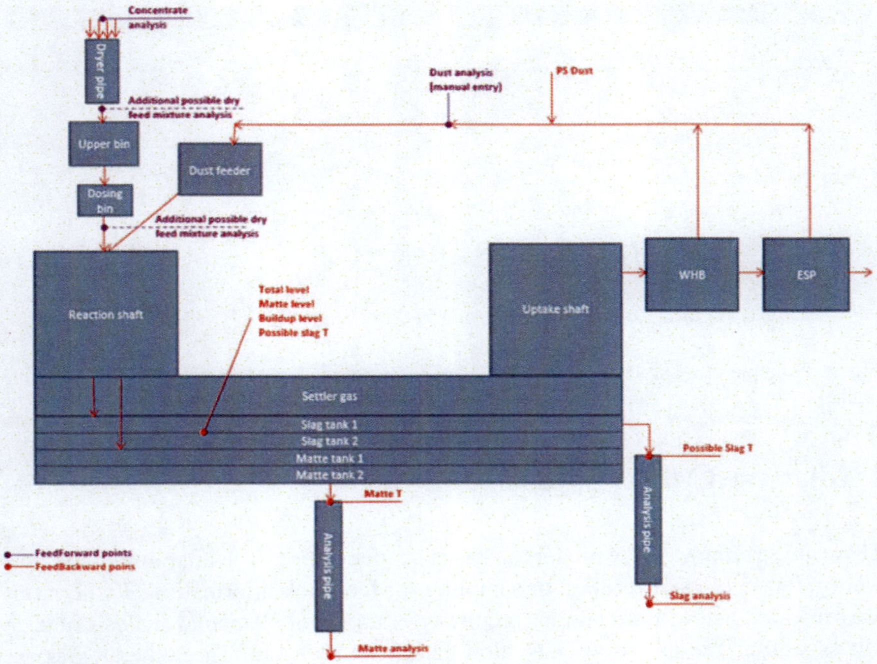

Fig. 2 Example typical model setup for a flash furnace. (Color figure online)

measurement or/and through the launder slag temperature measurement. The slag assay is taken either from the settler roof or from the slag launder. The slag assay is used for fluxing control and sometimes for matte oxidation control. The gas line model includes the waste heat boiler (WHB) and electrostatic precipitator (ESP).

The feedforward and feedback models are also used for estimation of the situation some hours ahead. This is shown in the main user screen as prediction trends. Suitable prediction trends can be matte level estimate, matte grade estimate or feed mixture heat value estimate.

Figure 3 left graph shows an example of a matte level prediction. It shows the estimated matte level rise on top of the measured bottom buildup level as well as the maximum allowed matte level before it starts to affect slag tapping.

Figure 3 right graph shows an example of a matte grade prediction after a feed blend change. It can be seen that the new matte grade (new matte formed from the reaction shaft), indicated with a blue color, changes before the actual matte, indicated with a brown color, from the analysis changes. This is due to both the sampling and analysis as well as the process delay.

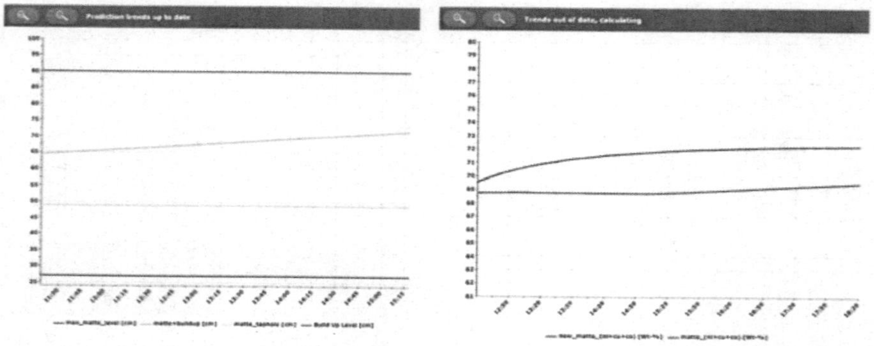

Fig. 3 Example of matte level prediction (left) and matte grade prediction (right). (Color figure online)

Feedforward Model Layouts

There are several ways the model can be set up, depending on availability of feedforward information and location of sampling for feedback information. Two options are presented below. Both options require information of elemental analysis of each bin as well as the discharge mass flow rate from each bin. The mineralogies are calculated based on the elemental assays as well as on the type of feed material.

One option is to base the feedforward information on the availability of commercial and/or larger batch feed assays, i.e. early feedforward information before the blend is in the blending plant. Figure 4 shows the model principle when based on larger commercial/larger batch assays.

This has a benefit of being able to calculate the feed assay throughout the feed system and early enough to make blend adjustments. It also links the online model with the actual raw material handling of the plant and enables an accurate offline model to be used in parallel for estimation purposes. The drawback of this method is the accuracy with short-term changes, it is important that the blend is stable in the short term. It is also required that the bin weights are known.

Another option is to regularly sample the feed assays after the wet or dry feed bins on-time. Figure 5 shows the model principle when based on regular sampling of feed assays.

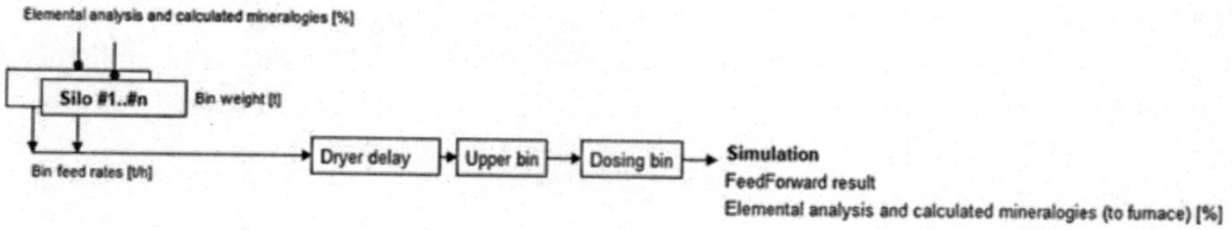

Fig. 4 Example setup of early feedforward information

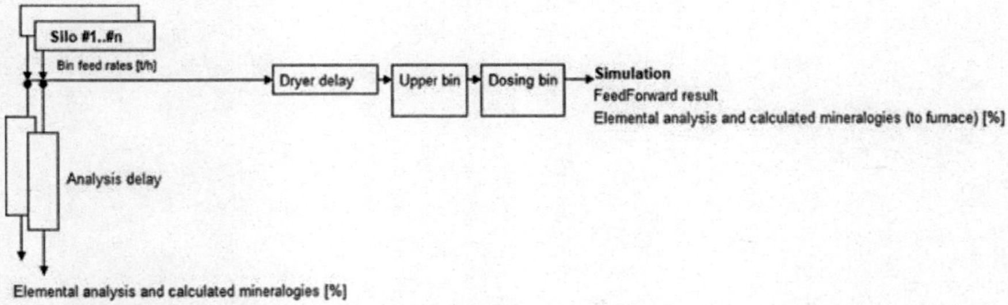

Fig. 5 Example setup of on-time feedforward information

This has a benefit of being able to quickly react to also short-term blend changes. It is not necessary to know the weight of the bins. The drawback of this method is the increased sampling requirement at the feed bin discharge points.

Feedforward Mineralogy

The feed assay is used to online estimate the mineralogy of the feed mixture. The mineralogy is based on separate mineralogy characterization of the different types of feed materials, including also different sludges, matte and slag reverts and other typically recyclable materials. Using this information a mineralogical algorithm determines the final mineralogy of the mixture.

Sulphidic minerals of the mineralogical algorithm include:
Pendlandite ($Ni_{4.5}Fe_{4.5}S_8$)
Millerite (NiS)
Pyrite/pyrrhotite (Fe_xS_y)
Additional minerals from characterization included in the mineralogical algorithm include:
Serpentine ($3MgO*2SiO_2*2H_2O$)
Sulphates (SO_4)
Elemental S
Carbonates (Ca, Mg)
Hydroxides (Ni, Cu, Fe)
Oxides (Ni, Cu, Fe)
Magnetite (Fe_3O_4)
Metallic Ni

Using mineralogical determination of an assay in combination with an online process model requires the resulting heat value of the feed mixture as a function of varying elemental analysis to be smooth. Otherwise unrealistic step changes in the heat value will occur during changes of the elemental assay, which would cause

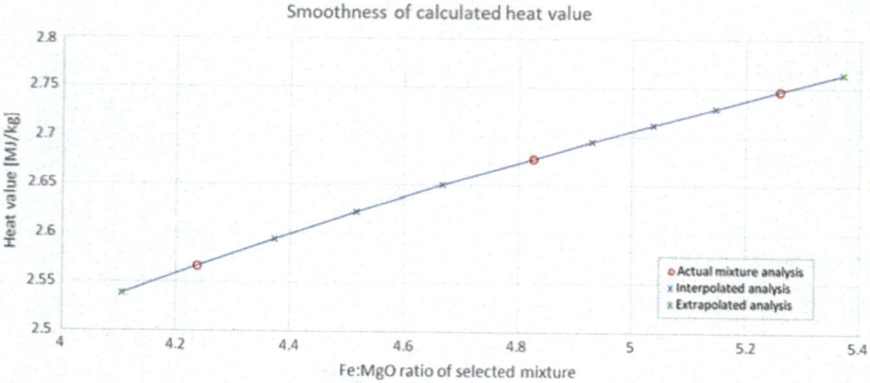

Fig. 6 Smoothness of concentrate mixture heat value. (Color figure online)

incorrect step changes in process control. Figure 6 shows an example resulting feed mixture heat value from resulting mineralogy calculations for varying S, Ni, Fe, MgO and Cu in the feed mixture while keeping other elements such as Co, Al2O3, CaO and Cr constant. Three representable mixture analysis points have been used, which are marked with red, and the other points are interpolated or extrapolated.

It can be seen that the resulting heat value based on the mineralogy estimation algorithm is smooth in the calculated intermediate points as well as the extrapolated points for the selected assays.

Optimizing Slag Liquidus Temperature

As an option, the Process Advisor includes a slag liquidus ternary diagram. The diagrams have been calculated with MTDATA software using thermodynamic data from MTOX database for oxide and sulphide developed in industrial cooperation. For a Ni FSF process with silica fluxing only, suitable axes have been selected as SiO_2, MgO+FeOx and FeOx. For a Ni FSF process with silica and lime fluxing, suitable axes have been selected as SiO_2, FeOx and CaO.

The optimal fluxing strategy using silica as a fluxing agent from a slag liquidus point of view can be visualized in the diagram so that for constant Fe_2O_3/FeO and Ni/(Ni+Cu) ratios an increasing MgO content of the slag is compensated with an increasing SiO_2 content in the direction of the arrow. This is shown in Fig. 7.

The optimal fluxing strategy using silica and lime as fluxing agents from a slag liquidus point of view can be visualized in the diagram so that when the silica is at its maximum lime can be used to avoid high slag liquidus temperatures by moving towards the CaO corner as indicated by the arrow. This is shown in Fig. 8. It must be noted that the slag viscosity change from adding lime is not taken into consideration here.

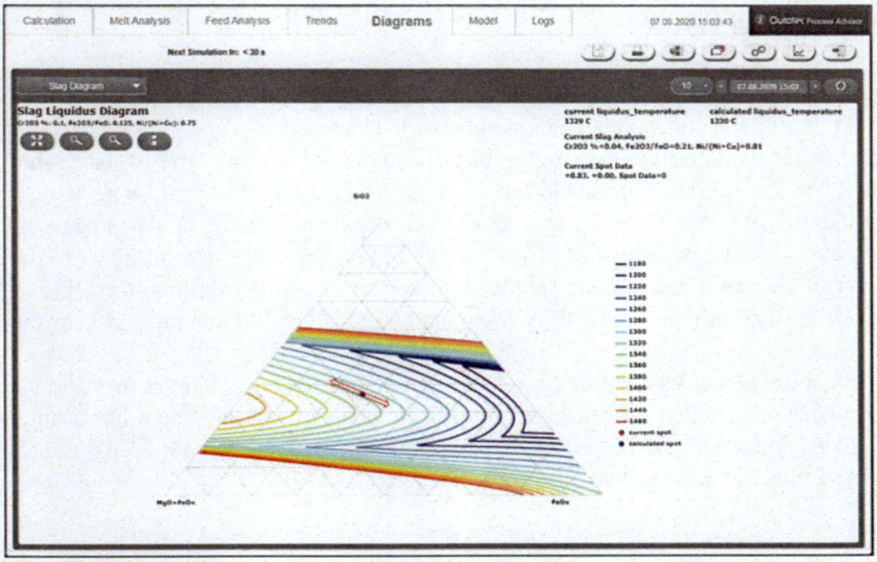

Fig. 7 Example typical slag ternary diagram for a Ni furnace showing the current operation point and liquidus temperature as well as the silica fluxing control strategy indicated by the arrow. (Color figure online)

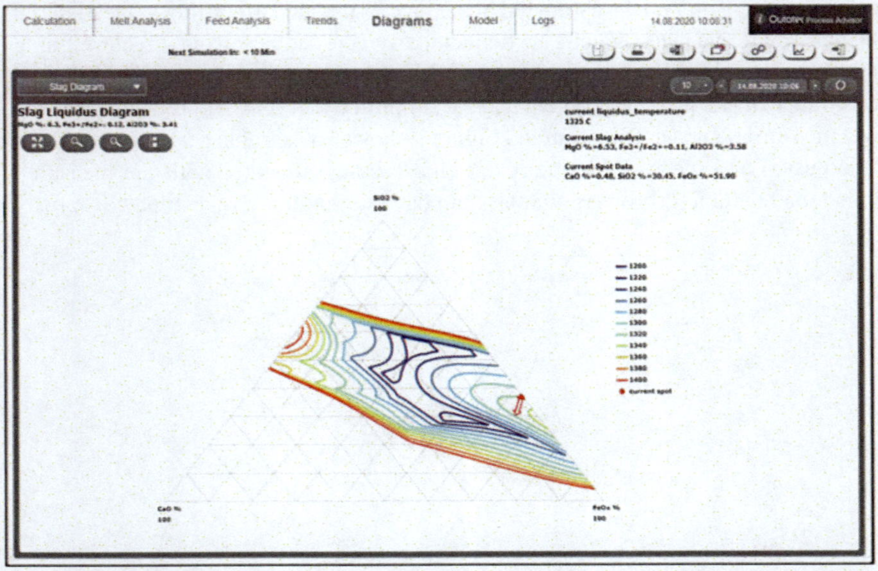

Fig. 8 Example typical slag ternary diagram for a Ni furnace showing the current operation point and liquidus temperature as well as the lime fluxing (in addition to silica) control strategy indicated by the arrow. [2]. (Color figure online)

Maximising Capacity

Since the process is simulated with a regular interval (typically every 5 or 10 min), it is also possible to connect other calculations to the simulation event. One such useful calculation is to calculate according to the online heat and material balance what would be the maximum possible feed rate at that moment. This can be performed by using different algorithms and restrictions depending on the actual process of bottlenecks and characteristics. The estimated maximum feed rate is shown in the Advisor UI, but it can also be sent to the automation system through the OPC or Historian link so that the value is shown continuously to the operator also in the DCS.

By continuously calculating the maximum possible feed rate, it helps the operator to always try to maximize the feed rate. Each small maximizing step has a huge influence in the long run, or vice versa each time the feed rate has not been optimal, a lot of possible lost revenue has been generated.

Furnace Inventory

Based on the online mass balance a matte level predictor was developed. During each simulation interval the estimated matte level is updated according to the material balance. Matte and slag amounts are deducted from the furnace using direct weighing of ladles or weighing the granulated material. The furnace level is slowly recalibrated according to the level measurement using a simplified soft sensor approach, since the level measurement is not so accurate. Figure 9 shows an example period of predicted (blue curve) and measured matte levels (red crosses) together with the measured tapped matte amount (brown points). It can be seen that the level estimate is mostly

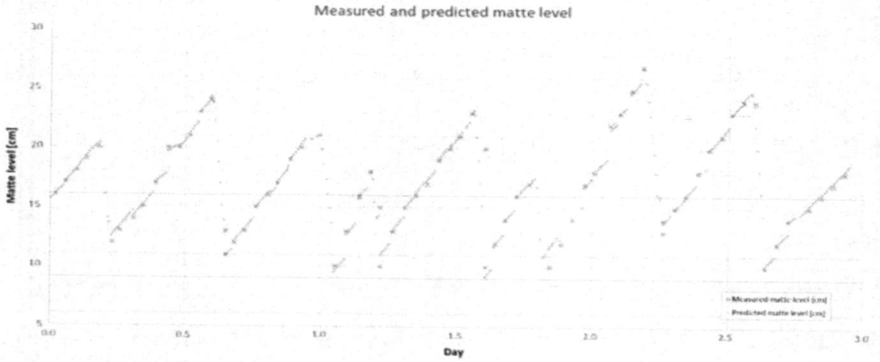

Fig. 9 Matte level predictor. (Color figure online)

good, although sometimes the level is higher than estimated, which can be seen where there is a vertical upwards jump between the predicted level and the measured level. This is probably due to an increased buildup formation at that time.

Offline Blend Planner Model

The latest development has been to separate the Blend Planner model into a complete offline model in order to be able to simulate and optimize different blends as well as coolant consumption. Coolant mixture may include different reverts, dusts and sludges. Pure theoretical models do not take into account the non-ideal situations that always occur in actual smelters and may thus produce unrealistic results, especially considering the heat balance. Since the Process Advisor includes error terms (or adapted parameters) the model will predict the actual situation very well. The error terms are selected depending on the process. The error terms bring an additional advantage as they can indicate issues with sampling, silica feeding or laboratory analysis.

Since the Process Advisor includes real-time error terms (or adapted parameters), which describe the difference between the theoretical and the real situation, and these are used in the offline model, the calculation results of the Process Advisor are highly accurate.

Achieved Benefits

Significant process benefits have been achieved using the Process Advisor. Some examples of achieved benefits are listed below.

Figure 10 shows process control improvements (matte grade) after installation and commissioning.

Figure 11 shows process control improvement (heat balance) after commissioning of fully automated closed loop control.

It can be seen from Fig. 11 that even though there was a drastic change in the concentrate mixture Fe:MgO ratio and the Advisor drastically removed the heat input to the furnace as a result, there were no spikes in matte temperature and the buckstay movement was minimal (no expansions).

Table 1 shows process control improvement (slag control) after improvement to target slag fluxing philosophy from changing from a concentrate assay based calculation to a combined model and slag assay based calculation.

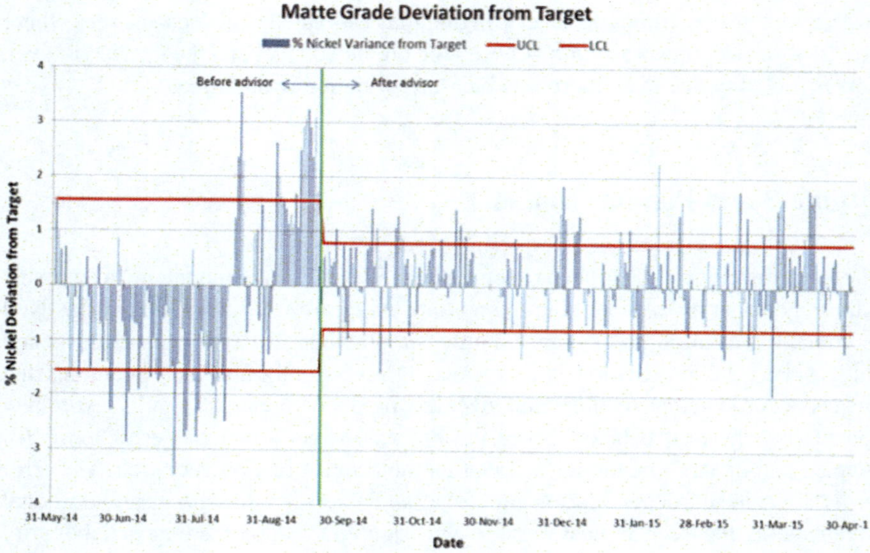

Fig. 10 Improvement in matte grade stability [3]. (Color figure online)

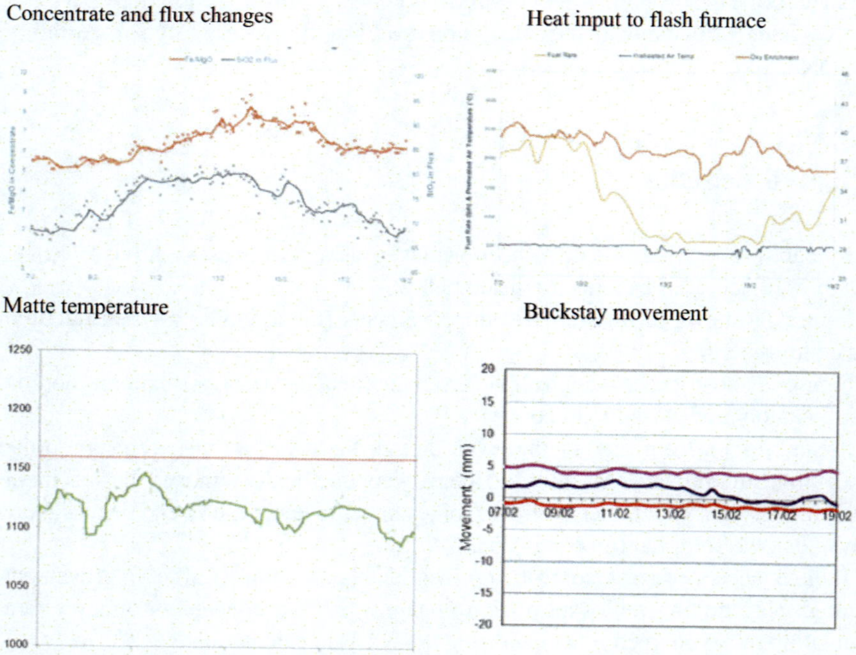

Fig. 11 Improvement in heat balance control [3]. (Color figure online)

Table 1 Improvement in slag stability [4]

	R^2 value of realised function: Slag Fe:SiO_2 = f(Slag Fe:MgO)	Standard deviation of Fe_3O_4% versus average Fe_3O_4%
Concentrate-based approach	0.24	29%
Combined model and slag-based approach	0.67	22%

Conclusions

The Process Advisor has successfully improved smelter profitability by stabilizing the process control through a fully automated process control, i.e. a closed loop system. The feedforward algorithm has proven to be smooth and prediction is good. The integrated slag liquidus ternary diagram has proven useful for calculation of the liquidus temperature and to visualize the optimal fluxing strategy. The matte level predictor has proven useful and can estimate the matte level with reasonable accuracy in combination with matte level measurements and weight feedback of tapped matte. The offline blend planner model and the maximum capacity calculation have proven useful.

Additionally, the actual process results have improved using the Process Advisor.

In the future the model development focuses on further fine tuning of the furnace inventory and additional optimizing calculations run as parallel threads to the model. Another intention is to increase the utilization of the Advisor to predicting the distribution of impurities.

Acknowledgements The authors acknowledge the support and enthusiasm of our customers and all people involved in the support and development process, especially BHP Nickel West Kalgoorlie Smelter and Boliden Harjavalta Smelter for the good cooperation throughout the projects.

References

1. Bjorklund P (2015) Outotec process advisor startup at Kalgoorlie nickel smelter in Kalgoorlie, Australia. (n.d.). In: Outotec Smelting Newsletter 3/2015. http://www.outotec.com/company/newsletters/smelting-newsletter/smelting-issue-3–2015/outotec-process-advisor/. Retrieved 25 Aug 2017
2. Gisby J, Taskinen P, Pihlasalo J, Li Z, Tyrer M, Pearce J, Avarmaa K, Björklund P, Davies RH, Korpi M, Martin S, Pesonen L, Robinson J (2017) MTDATA and the prediction of phase equilibria in oxide systems: 30 years of industrial collaboration. Metall Mater Trans B 48(1):91–98
3. Rich A, Terry M, Oswald M (2017) extending campaign life at kalgoorlie nickel smelter through process control and furnace binding improvement. In: 15th International Flash Smelting Congress September 2017. Helsinki, pp 3–8

4. Bjorklund P (2018) Continuing process improvements by further optimization of the Process Advisor at Kalgoorlie Nickel Smelter in Kalgoorlie, Australia Outotec Process Advisor Startup at Kalgoorlie Nickel Smelter in Kalgoorlie, Australia. (n.d.). In: Outotec Smelting Newsletter 1/2018. http://www.outotec.com/company/newsletters/smelting-newsletter/smelting-issue-1–2018/continuing-process-improvements-by-further-optimization-of-the-process-advisor-at-kalgoorlie-nickel-smelter-in-kalgoorlie-australia/. Retrieved 1 March 2019

Fluxing Optimisation and Control Improvements at the Kalgoorlie Nickel Smelter

David Grimsey, Eric Grimsey, and Peter Bjorklund

Abstract The Nickel West Kalgoorlie Smelter (NKS) treats a blended nickel sulphide concentrate from multiple ore bodies containing a high portion of magnesia in an integrated flash smelter. Fluxing and related control strategies have been an area of continuous improvement. Recent improvements in slag liquidus control have been achieved through the application of powerful thermodynamic modelling software (FACTSage) combined with improved process measurements and implementation of a highly automated control model (Outotec Process Advisor).

Keywords Slag chemistry · Fluxing · Nickel smelting · Process control · FACTSage · Outotec process advisor

Smelter Overview

NKS is located near Kalgoorlie-Boulder, Western Australia and forms part of BHP's Nickel West Asset, a vertically integrated nickel business including mines, concentrators, smelter and refinery. A brief overview is presented here with more detail published elsewhere [1].

NKS was commissioned in 1972 to treat concentrates from the Kambalda region, located 40 km from NKS. A series of upgrades have unlocked capacity from 30 to 100 ktpa Ni-in-matte, largely from concentrate produced in Western Australia's Mt Keith and Leinster regions. The process utilises an 'Integrated Flash Furnace' wherein the traditional flash and electric furnaces are combined. Matte converting

D. Grimsey (✉)
BHP–Nickel West Nickel West Kalgoorlie Smelter (NKS), Smelterman Dr, Feysville, WA 6431, Australia
e-mail: david.grimsey@bhp.com

E. Grimsey
WA School of Mines (Curtin University), Kalgoorlie, WA 6431, Australia

P. Bjorklund
Outotec (Finland) Oy, Rauhalanpuisto 9, FI-02231 Espoo, Finland

C. Anderson et al. (eds.), *Ni-Co 2021: The 5th International Symposium on Nickel and Cobalt*, The Minerals, Metals & Materials Series,
https://doi.org/10.1007/978-3-030-65647-8_23

is carried out in three Peirce-Smith Converters. Supporting unit processes include gas cleaning and acid production, oxygen production, flux preparation and power generation.

Fluxing of the Integrated Flash Smelter

The primary function of flash smelting is upgrading sulphide concentrate through oxidation then removal of impurities (primarily FeO, SO_2 and MgO) by separation of the slag and gas phases from the enriched sulphide stream (nickel matte) at temperatures in excess of 1300 °C. Flux must be added to allow the formation of a discrete molten oxide phase (slag) that can be separated from the matte.

The NKS Integrated Flash Furnace utilises a slag that solidifies with the initial formation of olivine, a Mg–Fe silicate solid solution with a low melting temperature Fe end-member (fayalite—Fe_2SiO_4, ~1200 °C) and a high melting temperature Mg end-member (forsterite—Mg_2SiO_4, ~1890 °C). The relative amounts of FeO and MgO are key drivers of the fluxing strategy, with concentrates characterised by their Fe/MgO ratio. Concentrates with lower Fe/MgO ratios require higher flux addition to maintain a suitable liquidus temperature and also more energy to smelt due to silicates replacing sulphides; concentrates with <5 Fe/MgO are generally unsuitable for flash smelting.

The NKS process was designed to treat high Fe/MgO concentrates, typical of those produced from Kambalda's massive nickel sulphide deposits. After start-up, NKS also treated concentrates with moderate MgO (5–7%). The detrimental impact of MgO on slag superheat wasn't anticipated, resulting in excessive accretions, operational challenges and reduced furnace life. A number of changes were implemented in response, including installation of electrodes in the settler (now removed), oxygen enrichment of the process air and addition of CaO as a slag modifier.

The ongoing operational challenges from processing higher MgO concentrates led to a maximum limit of 5% MgO in concentrate being imposed in 1975. In 1991, WMC (now BHP—Nickel West) acquired Mt Keith, a large, sulphur deficient disseminated orebody with high levels of MgO. Mine development began in 1993 and NKS again faced the challenge of processing higher MgO concentrates. An example of the fluxing strategy from 1994 to 1997 is shown in Fig. 1. The aim was to target a slag liquidus between 1330–1340 °C as MgO increased in feed, firstly by SiO_2 addition to maintain MgO in slag at 7% until Fe/SiO_2 decreased to 1.2, and then by addition of up to 9% CaO in slag, to prevent MgO increasing above 8%. The concentrate contained minor CaO, which explains the presence of 1% CaO in slag without lime addition. Figure 1 is taken directly from a NKS internal report [2] and is included here for historical reference.

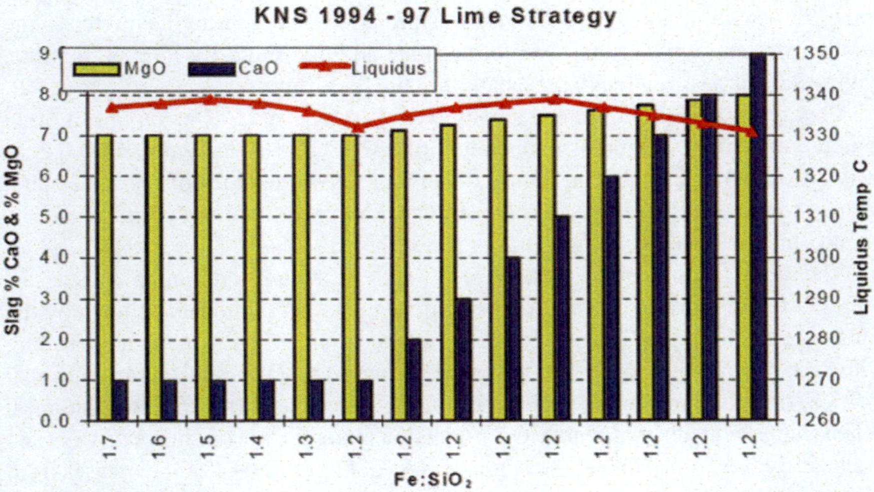

Fig. 1 NKS fluxing strategy (1994–1997). (Color figure online)

Strategy Development

The NKS fluxing strategy was reviewed in 1998 [2] to consider treatment of low Fe/MgO concentrates through extending the addition of SiO_2 down to 0.7 Fe/SiO_2, Al_2O_3 up to 12% and CaO up to 14%. Modification with Al_2O_3 was seen as promising but was not pursued due to inadequate data for the impacts on liquidus and recovery. Ultimately, CaO addition was preferred; it had already been used at NKS, and even though it was the highest cost, extrapolation of plant trial data indicated that addition of CaO would improve nickel recoveries compared to other fluxing options.

In 2007 fluxing strategies using SiO_2 with Al_2O_3 and CaO modifiers were again examined [3, 4] through calculation of slag liquidus temperatures for various fluxing combination using FACTSage thermodynamic modelling software. The KNS integrated flash furnace combines oxidising (flash smelting) and reducing (electric furnace) processes, with average slag oxygen pressure (pO_2) varying from 10^{-8} to 10^{-10} atm [5] due to variation of the ferric/ferrous iron ratio in slag. The FACTSage calculated liquidus temperatures were found to increase across this pO_2 range [4] to reach a maximum at ~10^{-9} atm; hence the slag liquidus will increase across the furnace as conditions become more reducing and cannot be controlled at a single temperature. In this assessment, the slag liquidus temperatures were primarily calculated at 10^{-9} atm oxygen pressure to provide maximum estimates for a margin of safety.

The optimum strategy was defined as that requiring the lowest mass of flux to achieve the target liquidus over the operating range, subject to acceptable slag fluidity and matte/slag immiscibility being achieved in the furnace. Operationally, a strategy using SiO_2 with minor Al_2O_3 (2–4%) was preferred as it appeared superior to the existing strategy without the need for separate addition of costly lime or alumina, with sufficient Al_2O_3 already present within the SiO_2 flux.

A comparison of the combined tonnes of SiO_2, Al_2O_3 and CaO per tonne of MgO required to achieve a liquidus temperature of 1325 °C for a slag with a Fe/MgO of 6, is illustrated in Fig. 2. The tonnes of flux are shown as lines of constant mass, relative to the resulting %Al_2O_3 and %CaO in slag. SiO_2 makes up the balance of the flux but the resulting %SiO_2 in slag is not shown.

When only SiO_2 is added, both %Al_2O_3 and %CaO are zero, and 5.3 tonnes of SiO_2 per tonne of MgO are required to achieve the target liquidus, as shown by the constant mass line at the intersection of the x and y axes. When Al_2O_3 is added along with SiO_2, the change can be seen along the y-axis; as Al_2O_3 in slag increases from 0 to 7.3%, the required mass of flux ($SiO_2 + Al_2O_3$) decreases from 5.3 to 3.7 tonnes. When CaO is added along with SiO_2, projection along the x-axis shows that as CaO in slag increases from 0 to 6%, the required mass of flux ($SiO_2 +$ CaO) increases from 5.3 to 6.4 tonnes, before decreasing to 6 tonnes at 9% CaO and then falling sharply. An important observation is that a minor increase in %Al_2O_3 in slag meaningfully reduces the total flux requirement, whereas over 9% CaO in slag is required before reduction in flux mass is achieved relative to using SiO_2 alone. Further, operating with 9% CaO in slag would significantly enhance the risk of solid monoxide or spinel precipitation within the slag [3, 4].

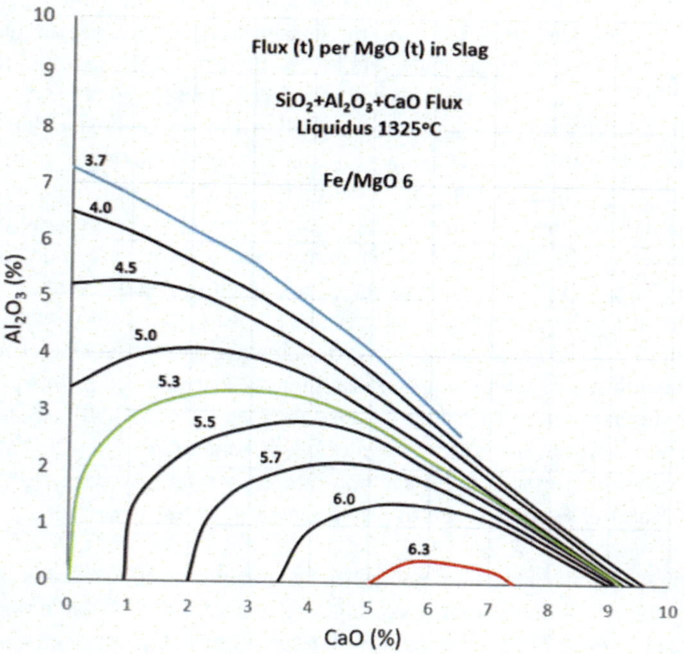

Fig. 2 Flux requirement (SiO_2, Al_2O_3 and CaO) to achieve a 1325 °C liquidus for a 6 Fe/MgO slag. (Color figure online)

The FACTSage study showed that, in general, to achieve a 1325 °C liquidus the addition of Al_2O_3 decreased the total flux required while addition of CaO increased it. SiO_2 is the primary flux for slag liquidus control at NKS and is significantly lower cost than CaO and Al_2O_3 additives; thus, these should only be used from a liquidus control perspective when SiO_2 is no longer effective. As discussed, CaO was previously added at NKS due to an expected nickel recovery benefit but review of plant data since that time did not support this. Metal recovery is influenced by many competing factors such as feed rate, concentrate burner performance, electrode operation and slag superheat control in addition to slag composition. These competing factors make it difficult to reliably predict trends from plant data.

A review of the fluxing control strategy was also completed as part of the 2007 study [3, 4]. The existing strategy related the Fe/MgO in concentrate to a target Fe/SiO_2 in slag to set the SiO_2 flux requirement. The preferred method was to more directly target SiO_2/MgO in slag, which is relatively constant (4.5–4.7) over the NKS operating range when compared to the Fe/SiO_2 ratio (1–1.7), as illustrated on the MgO-'FeO'-SiO_2 ternary phase diagram shown in Fig. 3. Lines representing constant SiO_2/MgO ratios lie parallel to lines of constant liquidus temperature within the NKS operating range. The 4.7 line shown in Fig. 3, for example, lies approximately parallel

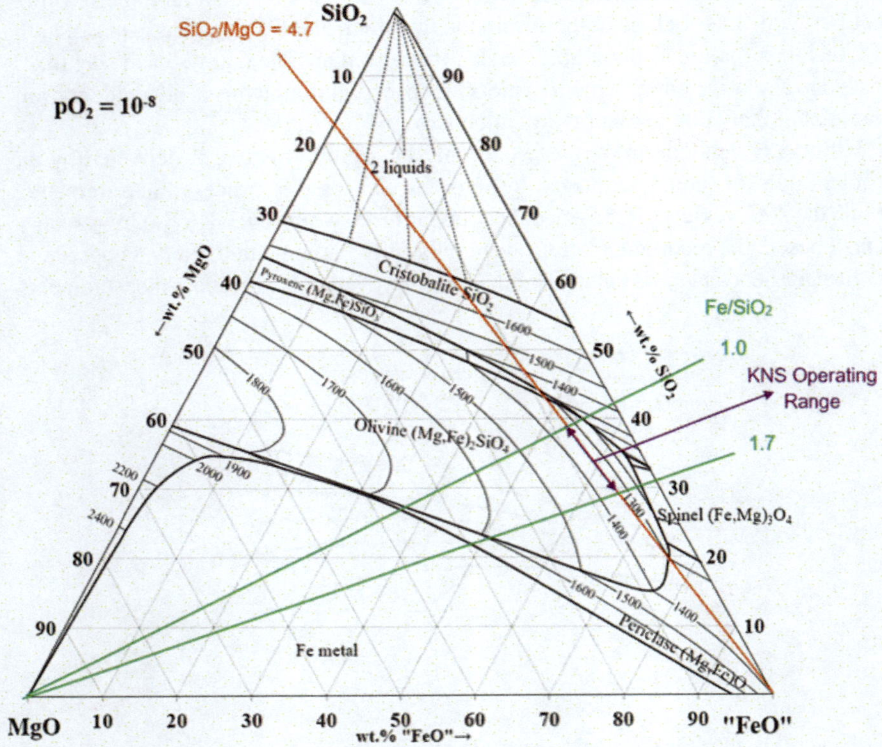

Fig. 3 MgO-'FeO'-SiO_2 phase diagram. (Color figure online)

to the 1300 °C liquidus line. Additionally, the narrower set point range for SiO_2/MgO makes it easier for an operator to quickly verify that slag chemistry is under control, compared to looking at the more variable Fe/SiO_2 ratio.

Current NKS Fluxing Strategy

The current fluxing strategy is based on the theory from the 2007 study [4] taking into consideration unique plant conditions and operational experience. The strategy uses SiO_2 along with the Al_2O_3 contained within the flux to target a slag liquidus of 1325 °C. The method accounts for the major liquidus influencing components in the NKS slag system (FeO, MgO and Al_2O_3) by using the predicted Fe/MgO and Al_2O_3/MgO in slag to set the SiO_2/MgO flux requirement, based on the relationships determined using FACTSage as shown in Fig. 4 [4]. Background CaO in slag may also need to be taken into account.

A significant change with the new fluxing strategy was the discontinuation in 2015 of the addition of CaO when treating low Fe/MgO concentrates. This change did not necessarily improve liquidus control, but had other key benefits. The replacement of CaO with SiO_2 flux not only reduced costs, but also reduced the total tonnes of flux required which has helped de-constrain the flux plant and furnace heat balance, both of which can feature as plant bottlenecks. It is also plausible that reduced slag mass has led to a relative improvement in nickel recovery, although this is not evident from plant data, likely due to competing influences.

Ultimately, any control strategy can only be highly effective if delivered by an efficient control model supported by relevant and reliable process measurements. While the 2007 review [3] provided the foundation on which the NKS fluxing strategy is now based, the recommendations were only fully implemented after an upgrade to the furnace heat and mass balance model and improvements in process measurements

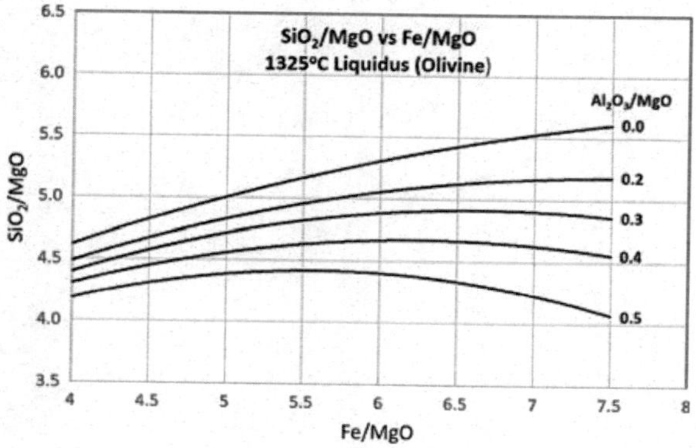

Fig. 4 SiO_2/MgO for 1325 °C liquidus temperature from FACTSage

Outotec Process Advisor

In late 2014, the NKS furnace control model, FFMod32 [6], was replaced by Outotec Process Advisor which is a fundamentally based heat and mass balance flash furnace control model incorporating feedforward and feedback components. The model runs every ten minutes, automatically adjusting inputs to achieve process set points and calibrates error terms whenever new assay or temperature feedback data become available. The previous control model was also fundamentally based but was reliant on operator response to process instability; this could be subjective and typically resulted in a delayed response to process variation. Overall, the higher degree of automation and simulation frequency of Process Advisor resulted in significant improvements in NKS furnace process control.

Improvements in process measurements accompanied the implementation of Process Advisor. These included increased sampling frequency, relocation of feed sampling further upstream and improved slag temperature measurement. As an example, the settler temperature was previously taken using a roof pyrometer to measure gas temperature. Direct settler slag temperature measurement using an immersion thermocouple, essential for slag superheat control, was implemented for Process Advisor. Further details of the implementation at NKS are available elsewhere [7–9].

Fluxing Optimisation and Control Improvements

The impact of the fluxing optimisation and control improvements can be assessed by their effect on the variation of the SiO_2/MgO ratio in slag, since as discussed, this is the key indicator of slag liquidus temperature. Figure 5 shows the SiO_2/MgO ratio in slag when targeting a 1325 °C liquidus temperature over three periods; namely, (A) when using the previous fluxing strategy without Process Advisor (Jan 2012 to Oct 2014), (B) with Process Advisor (Oct 2014 to Aril 2017) and (C) when using Process Advisor with the new SiO_2/MgO fluxing strategy and a slag Fe/MgO error correction, discussed subsequently (April 2014 to July 2020). The average and standard deviation of SiO_2/MgO in slag within each range is also shown along with the average liquidus (calculated in FACTSage from the average slag assay); however given that the SiO_2/MgO set point can vary within each range from ~4.5 to 4.7, the standard deviation shown in Fig. 5, while useful for comparative purposes, will exaggerate the true variation. A more accurate analysis would directly compare the liquidus deviation from the target.

Figure 5 shows a noticeable decrease in SiO_2/MgO variation with progression through time periods A, B and C. Comparison of A to B shows that implementation of Process Advisor and the associated improved process measurements did improve liquidus control using the original fluxing strategy, with a reduction in the absolute

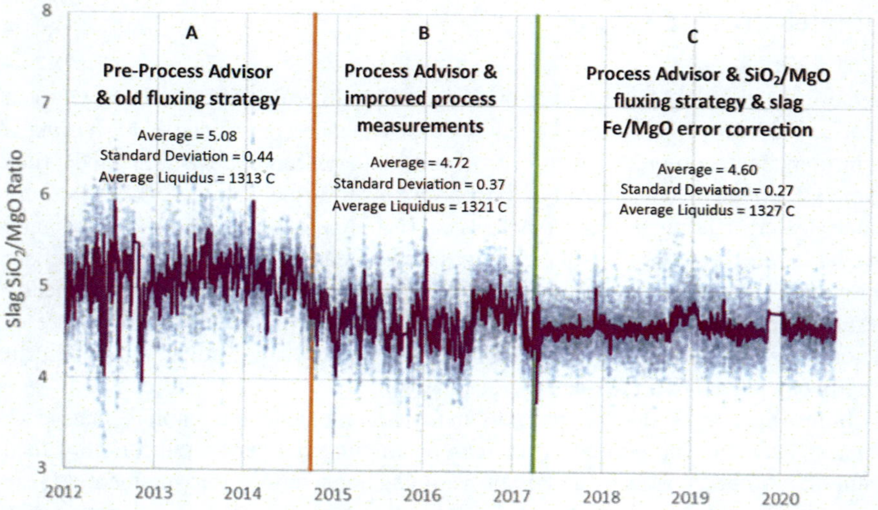

Fig. 5 Slag SiO_2/MgO from 2012 to 2020; trend line is a 50-point moving average of 2-hourly slag assays. (Color figure online)

standard deviation of SiO_2/MgO from 0.44 to 0.37. However, improvements in other important heat and mass balance control parameters were more noticeable, such as improved control of matte grade [9] and heat balance.

Notably, the average SiO_2/MgO decreased with implementation of Process Advisor, from 5.08 in period A to 4.72 in period B, corresponding to an average liquidus temperature increase from 1313 to 1321 °C. This better aligned the process to the 1325 °C target but was unexpected since the slag fluxing strategy had not changed, only the controller and some measurements. Later it was determined that the pre-Advisor concentrate sampling method led to an overestimation of MgO relative to the new method, resulting in excess SiO_2 addition, which explained the lower average liquidus temperature during period A.

Comparison of periods B and C in Fig. 5 shows a further improvement in the stability of SiO_2/MgO in slag when the new fluxing strategy was implemented with Process Advisor. A new feedback correction was also implemented to reduce systemic errors in the MgO and Fe balances. The correction compares the Fe/MgO in slag with that predicted by Advisor using the Fe/MgO feed assay, and then adjusts the overall balance accordingly. The assumption is that the well-mixed slag, after taking into account Fe in matte, is more representative of the Fe/MgO input than the feed assays. These changes further reduced the slag SiO_2/MgO standard deviation from 0.37 to 0.27, while increasing the average liquidus from 1321 °C to 1327 °C, closer to the target of 1325 °C. Overall, this resulted in a significant improvement in liquidus temperature control.

Considerations for Future Improvement

The recent changes in fluxing strategy have, as noted, increased the average slag liquidus by 14 °C, from 1313 to 1327 °C (Fig. 5). Since this likely reflects a decrease in average slag superheat, an increase in below bath accretion may be expected, although the superheat change had not been monitored prior to the recent introduction of a representative slag temperature measurement. A noticeable increase in accretions was confirmed after a furnace drain as part of a major outage in 2019. These accretions likely played a role in extending furnace refractory and hearth life [10] while negatively impacting furnace operability and metal recovery.

Appropriate superheat control is a balance between the above factors and will be a consideration for future fluxing optimisation and slag temperature control changes. However, superheat control across the furnace can be problematic because the slag temperature can increase up to 80 °C as it flows from the settler and past the electrodes in the electric furnace appendage. The slag superheat can therefore vary significantly across the furnace due to changes in slag temperature and the liquidus temperature can also vary with change in oxygen pressure, as discussed. Currently, the slag temperature is controlled against a single point measurement in the settler and the slag liquidus is targeted at a single maximum temperature of 1325 °C. While the new fluxing strategy and control have provided significant improvements, further control optimisation will need to consider how variation in slag temperature and liquidus across the furnace can be accounted for to improve operational outcomes.

Any errors associated with model inputs result in a period of instability before the model can appropriately adjust error terms from process feedback. Improvement to model inputs is therefore a consideration for future control optimisation. An example is converter slag at NKS; it contains iron oxide and SiO_2, and is recycled to the flash furnace. Process Advisor does not currently account for converter slag in a feedforward manner and is reliant on adjustment of the Fe and SiO_2 mass balance error terms from slag assay feedback. Although the tonnes of recycled converter slag are relatively small, converting is a batch process while and the flash furnace is continuous; the sole reliance on feedback in this situation may introduce control error.

Conclusions

NKS treats a blended concentrate from multiple orebodies with varying compositions containing a high portion of MgO. Fluxing strategy and control methodology has been an area of continuous improvement over NKS's 48 year history, with the development and implementation of appropriate fluxing strategies enabling successful treatment of variable feed types. Significant recent improvements in slag chemistry control have been achieved through a combination of:

1. Application of powerful thermodynamic modelling software (FACTSage) to improve fundamental understanding of the slag system and to compare potential fluxing strategies.
2. Improved process measurements to provide valid and timely inputs for process control.
3. Implementation of a highly automated heat and mass balance control model (Outotec Process Advisor).

This work has highlighted some more generally applicable learnings that are industrially relevant. Firstly, thermodynamic modelling is a valuable tool to understand and optimise chemical processes. However, to add business value from thermodynamic modelling, outcomes such as process control strategies must be matched with a suitably capable control model to be practically implemented.

Secondly, while representative and timely process measurements are required for process control, they are also essential to enhance the practical process understanding, especially for pyrometallurgical processes, which are difficult to visually observe and can be very different from smelter to smelter. Improved measurements allow for previous assumptions to be validated or corrected, and for further improvements to be more easily identified and implemented. With continuous improvement in technology, measurement of key process parameters should be an enduring priority for any industrial operation.

Thirdly, any industrial plant can be susceptible to measurement error that can result in process instability. Some error is unavoidable and can result from multiple sources, such as sampling and analysis methods, and feeder calibration. A key lesson is that measurements for important process control parameters should include error correction using process feedback to help mitigate systemic errors.

Fourthly, heat and mass balance models, such as those within Outotec Process Advisor, are powerful tools with wide ranging potential applications. Maximising value requires engagement and ownership of the model from site technical staff; site staff understand operational problems and are best positioned to identify improvement opportunities. The strong collaboration between NKS and Outotec has been fundamental to the ongoing process improvements implemented through Process Advisor.

Lastly, slag chemistry has a competing influence on a number of operational priorities, such as recovery, furnace build-up and operability. As fundamental and practical understanding of a process improves, the goal of the fluxing strategy should be continually reviewed to ensure alignment with the best overall process outcome.

Acknowledgements The authors acknowledge NKS and Outotec staff for their support through the identification and implementation of these improvements; it has been a team effort. The authors also acknowledge University of Queensland PYROSEARCH staff for their input on some of the fundamental aspects discussed in the paper.

References

1. Palmer J, Malone J, Loth D (2005) Kalgoorlie Nickel Smelter operations overview 1972–2005. In: Donald J, Schonewille R (eds) Nickel and cobalt 2005: challenges in extraction and production, Met Soc of CIM (Montreal), pp 441–455
2. Elliot B (2001) The effect of lime addition to the flash furnace for processing of high magnesia concentrates, Nickel West Internal Report, KNS Technical Report #565
3. Grimsey EJ (2007) KNS Furnace rebuild definition phase study: slag systems and fluxing option, Nickel West Internal Report, Document Number PREP-NS-03-0231
4. Grimsey EJ, Tran BP (2009). Fluxing strategy for nickel slags containing magnesia, In: Liu J et al (eds) Pyrometallurgy for nickel and cobalt 2009, Met Soc of CIM (Montreal), pp 111–129
5. Thay PK et al (1991). Oxygen potentials and slag-matte reactions in the Kalgoorlie Nickel Smelter, In: 5th AusIMM Extractive Metallurgy Conference, AusIMM (Parkville, Vic), pp 277–284
6. Hunt A et al (1993) Modelling for design and control at the Kalgoorlie Nickel Smelter, In: Reddy RG, Weizenbach RN (eds) Proceeding of the Paul E. Queneau symposium extractive metallurgy of copper, nickel and cobalt, vol 1: fundamental aspects. TMS—AIME (Warrendale, PA), pp 489–499
7. Bjorklund P (2015) Outotec process advisor startup at Kalgoorlie Nickel Smelter in Kalgoorlie, Australia. (n.d.). In: Outotec Smelting Newsletter 3/2015. http://www.outotec.com/company/newsletters/smelting-newsletter/smelting-issue-3–2015/outotec-process-advisor/. Retrieved 25 Aug 2017
8. Bjorklund P (2018) Continuing process improvements by further optimization of the process advisor at Kalgoorlie Nickel Smelter in Kalgoorlie, Australia Outotec Process Advisor Startup at Kalgoorlie Nickel Smelter in Kalgoorlie, Australia. (n.d.). In: Outotec Smelting Newsletter 1/2018. http://www.outotec.com/company/newsletters/smelting-newsletter/smelting-issue-1–2018/continuing-process-improvements-by-further-optimization-of-the-process-advisor-at-kalgoorlie-nickel-smelter-in-kalgoorlie-australia/. Retrieved 1 March 2019
9. Bjorklund P, Grimsey D, Marakoski, M (2021) Continuous improvement of process advisor optimizing furnace model. Paper presented at the 5th international symposium on nickel and cobalt, March 14–18, 2021
10. Rich A, Terry M, Oswald M (2017) Extending campaign life at the Kalgoorlie Nickel Smelter through process control and furnace binding improvement. Paper presented at the 15th international flash smelting congress, Helsinki, Finland, 3–8 September 2017

Preparation of Refractory Materials by Co-sintering of Ferronickel Slag and Ferrochromium Slag: Thermodynamic Analysis

Foquan Gu, Yuanbo Zhang, Zhiwei Peng, Huimin Tang, Zijian Su, and Tao Jiang

Abstract During the past ten years, the stainless-steel industry has undergone rapid development around the world. At the same time, a great quantity of stainless-steel slag (mainly ferronickel slag and ferrochromium slag) was co-generated. It is urgent to seek an efficient method for the comprehensive utilization of ferronickel slag and ferrochromium slag. In this study, the feasibility of preparing refractory materials by co-sintering of ferronickel slag and ferrochromium slag was evaluated based on thermodynamic analysis. The thermodynamic results showed that by co-sintering of ferronickel slag and ferrochromium slag in the presence of magnesia, the phase composition of the system is forsterite, spinel, and perclase, which is beneficial to prepare high refractoriness of the refractory materials. The experimental results verified those findings, a refractory material with refractoriness of 1780 °C, bulk density of 2.30 g/cm^3, apparent porosity of 16.19%, and compressive strength of 89.18 MPa could be obtained.

Keywords Ferronickel slag · Ferrochromium slag · Refractory materials · Comprehensive utilization · Co-sintering

Introduction

Ferronickel slag and ferrochromium slag are two main slags generated during stainless-steel production. Ferronickel slag is a waste of ferronickel smelting process with the amount of almost four to six times that of metal production [1, 2]. As the rotary kiln-electric furnace (RKEF) process has undergone a rapid development during the past decade, a significant quantity of ferronickel slag was co-generated, and the ferronickel slag has become the fourth largest industry solid waste in China and increased with exceeds 40 Mt each year [3]. Ferrochromium slag is generated from ferrochromium smelting process, when production of 1 ton ferrochromium

F. Gu · Y. Zhang (✉) · Z. Peng · H. Tang · Z. Su · T. Jiang
School of Minerals Processing and Bioengineering, Central South University, Changsha 410083, Hunan, China
e-mail: sintering@csu.edu.cn

C. Anderson et al. (eds.), *Ni-Co 2021: The 5th International Symposium on Nickel and Cobalt*, The Minerals, Metals & Materials Series,
https://doi.org/10.1007/978-3-030-65647-8_24

alloy, about 1.1–1.6 tons of ferrochromium slag is co-generated [4]. Compared with other metallurgical slags, the ferronickel slag and ferrochromium slag have low activity and low recovery value. Nowadays, only about 10 wt% of ferronickel slag and ferrochromium slag are applied in construction and building materials, and producing glass ceramics and geopolymers [5–12]. A majority of ferronickel slag and ferrochromium slag are discarded and sent for landfilling or dumping. However, those two slags, especially ferrochromium slag, contain high content of hazardous elements (e.g., Cr), which may lead to environmental issues, such as contaminating soil and water. Thus handling such a huge quantity of slag is a huge challenge to the stainless-steel industry. It is urgent to seek an efficient method for utilization of ferronickel slag and ferrochromium slag.

In our previous work, a refractory material with the refractoriness of 1660 °C, bulk density of 2.92 g/cm^3, apparent porosity of 1.82%, and compressive strength of 100.61 MPa could be obtained after sintering ferronickel slag at 1350 °C for 3 h in the presence of 20 wt% sintered magnesia [12, 13]. In addition, chromium components in the system were indicated stabilized as trivalent chromium in the chromate spinel phases, which were beneficial to promote the refractory material refractoriness. After sintering ferronickel slag at 1350 °C for 3 h in the presence of 20 wt% sintered magnesia and 6 wt% chromium oxide, a great refractory material with the refractoriness of 1840 °C, bulk density of 2.68 g/cm^3, apparent porosity of 15.19%, and compressive strength of 96.28 MPa was obtained [14, 15]. It was obvious that the addition of chromium oxide possessed a much higher refractoriness, however, the cost of analytically pure chromium oxide is high. Due to the main compositions ferrochromium slag is similar to ferronickel slag, and the content of chromium oxide in the ferrochromium slag is high. The ferrochromium oxide may use chromium oxide additive for preparing refractory material.

In this study, the feasibility of preparing refractory material by co-sintering ferronickel slag and ferrochromium slag was performed based on thermodynamic analysis.

Experimental

Materials

The raw materials used in this study included ferronickel slag and ferrochromium slag. The ferronickel slag was obtained from the RKEF process of ferronickel smelting. The ferrochromium slag was obtained from the electric arc furnace process of ferrochromium smelting. The chemical compositions of ferronickel slag and ferrochromium slag are shown in Table 1. The ferronickel slag contained 48.29 wt% SiO_2, 30.95 wt% MgO, 7.39 wt% FeO, 4.04 wt% Al_2O_3, 2.40 wt% CaO, and 2.11 wt% Cr_2O_3. The ferrochromium slag contained 28.99 wt% SiO_2, 25.64 wt% MgO, 15.37 wt% Al_2O_3, 10.04 wt% Cr_2O_3, 4.28 wt% CaO, and 4.06 wt% FeO. The main

Table 1 Main chemical compositions of the ferronickel slag and ferrochromium slag

	SiO_2	MgO	Al_2O_3	Cr_2O_3	CaO	TFe	FeO
Ferronickel salg	48.29	30.95	4.04	2.11	2.40	8.67	7.39
Ferrochromium slag	28.99	25.64	15.37	10.04	4.28	3.90	4.06

phase composition of the ferronickel slag was olivine, the main phase composition of the ferrochromium slag was forsterite, magnesium aluminate spinel, and aluminum iron oxide. Magnesia of analytical grade was used in the preparation of refractory materials from ferronickel slag and ferrochromium slag.

Methods

The thermodynamic analysis of relevant transforms during sintering of the ferronickel slag and ferrochromium slag with the addition of magnesia was performed by establishing corresponding equilibrium and phase diagrams using the software FactSage 7.3 (Thermfact/CRCT and GTT-Technologies, Montreal, Canada, and Herzogenrath, Germany).

In the refractory materials preparation process, the mixture of grounded ferronickel slag, ferrochromium slag (86 wt% particles passing 74 μm sieve), and magnesia was carefully mixed in a planetary ball mill for 20 min. Briquettes of 20 mm in diameter and 20 mm in height were produced by manual hydraulic pressing at 100 MPa. After drying, the briquettes were placed in a muffle furnace before sintering. After sintering at a designated temperature for a given period of time, the briquettes were cooled to room temperature and then taken out for the subsequent characterization tests. The characterization tests method is consistent with our previous research [13].

Results and Discussion

Phase Diagram Analysis

In the ferronickel slag and ferrochromium slag, the main chemical compositions are MgO, SiO_2, Cr_2O_3, and Al_2O_3. In order to reveal the phase transformations of refractory materials obtained by co-sintering ferronickel slag and ferrochromium slag, the MgO-SiO_2-Cr_2O_3 and MgO-SiO_2-Al_2O_3 ternary phase diagrams have been calculated.

The phase diagrams of MgO-SiO_2-Cr_2O_3 and MgO-SiO_2-Al_2O_3 are shown in Figs. 1 and 2, respectively. The chemical composition of the triple invariable point

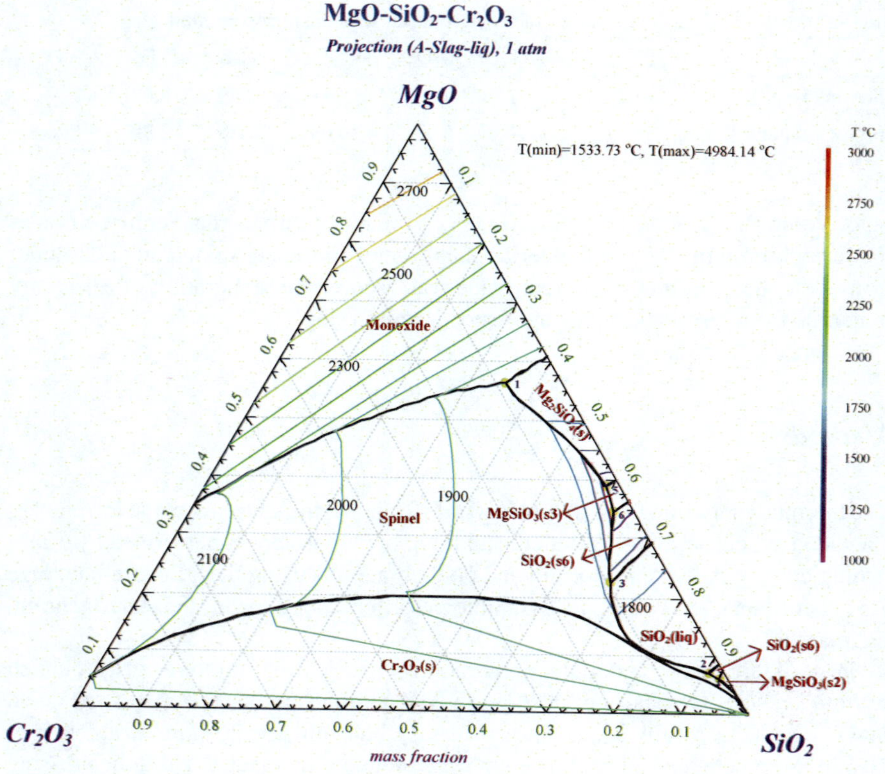

Fig. 1 MgO-SiO_2-Cr_2O_3 ternary phase diagram. (Color figure online)

and the corresponding eutectic point in the MgO-SiO_2-Cr_2O_3 system and MgO-SiO_2-Al_2O_3 are shown in Tables 2 and 3, respectively. As shown in Fig. 2, in the MgO-SiO_2-Cr_2O_3 system, the possible forming phases include periclase (monoxide, MgO), spinel ($MgCr_2O_4$), chromium oxide (Cr_2O_3), forsterite (Mg_2SiO_4), enstatite ($MgSiO_3$), cristobalite (SiO_2), and liquid (SiO_2). The periclase, spinel and chromium oxides are the main phases in the MgO-SiO_2-Cr_2O_3 system. As shown in Table 2, The temperature range of the eutectic phase formed by solid solution between different phases is 1534–1814 °C, exhibit high eutectic point. It is obvious that the presence of the enstatite phase will lower the eutectic point. The eutectic phase (point 1 in Fig. 1) formed by perclase, forsterite, and spinel shows the highest eutectic point (1814 °C.), the corresponding component contents of MgO, SiO_2, and Cr_2O_3 are 56.27 wt%, 35.14 wt%, and 8.59 wt%, respectively. Based on the above analysis, it could be concluded that it is possible to obtain high melting point forsterite, spinel, and perclase phases by increasing the contents of MgO and Cr_2O_3 in the MgO-SiO_2-Cr_2O_3 system.

As shown in Fig. 2, in the MgO-SiO_2-Al_2O_3 system, the possible forming phases include periclase (MgO), spinel ($MgAl_2O_4$), forsterite (Mg_2SiO_4), cordierite

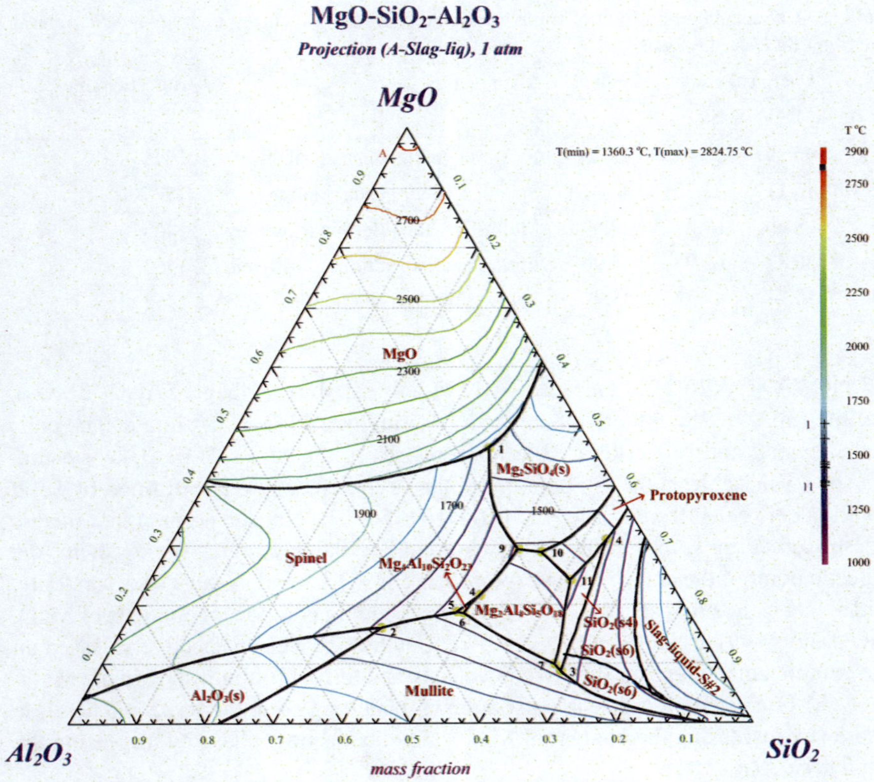

Fig. 2 MgO-SiO$_2$-Al$_2$O$_3$ ternary phase diagram. (Color figure online)

Table 2 Chemical composition of triple invariable point and the corresponding eutectic point in the MgO-SiO$_2$-Cr$_2$O$_3$ system

No.	Chemical composition/wt%			Main phases	Eutectic point/°C
	MgO	SiO$_2$	Cr$_2$O$_3$		
1	56.27	35.14	8.59	Forsterite, periclase, and spinel	1814
2	7.12	90.54	2.34	Cristobalite, liquid, and spinel	1723
3	22.52	68.03	9.45	Cristobalite, liquid, and spinel	1723
4	6.11	92.55	1.35	Enstatite, cristobalite, and spinel	1697
5	38.90	58.99	2.12	Enstatite, forsterite, and spinel	1547
6	34.56	62.32	3.11	Enstatite, cristobalite, and spinel	1534

Table 3 Chemical composition of triple invariable point and the corresponding eutectic point in the MgO-SiO_2-Al_2O_3 system

No.	Chemical composition/wt%			Main phases	Eutectic point/°C
	MgO	SiO_2	Al_2O_3		
1	46.39	38.41	15.20	Forsterite, periclase, and spinel	1646
2	16.11	37.82	46.07	Spinel, corundum, and mullite	1575
3	8.66	67.67	23.67	Mullite, cristobalite, and tridymite	1465
4	30.82	63.28	5.90	Enstatite, cristobalite, and tridymite	1465
5	18.74	47.44	33.82	Spinel, sapphire, and mullite	1464

($2MgO{\cdot}2Al_2O_3{\cdot}5SiO_2$), enstatite ($MgSiO_3$), sapphire ($4MgO{\cdot}5Al_2O_3{\cdot}2SiO_2$), mullite ($3Al_2O_3{\cdot}2SiO_2$), cristobalite ($SiO_2$), tridymite ($SiO_2$), and liquid. The periclase, spinel, and forsterite are the main phases in the MgO-SiO_2-Al_2O_3 system. As shown in Table 3, the temperature range of the eutectic phase formed by solid solution between different phases is 1464–1646 °C. The eutectic point of the eutectic phase formed by spinel in the ternary system is relatively high. Among them, the eutectic point of the eutectic phase (point 1 in Fig. 2) formed by periclase, forsterite, and spinel is the highest at 1646 °C, the corresponding component contents of MgO, SiO_2, and Al_2O_3 are 46.39 wt%, 38.41 wt%, and 15.20 wt%, respectively. Based on the component contents of the ferronickel slag and ferrochromium slag, in the MgO-SiO_2-Al_2O_3 system, it is possible to obtain high melting point forsterite and perclase phases by increasing the content of MgO, and obtain spinel phase by increasing the content of Al_2O_3.

Thermodynamic Equilibrium Phases Analysis

In order to further identify the thermodynamic characteristics of ferronickel slag and ferrochromium slag by co-sintering in the presence of magnesia. The effect of sintering temperature, and oxygen concentration on the contents of thermodynamic equilibrium phases of the ferronickel slag and ferrochromium slag have been researched. It should be noted that based on our previous studies [3, 12–15], during the thermodynamic equilibrium calculation, the mass ratio of ferronickel slag and ferrochromium slag was fixed at 1, and the addition of MgO was fixed at 30%.

As shown in Fig. 3, when the sintering temperature is less than 1400 °C, the slags system including forsterite, spinel, and perclase, the contents of forsterite, spinel, and perclase are about 76%, 17%, and 5%, respectively, and those composition contents have a slight change with the sintering temperature increases. However, when the sintering temperature exceeds 1400 °C, the quantity of the liquid phase increases rapidly, and reaches 37.48% at 1600 °C. It is obvious that it is easy to obtain high melting point phases (including forsterite, spinel, and perclase) by controlling the sintering temperature.

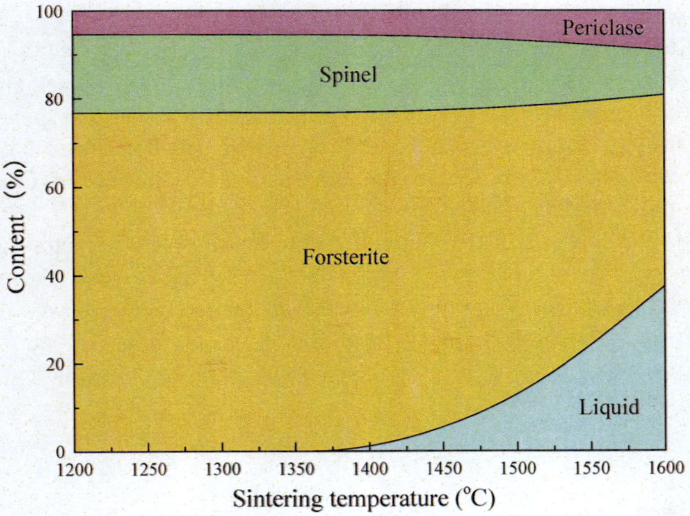

Fig. 3 Calculated contents of thermodynamic equilibrium phases in the ferronickel slag and ferrochromium slag system at different temperatures. (Color figure online)

The thermodynamic equilibrium phases in the ferronickel slag and ferrochromium slag system at different oxygen concentrations were calculated by fixing the addition of MgO of 30 wt% and sintering temperature of 1350 °C. As shown in Fig. 4, when the oxygen concentration of 0 vol.%, the phase composition in the slags is

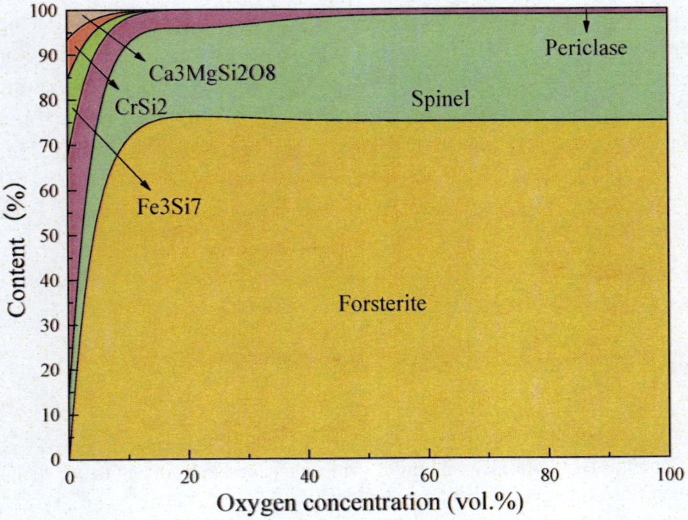

Fig. 4 Calculated contents of thermodynamic equilibrium phases in the ferronickel slag and ferrochromium slag system at different oxygen concentration. (Color figure online)

complex, with 13.72% spinel, 53.79% perclase, 17.14% Fe_3Si_7, 8.76% $CrSi_2$, and 6.59% $Ca_3MgSi_2O_8$. With the oxygen concentration increased to 20 vol.%, the phase composition contents in the slags are 76.81% forsterite, 17.66% spinel, and 5.52% perclase. As the oxygen concentration further increased, the content of spinel slightly increased and the content of perclase slightly decreased, which indicated that increase in oxygen concentration contribute to the complete reaction between MgO and slags, improving the formation of spinel.

To verify the findings of the thermodynamic analysis, the refractory materials preparation experiments of slag were carried out by sintering the slags (the mass ratio of ferronickel slag and ferrochromium slag of 1) in the presence of 30 wt% magnesia at 1350 °C for 3 h with the oxygen concentration of 20 vol.%. A good refractory material with refractoriness of 1780 °C, bulk density of 2.30 g/cm^3, apparent porosity of 16.19%, and compressive strength of 89.18 MPa could be obtained.

Conclusions

In the present study, the thermodynamic analysis of ferronickel slag and ferrochromium slag co-sintered in the presence of magnesia was carried out. It was shown that by co-sintering ferronickel slag with the addition of magnesia at appropriate temperatures (1200–1400 °C) and oxygen concentration (20–100 vol.%), it is possible to promote the formation of high melting point phases, including forsterite, spinel, and perclase, which would improve refractoriness of the refractory materials derived from the slags. The verification experiment results showed that under the conditions of mass ratio of ferronickel slag and ferrochromium slag of 1, magnesia addition of 30 wt%, sintering temperature of 1350 °C, sintering time of 3 h, oxygen concentration of 20 vol.%, a refractory material with refractoriness of 1780 °C, bulk density of 2.30 g/cm^3, apparent porosity of 16.19%, and compressive strength of 89.18 MPa could be obtained. Overall, co-sintering of ferronickel slag and ferrochromium slag to prepare refractory has theoretical and practical feasibility.

Acknowledgements This work was supported by the National Natural Science Foundation of China under Grants U1960114.

References

1. Dourdounis E, Stivanakis V, Angelopoulos GN, Chaniotakis E, Frogoudakis E, Papanastasiou D, Papamantellos DC (2004) High-alumina cement production from FeNi-ERF slag, limestone and diasporic bauxite. Cem Concr Res 34(6):941–947
2. Balomenos E, Panias D (2013) Iron recovery and production of high added value products from the metallurgical by-products of primary aluminum and ferronickel industries. In: Proceedings of the 3rd international slag valorisation symposium, Leuven, Belgium, pp 72–161

3. Gu F, Peng Z, Zhang Y, Tang H, Ye L, Tian W, Liang G, Lee J, Rao M, Li G, Jiang T (2019) Valorization of ferronickel slag into refractory materials: Effect of sintering temperature. JOM 71:1024–1032
4. Spooren J, Kim E, Horckmans L, Broos K, Nielsen P, Quaghebeur M (2016) In-situ chromium and vanadium recovery of landfilled ferrochromium and stainless steel slags. Chem Eng J 303359–368
5. Zhang Z, Zhu Y, Yang T, Li L, Zhu H, Wang H (2017) Conversion of local industrial wastes into greener cement through geopolymer technology: a case study of high-magnesium nickel slag. J Clean Prod 141:463–471
6. Lemonis N, Tsakiridis PE, Katsiotis NS, Antiohos S, Papageorgiou D, Katsiotis MS, BeaziKatsioti M (2015) Hydration study of ternary blended cements containing ferronickel slag and natural pozzolan. Constr Build Mater 81:130–139
7. Saha AK, Sarker PK (2016) Expansion due to alkali-silica reaction of ferronickel slag fine aggregate in OPC and blended cement mortars. Constr Build Mater 123:135–142
8. Maragkos I, Giannopoulou I, Panias D (2009) Synthesis of ferronickel slag-based geopolymers. Miner Eng 22(2):196–203
9. Komnitsas K, Zaharaki D, Perdikatsis V (2009) Effect of synthesis parameters on the compressive strength of low-calcium ferronickel slag inorganic polymers. J Hazard Mater 161(2–3):760–768
10. Rawlings RD, Wu JP, Boccaccini AR (2006) Glass-ceramics: their production from wastes-a review. J Mater Sci 41:733–761
11. Karamanov A, Paunovi´ c P, Ranguelov B, Ljatifi E, Kamusheva A, Nacevski G, Karamanova E, Grozdanov A (2017) Vitrification of hazardous Fe-Ni wastes into glass-ceramic with fine crystalline structure and elevated exploitation characteristics. J Environ Chem Eng 5(1):432–441
12. Gu F, Peng Z, Zhang Y, Tang H, Su Z, Tang H, Tian W, Liang G, Lee J, Rao M, Li G, Jiang T (2019) Selective recovery of chromium from ferronickel slag via alkaline roasting followed by water leaching. J Hazard Mater 374:83–91
13. Gu F, Peng Z, Zhang Y, Tang H, Ye L, Tian W, Liang G, Rao M, Li G, Jiang T (2018) Facile route for preparing refractory materials from ferronickel slag with addition of magnesia. ACS Sustain Chem Eng 6:4880–4889
14. Peng Z, Gu F, Zhang Y, Tang H, Ye L, Tian W, Liang G, Rao M, Li G, Jiang T (2018) Chromium: a double-edged sword in preparation of refractory materials from ferronickel slag. ACS Sustain Chem Eng 6:10536–10544
15. Gu F, Peng Z, Zhang Y, Tang H, Tian W, Lee J, Rao M, Li G, Jiang T (2020) Promoting spinel formation and growth for preparation of refractory materials from ferronickel slag. Int J Appl Ceram Technol 17:1701–1712

PGM Furnace Design, Construction, Improvement, and Performance Optimisation

Isobel Mc Dougall, Gerrit de Villiers, Hugo Joubert, Burger van Beek, John Davis, and Trevor Goff

Abstract Tenova Pyromet designed, supplied, and commissioned a 12 MVA circular electric furnace for Sibanye-Stillwater (previously Lonmin) in South Africa. The furnace has been operating successfully since its commissioning in 2012. The paper provides an overview of the furnace performance and campaign life history, and explores the innovative design concepts and improvements implemented. Of particular interest is the use of a novel sidewall lining design that utilizes indirect cooling to remove excess heat from the matte/slag tidal zone without the need to use water-cooled copper. Furthermore, the performance of Tenova's Söderberg electrode column designed for base metal applications, limiting water use above the furnace, is discussed. In addition, the joint development and performance of an uncooled cast iron matte taphole faceplate is presented. Potential future design innovations to further improve smelter operating safety are discussed.

Keywords Furnace cooling · Refractory lining · Graphite · Söderberg electrode · Taphole · Cast iron

I. Mc Dougall · G. de Villiers · H. Joubert (✉)
Tenova Pyromet, Midrand Business Park, 563 Old Pretoria Main Road, Halfway House, Midrand 1685, South Africa
e-mail: hugo.joubert@tenova.com

I. Mc Dougall
e-mail: Isobel.mcdougall@tenova.com

G. de Villiers
e-mail: Gerrit.devilliers@tenova.com

B. van Beek · J. Davis · T. Goff
Sibanye-Stillwater Smelter Marikana SA PGM Operations, Johannesburg, South Africa
e-mail: Burger.VanBeek@sibanyestillwater.com

J. Davis
e-mail: John.Davis@sibanyestillwater.com

T. Goff
e-mail: Trevor.Goff@sibanyestillwater.com

C. Anderson et al. (eds.), *Ni-Co 2021: The 5th International Symposium on Nickel and Cobalt*, The Minerals, Metals & Materials Series,
https://doi.org/10.1007/978-3-030-65647-8_25

Introduction

The Marikana smelter complex currently consists of five electric smelting furnaces and three Peirce Smith type converters.

In April 2010 Tenova Pyromet was contracted to design and build a new circular 12 MVA primary platinum smelting AC electric furnace (Furnace 2). The furnace replaced the decommissioned rectangular Merensky six-in-line electric smelting furnace. At the time of project initiation in 2010 the smelter complex operated with the primary existing circular 28 MW Furnace 1 as well as the three Pyromet circular 5 MW furnaces (Furnaces 3, 4, & 5). In 2009 management estimated the smelter plant would soon run at full capacity and additional furnace capacity will be required to recover any concentrate stock accumulated during furnace planned and unplanned outages. It was decided that the upgrade and modernisation of the existing six-in-line furnace would not be suited for the high operating temperatures and that a new circular furnace would be better suited to smelting feed blends containing chromite rich UG2 concentrate [1].

Tenova Pyromet was selected to design, supply, construct, and commission the new Furnace 2 following the successful completion of a pre-feasibility study commissioned in 2009. In selecting Tenova Pyromet for the project, the then Lonmin recognised Tenova's experience with circular bath smelting furnaces in the platinum and base metal industries, including the three Pyromet furnaces at the Marikana smelter complex that have been operating for more than 20 years at that time. Tenova Pyromet is part of the Tenova group of companies with its headquarters in Castellanza, Italy. Based in Johannesburg South Africa, Tenova Pyromet is the electric submerged arc furnace centre-of-excellence within the Tenova group, and supplies furnace technology and equipment to clients all around the world. The project was completed within 28 months and the first matte tap took place during July 2012.

This paper provides an overview of the performance and campaign life history of Furnace 2 since its start-up in 2012. Some of the key design concepts and innovations implemented on the furnace during the original project as well as subsequent design improvements are discussed. These include the use of graphite in the critical matte–slag tidal zone lining design and less water cooling above the furnace roof, as well as the successful development and implementation of an uncooled cast iron matte taphole faceplate.

Furnace Operation and Performance

Following the commissioning of Furnace 2 in July 2012 a performance test was successfully conducted, and the Furnace was operated at the design power set point of 10 MW for the duration of the performance test. Subsequently the furnace has been operated between 8 and 9 MW depending on the production throughout requirements and has cumulatively processed over a half million tonnes of concentrate.

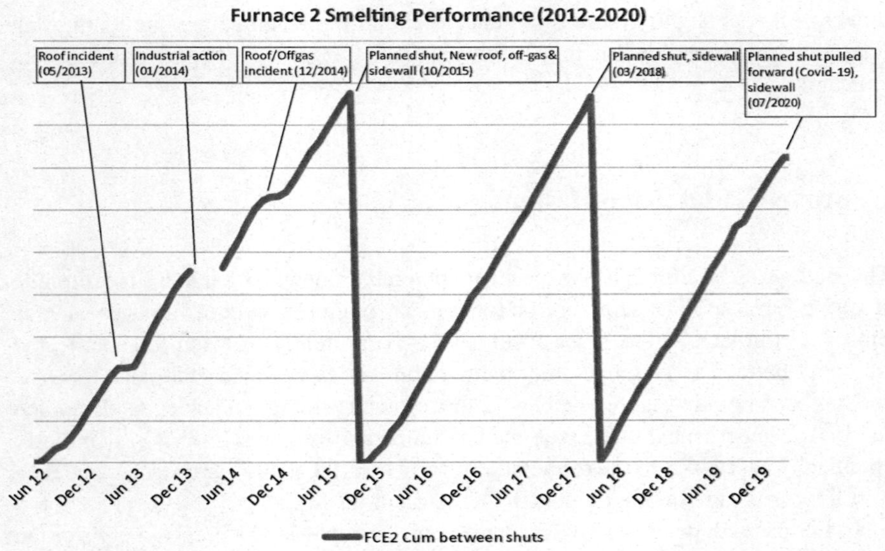

Fig. 1 Furnace 2 cumulative tonnes per sidewall campaign. (Color figure online)

The refractory crucible design performed as per design, and no slag or matte leak occurred during the last eight years of operation. The hearth and skew design performed well with limited hearth ratcheting recorded. The sidewall wear was as expected for a shallow cooled plate cooler sidewall and good structural stability of the sidewall was noticed as the campaigns progressed. The matte tap block design performed well with almost no wear experienced on the matte tap block lintel cooler. The slag tap-block and granulation system performed as per design with no safety-related incidents reported.

In Fig. 1 the cumulative tonnes smelted can be seen for the three sidewall campaigns for Furnace 2.

During the first sidewall campaign off-gas and roof-related problems where experienced, resulting in high freeboard temperatures and off-gas blockages. This was addressed during the first sidewall repair in October 2015 by installing a new roof and off-gas uptake design similar to what is installed at Furnace 1. An additional row of plate lintel coolers was also installed in the freeboard area to assist in supporting the upper sidewall as undercutting is experienced throughout the campaign.

During the second sidewall campaign the addition of lime flux was stopped, and the sidewall monitoring was used to schedule the replacement of the sidewall. The sidewall wear profile was more pronounced compared to the first campaign. From the start of the third sidewall campaign in April 2018 stable Furnace operation was experienced during the campaign. A strategic decision was taken to move the planned sidewall rebuild three months forward due to the lower production volumes experienced as a result of the impact from the Covid-19 pandemic. The sidewall wear profile was similar compared to the first two campaigns although the sidewall upward tilt was slightly more pronounced. On all three campaigns the matte–slag tidal zone performed well with a 13% wear reduction in wall thickness of the tidal zone area.

Improved Tidal Zone Design

The matte–slag tidal zone is one the most vulnerable sidewall wear areas for sulphide smelting furnaces. The same is true for primary platinum smelter furnaces such as Sibanye's Furnace 2. The sidewall lining is intermittently exposed to superheated slag and matte. The slag operating temperature varies between 1550 and 1680 °C and has been found to aggressively wear the refractory lining if not adequately cooled [2, 3]. The superheated matte, operated at temperatures above 1500 °C, is similarly chemically aggressive to the refractory. In particular, the matte has a strong tendency to sulphidize mag-chrome (MgO-Cr_2O_3) refractories [4].

In the past copper plate coolers have been installed intermittently between the bricklayers to improve the campaign of the matte–slag tidal zone refractory lining [2, 5]. The theory is for the copper plate coolers to remove excess heat from the lining and reduce the operating hot face temperature. This in turn will slow the brick-wear and potentially stop wear through the formation of a slag freeze lining. The slag freeze lining will prevent direct contact between the molten slag or matte and the brick hot face, and therefore limit chemical attack. The downside to this solution is the potential contact between matte and the copper plate cooler. This could happen when, usually due to abnormal operating conditions that do occur from time to time, the remaining working lining in front of the copper cooler is worn away and no slag freeze lining is maintained. A copper cooler is likely to rapidly melt or dissolve in contact with the superheated matte, resulting in a containment failure with all the associated safety risks and prematurely ending the lining campaign life.

During the engineering phase for Furnace 2 it was decided to investigate alternative ways to remove excess heat from the matte–slag tidal zone without the use of water-cooled copper cooling elements. Graphite was identified as a potential material to use due to its high thermal conductivity enabling the removal of excess heat from the lining hot face. Furthermore, graphite has been used successfully on the hot face of copper coolers adjacent to the slag bath in other platinum smelting furnaces, showing that it can withstand the attack mechanisms if its operating temperature is kept low enough [6, 7]. Nonetheless, as part of the investigation graphite's resistance to chemical attack by both superheated slag and matte was tested [8]. It was found to provide better resistance to both slag and mate compared to mag-chrome refractory material. Mag-chrome is typically used as the working lining in platinum and other sulphide smelting furnaces.

Following careful consideration of the risk of graphite oxidising in the long term while in contact with the superheated slag, it was decided to focus on graphite as a potential back lining in the matte–slag tidal zone [9]. In the concept mag-chrome is maintained as the working lining with the graphite back lining, replacing the typically

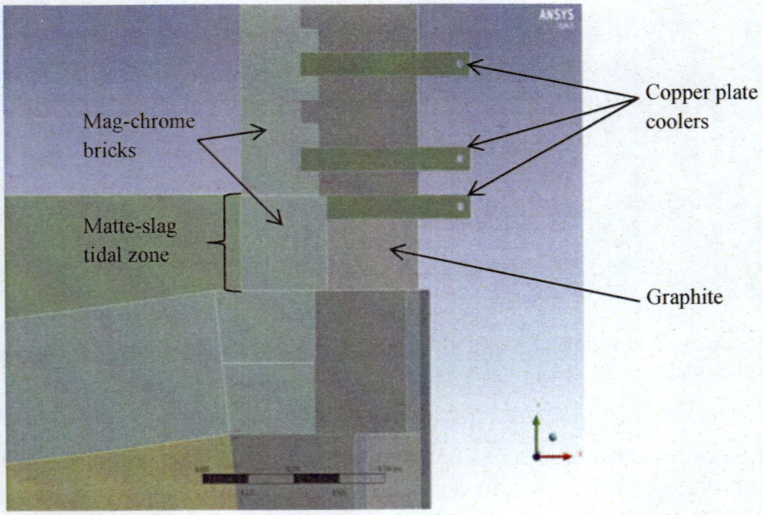

Fig. 2 Matte–slag tidal zone sidewall design concept with a graphite backing lining. (Color figure online)

used alumina-chrome refractory, enabling the more effective removal of excess heat from the working lining. A water-cooled copper plate cooler is placed directly on top of the graphite back lining. The proposed concept tidal zone design is shown in Fig. 2.

Thermal FEA (Finite Element Analysis) modelling was employed to evaluate the concept design compared to typical tidal zone lining design arrangements used in industry [8, 9]. It was compared to a design without any cooling, copper or otherwise, in the tidal zone, as well as an arrangement with a water-cooled copper plate cooler installed one brick above the hearth skewback level, directly extracting heat from the tidal zone. The results indicated that the graphite backing lining enabled the removal of enough excess heat from the working lining to ensure similar working lining temperatures achieved with the use of a copper plate cooler installed one brick above the hearth skewback. The use of the graphite backing lining with the copper plate cooler on top further ensured that the 800 °C isotherm, the temperature at which matte will be completely solid, is kept closer to the hot face of the working lining. In addition, the 450 °C isotherm, above which oxidation of the low porosity graphite material becomes possible, is kept well in front of the graphite hot face. It was concluded that the same benefits to the critical tidal zone working lining can be achieved through the use of a graphite backing lining as compared to the use of direct copper cooling without the downside risk associated with water-cooled copper coming in contact with molten matte. The comparative thermal FEA results, as well as the location of respectively the 450–800 °C isotherms are shown in Fig. 3 for the case with no cooling and the case with the graphite backing lining in the tidal zone.

Based on the results of this investigation, Sibanye-Stillwater and Tenova jointly agreed to implement the use of the graphite backing lining in Furnace 2. The original

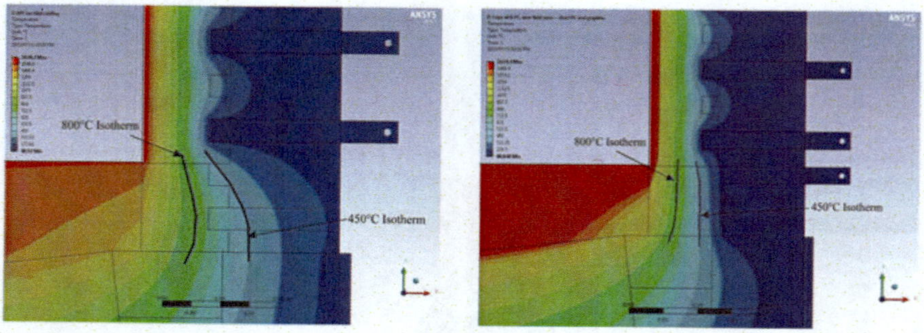

Fig. 3 Comparative thermal FEA results between an uncooled tidal zone and one with the proposed graphite backing lining [9]. (Color figure online)

graphite backing lining has been in operation since the start-up of Furnace 2 in 2011, and was replaced for the first time during the 2020 furnace reline. During the reline in 2018, the sidewall demolition damaged a few of the graphite blocks mechanically. However, no operating process-related wear or oxidation was noted. The machining marks on the blocks were still visible. To do repairs to the upper skewback ring the complete ring of graphite blocks were removed, cleaned, and reinstalled and only the mechanically damaged graphite blocks were replaced. The graphite blocks are shown in Fig. 4 during the 2018 Furnace 2 reline. The same matte–slag tidal zone lining design was installed during the 2020 reline.

Fig. 4 Photos showing the graphite blocks during the **a** demolition and **b** rebuilding phases of the most recent Furnace 2 reline. (Color figure online)

Moving Towards Waterless Electrodes

The presence of cooling water above a smelting furnace has been a risk many furnace operators had to live with, and still has to live with. Examples of explosions due to water coming into contact with molten metal, matte, or slag are unfortunately numerous [10]. Cooling water is typically used to cool the electrode components, and in some industries the furnace roof, all equipment located directly above the furnace bath. In the platinum industry most furnace roofs are not water-cooled, and consist of refractory bricks supported by an anchoring system, commonly referred to as a hanging brick roof design. However, on all operating platinum furnaces known to the authors water is used to cool the electrode components. This is the case for Söderberg type self-baking electrodes as used on typical platinum smelting furnaces, for example Furnace 2, as well as for prebaked graphite electrodes as used on some smaller and older furnaces, for example Furnaces 3, 4, and 5 at Sibanye-Stillwater's smelter complex at Marikana (originally designed and built by Tenova Pyromet). Explosions due to water leakages on furnace roofs have occurred in the past. For example, a large explosion occurred on a silicon smelting furnace in Norway in 2006 due to water leaking from electrode components [11]. In 2004 Sibanye-Stillwater experienced an explosion in Furnace 1 due to a water leak from a flexible hose that blew of the furnace roof.

Due to the relatively low electrode current densities present in Söderberg electrodes used in the platinum industry, and the base metals industry in general, the electrode casing remains intact for a relatively long distance below the contact shoes and provides structural strength to the electrode. This allows the water-cooled lower electrode components, such as the contact shoes and pressure rings, to be located above the furnace roof. This is in contrast to other industries, such as ferroalloys, where the water-cooled contact shoe and pressure ring arrangement has to be located close to the feed burden in the furnace freeboard to limit the free electrode length. In addition, as the casing remains intact below the contact shoes, sealing directly against the electrode casing at the furnace roof level is possible. There is thus no need for water-cooled heat shields extending through the seal into the furnace freeboard as is the case in most other furnace operations.

Notwithstanding that this arrangement is typical of most base metal furnaces, these furnaces tend to still use extensive water cooling on the lower electrode components, including water-cooled pressure rings, contacts shoes, bus tubes, and in some cases water-cooled heat shields. Tenova Pyromet realised this arrangement presented an opportunity to significantly reduce the use of cooling water above the furnace roof for base metal furnaces. Starting with Sibanye-Stillwater's Furnace 2, the water-cooled electrode pressure ring was changed to an uncooled stainless steel design and the water-cooled heat shields were removed. The water-cooled heat shields can be replaced with uncooled dust shields if excessive dust build-up is a concern above the furnace roof. The only water-cooled electrode components remaining above the furnace roof are the contact shoes and the bus tubes supplying electrical current and cooling water to the contact shoes, as well as the pressure below between the

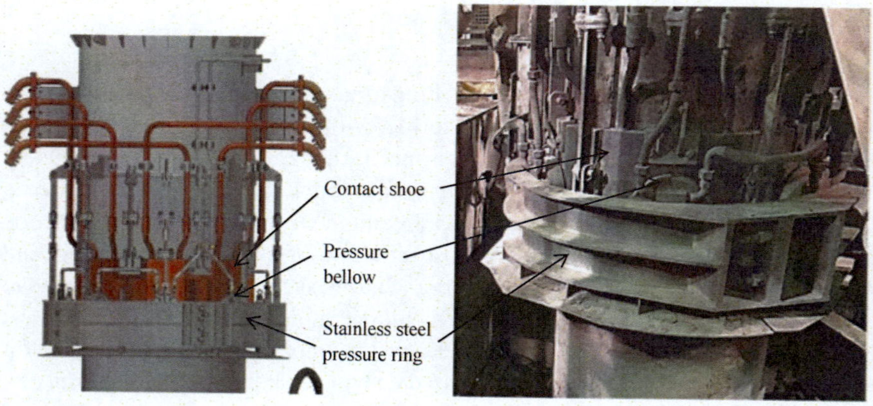

Fig. 5 Furnace uncooled lower Söderberg electrode arrangement and components. (Color figure online)

contact shoes and the pressure ring. A 3D model of the Furnace 2 lower electrode arrangement as well as a photo of the actual installation is shown in Fig. 5. The design of Tenova Pyromet's electrode for base metal furnace applications is discussed in more detail by Bantubanie et al. [12].

The new lower electrode design has been in operation on Furnace 2 since start-up in 2012. The maintenance required on the electrode components has been minimal apart from a single contact shoe that was replaced and inspected during 2015 scheduled reline as a precautionary measure.

Tenova Pyromet subsequently implemented the base metal style lower electrode design on another project with similar success. The design was expanded to include a new innovative, robust, and easy maintainable seal design for sealing directly onto the electrode casing [13]. Work is currently underway to develop a completely waterless electrode for base metal furnaces, removing all cooling water from above the furnace roof.

Matte Taphole Faceplate

Tenova Pyromet's standard matte taphole design consists of water-cooled copper elements split between the inner taphole block, the outer taphole block, and the taphole faceplate. The same design was employed for the original matte tapholes on Furnace 2 as shown in Fig. 6. Both the conical faceplate insert and the taphole blocks are manufactured from magnesite-chrome refractory material.

Based on the positive experience with the cast iron taphole faceplates on the smaller Pyromet furnaces (designed and supplied by Tenova Pyromet in 1989), Sibanye-Stillwater requested Tenova Pyromet to jointly investigate the use of uncooled cast iron faceplates for Furnace 2. The use of uncooled cast iron faceplates

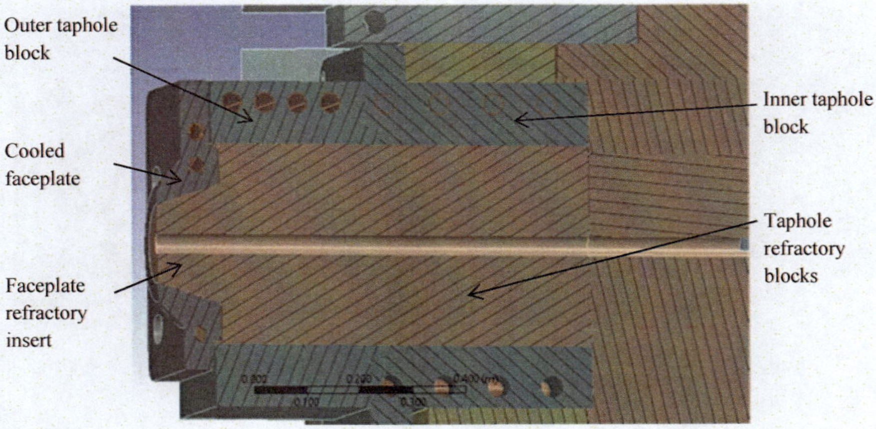

Fig. 6 Model showing section through the Furnace 2 matte taphole. (Color figure online)

assists in the removal of water from close proximity to the molten matte stream and would reduce the safety risk. In addition, the existing water-cooled copper faceplate is physically damaged from time-to-time during tapping operations, and pre-mature cracking of the refractory insert is attributed to overcooling by the faceplate resulting in a high radial temperature gradient across the insert.

Three material options were selected for further investigation: A typical ductile cast iron, a high-temperature Meehanite grade cast iron, and a high-temperature stainless steel casting grade. The three materials are described in Table 1.

Thermal finite element analysis (FEA) was performed to evaluate the different material options. A matte tapping temperature of 1550 °C was assumed in line with normal operating conditions. In the lead up to a taphole repair or a furnace rebrick, the furnace is emptied through one of the matte tapholes. Therefore, another load case considered was the tapping of slag at 1650 °C through the matte taphole. Both steady

Table 1 Alternative faceplate materials investigated [14, 15]

Material type	Max operating temperature	Material	Main attributes
Ductile cast iron	Up to 500 °C	Nodular graphite, according to DIN EN 1563 Grade EN-GJS-700-2	Readily available, good machinability.
Meehanite HE	Up to approximately 900 °C	Flake graphite, pearlitic	Best for cyclical heating applications, involving thermal shock
ASTM A297 HH	Scaling resistance up to 1093 °C Long term strength up to 954 °C	Fe–Cr-Ni alloy. 25% Cr 12% Ni	Excellent resistance to hot sulphur bearing gasses.

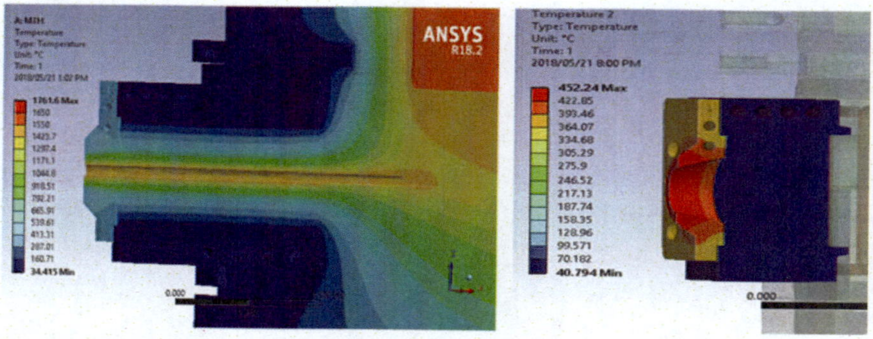

Fig. 7 Steady state temperature plot for uncooled Meehanite HE - Load case no.3. (Color figure online)

state and transient analyses were performed. The steady state analysis, which implies continuous tapping, was performed to represent the worst case. The transient analysis was limited to 1 h, the maximum tap duration for matte on Furnace 2. For comparison, analyses were performed for the current water-cooled copper faceplate design. The main purpose of the FEA modelling was to determine the maximum expected operating temperature for the faceplate. For example, the steady state temperature plot results for the uncooled Meehanite HE grade cast iron as shown in Fig. 7 indicates a maximum faceplate temperature of 452 °C. This represents the worst case for Meehanite HE and is well below the maximum allowable operating temperature of 900 °C as referenced in Table 1.

The maximum faceplate temperature for a number of load cases are listed in Table 2. The maximum temperature for the existing water-cooled copper faceplate remains low, even under steady state conditions. This is positive for the faceplate itself, but results in the expected high radial temperature gradient through the refractory insert (see Fig. 8a). Of the three new materials tested, Meehanite HE has the lowest maximum operating temperature, both under steady state and transient conditions, as well as during slag tapping. Next is ductile cast iron followed by the stainless steel cast material. The maximum faceplate operating temperature tends to be higher relative to a lower material thermal conductivity. Of interest for Meehanite HE the maximum operating temperature is calculated to be 177 °C after 1 h of matte tapping and 315 °C after 3 h of slag tapping, well below the steady state maximum temperature, and significantly lower than the Meehanite HE operating temperature limit. Ductile cast iron operates at temperatures close to its maximum operating limit. From Fig. 8b the maximum radial temperature gradient under steady state conditions for the Meehanite HE faceplate is estimated to be 20 °C/mm compared to 28 °C/mm for the water-cooled copper faceplate. The refractory insert should therefore experience less thermal stress and thermal fatigue when used in combination with the uncooled Meehanite HE faceplate.

It was decided to proceed with Meehanite HE and manufacture two uncooled faceplates for testing on Furnace 2 for the following reasons:

Table 2 Maximum calculated faceplate temperature for selected load cases

Load case no.	Steady state or transient	Face plate material	Faceplate cooling	Maximum faceplate temperature
1	SS	Copper	Yes	62 °C
3	SS	Meehanite HE	No	452 °C
6	SS	Ductile Cast Iron	No	498 °C
7	SS	Ductile Cast Iron slag tapping	No	512 °C
8	SS	Meehanite HE slag tapping	No	464 °C
9	SS	ASTM A297 HH	No	542 °C
10	SS	ASTM A297 HH slag tapping	No	554 °C
2T	T 1 h	Copper	Yes	59 °C
3T	T 1 h	Meehanite HE	No	177 °C
6T	T 1 h	Ductile Cast Iron	No	222 °C
8T	T 3 h	Meehanite HE slag tapping	No	315 °C

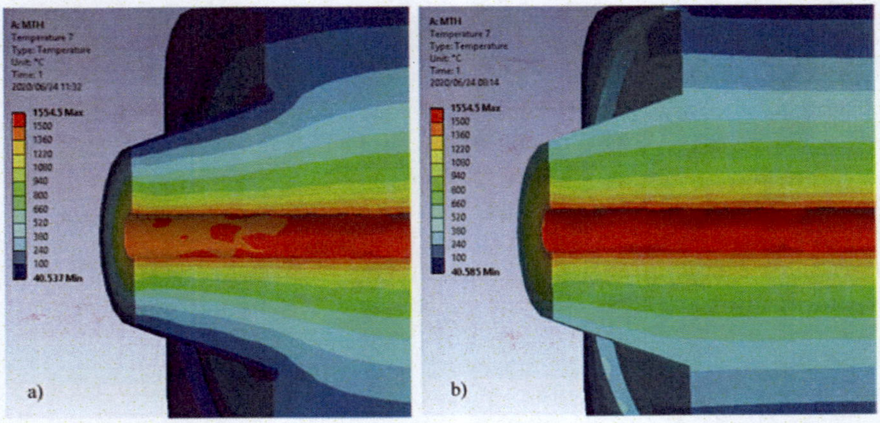

Fig. 8 Temperature gradient through faceplate refractory insert: **a** Water-cooled copper faceplate; **b** Meehanite HE faceplate. (Color figure online)

- Meehanite HE's calculated maximum operating temperature relative to its allowable maximum operating temperature is significantly lower compared to that for the Ductile Cast Iron. The cost for Meehanite HE is very similar to that for Ductile Cast Iron.

Fig. 9 Cast and machined Meehanite HE faceplates. (Color figure online)

- Meehanite HE's calculated maximum operating temperature relative to its allowable maximum operating temperature is very similar compared to that for the cast stainless steel option. Meehanite HE is less than half the cost of ASTM A297.
- Meehanite HE is less than half the cost of the existing copper faceplate, is anticipated to reduce the thermal stress experienced by the refractory insert, and reduces the overall safety risk associated with water in close proximity to the molten matte.

The first two Meehanite HE faceplates were manufactured in 2018. The cast and machined faceplates are shown in Fig. 9. The first Meehanite HE faceplate was commissioned in December 2018 on one of the matte tapholes and was inspected and replaced in January 2019 due to non-related oxygen lancing damage. Subsequently both matte tapholes were equipped with Meehanite HE faceplates that lasted for 12 months of operation before being replaced. It was decided to continue with the use of the Meehanite HE faceplates on Furnace 2 and subsequently on Furnace 1 as well. To date, the average campaign life for a Meehanite HE faceplate is six months, or the equivalent of approximately 270 matte taps. In most cases the campaign was terminated after a small hairline crack has been noticed at the top of the faceplate. An investigation is currently underway to determine the cause of this crack.

Conclusions

Furnace 2 has been in operation since July 2012 and the refractory crucible design performed very well. Improvements to the roof and off-gas design assisted in achieving uninterrupted sidewall campaigns. Three successful sidewall campaigns have been completed since start-up of the furnace. The tidal zone refractory design, including the graphite back lining, assisted in limiting refractory wear in this critical area of the Furnace. The graphite back lining has been replaced for the first time during the July 2020 furnace reline.

The uncooled pressure ring design on the Furnace 2 electrodes has been performing well with only minor maintenance required since start-up in 2012. Tenova Pyromet has installed the same design on other base metal furnaces with similar success. Future developments include the removal of all cooling water from above the furnace roof.

3D thermal FEA modelling was performed to determine whether the current water-cooled copper matte taphole faceplate could be replaced by an uncooled cast iron faceplate. Results indicated that the maximum temperature after 1 h of matte tapping was well within the limits of operation of grey cast iron, and that the maximum steady state temperature attained by the cast iron was well within the limits of high temperature cast iron material. It was concluded that the water-cooled copper matte taphole faceplate could safely be replaced by an uncooled faceplate made of Meehanite HE grade cast iron. A six-month campaign life has been averaged since its implementation on Furnace 2.

References

1. Eksteen JJ, Van Beek B, Bezuidenhout, GA (2011) Cracking a hard nut: An overview of Lonmin's operations directed at smelting of UG2-rich concentrate blends. Paper presented at Southern African Pyrometallurgy 2011 International Conference, Cradle of Humankind, South Africa, pp 231–251
2. Joubert H, Nourse RB, Masters B, Hundermark R (2005) Copper Cooling Design, Installation and Operational Results for the Slag Cleaning Furnace at Waterval Smelter, Rustenburg Platinum, South Africa. Paper presented at COM2005 44th Conf of Metallurgists, International Symposium on Nickel and Cobalt Production, Calgary, Alberta, Canada, pp 21–25
3. Nelson LR, Geldenhuis JMA, Emery B, de Vries M, Joiner K, Ma T, Sarvinis J, Stober FA, Sullivan R, Voermann N, Walker C, Wasmund BH (2006) Developments in furnace design in conjunction with smelting plants in Africa. In: Jones RT (ed) Presented at Southern African Pyrometallurgy 2006, SAIMM, Johannesburg, 5–8 March 2006
4. Eksteen JJ (2011) A mechanistic model to predict matte temperatures during smelting of ug2-rich blends of platinum group metal concentrates, minerals engineering special issue. In: Processing Nickel Ores and Concentrates, vol 24 (7), pp 675–687
5. Nelson LR, Geldenhuis JMA, Emery B, de Vries M, Joiner K, Ma T, Sarvinis J, Stober FA, Sullivan R, Voermann N, Walker C, Wasmund BH (2006) Developments in furnace design in conjunction with smelting plants in africa. In: Jones RT (ed) Southern African Pyrometallurgy 2006, SAIMM, Johannesburg, pp 5–8
6. Thethwayo BM, Garbers-Craig AM (2012) Corrosion of Copper Coolers in PGM Smelters". Paper presented at the 4th International Platinum Conf, Platinum in Transition "Boom or Bust", SAIMM., 2010
7. Shaw A, De Villiers LPVS, Hundermark RJ, Ndlovu J, Nelson LR, Pieterse B, Sullivan R, Voermann N, Walker C, Stober F, McKenzie AD (2013) Challenges and solutions in PGM furnace operation: high matte temperature and copper cooler corrosion. J South Afr Inst Min Metall 113(3):10–25
8. McDougall I, Eksteen JJ (2012) Sidewall Design to Improve Lining Life in a Platinum Smelting Furnace. Paper presented at the TMS 2012 International Smelting Technology Symposium, Orlando Florida, pp 11–15
9. Mc Dougall I (2013) Sidewall design for improved lining life in a PGM smelting furnace. J Southern Afr Inst Mining Metall 113:631–636

10. Kennedy MW, Nos P, Bratt M, Weaver M (2013) Alternative coolants and cooling system designs for safer freeze lined furnace operation. Present Ni-Co 2013:299–314
11. Tveit HM, Garcia H, Delbeck AT, Haug B, Saugestad, and Eikeland IJ (2006) Water leakages in ferroalloy and silicon reduction furnaces—experience gained from a severe accident in 2006, presented at the Silicon for the Chemical and Solar Industry IX, Oslo, Norway, pp 1–16
12. Bantubani S, Hansraj R, Arjun S, Mc Dougall I, Joubert H (2019) Electrode and electrode management technology for use on base metal electric furnaces. The 68th Annual Conference of Metallurgists (COM) hosting the 10th International Copper Conference 2019
13. Esterhuizen A, Joubert H, Essack S, Fowler N (2019) Innovative technologies for copper smelting and electric slag cleaning and matte settling furnaces. The 68th Annual Conference of Metallurgists (COM) hosting the 10th International Copper Conference 2019
14. Steel Casting Handbook, Supplement 8-High Alloy Data Sheets-Corrosion Series, Steel Founders' Society of America (2004). http://www.sfsa.org/sfsa/pubs/hbk/s8.pdf
15. Meehanite Specification Handbook, Meehanite Worldwide Corporation (2013). http://www.meehanite.co.uk. Accessed xxx 2013

A Novel Process to Reduce SO_2 Emissions During Electric Furnace Smelting of Sulphides

D. G. Tisdale, S. J. Muinonen, M. D. Molinski, and A. G. Stokreef

Abstract A novel process was developed and implemented to reduce sulphur dioxide generation during the electric furnace smelting stage at Glencore's Sudbury Integrated Nickel Operations, located in Ontario, Canada. The new process reduced the oxygen concentration in the furnace freeboard, thereby reducing unwanted roasting of residual sulphides in the furnace feed. A program of increasing-scale testing verified and demonstrated the viability of what was dubbed the Controlled Furnace Atmosphere (CFA) process. A significant reduction of site annual emissions of sulphur dioxide has been achieved since the commercial installation in 2015. Capital cost per tonne of emission reduction was lower than market-proven alternatives, and operating cost was minimal. This paper will describe the development and subsequent commercial operation of the new CFA process.

Keywords Sulphur dioxide · Emissions · Electric furnace

Introduction

There have been many process modifications throughout the history of the smelter at Glencore's Sudbury Integrated Nickel Operations. Nickel operations began at the site in the town of Falconbridge, Ontario, Canada in 1929. Motivating factors for

D. G. Tisdale (✉) · S. J. Muinonen · M. D. Molinski · A. G. Stokreef
Glencore Sudbury Integrated Nickel Operations, 2 Longyear Drive, Falconbridge, ON, Canada
e-mail: Dave.Tisdale@Glencore.ca

S. J. Muinonen
e-mail: Sari.Muinonen@Glencore.ca

M. D. Molinski
e-mail: Michael.Molinski@Glencore.ca

A. G. Stokreef
e-mail: Arthur.Stokreef@Glencore.ca

C. Anderson et al. (eds.), *Ni-Co 2021: The 5th International Symposium on Nickel and Cobalt*, The Minerals, Metals & Materials Series,
https://doi.org/10.1007/978-3-030-65647-8_26

modifications have been many, including increased throughput, increased productivity, changes in concentrate supply, and the need for improvements in environmental performance. Feeds to the smelter include sulphide concentrates from several sources. The need to reduce sulphur dioxide emissions, in particular, drove a number of changes, including the adoption of the Controlled Furnace Atmosphere (CFA) process described in this paper.

The current smelter flowsheet includes two fluid-bed roasters, which produce calcine fed to a six-in-line electric smelting furnace. A modified Pierce-Smith converter aisle treats matte tapped from the furnace. Sulphur capture at the Smelter has reached around 92%. The main streams contributing to the remaining emissions include acid plant tail gas, furnace off-gas, and converter aisle gases. Descriptions of these streams follow.

The roasters are fluidized with air, and deliberately restricted to a partial roast. Air flow is carefully controlled to match feed chemistry so that pay metals (nickel, copper and cobalt) remain as sulphides, allowing collection in a matte phase in the subsequent smelting stage. Iron sulphides in the feed are selectively roasted. Increasing the overall roast to nearly eliminate iron sulphide in the product calcine has been implemented to reduce sulphur emissions in processing downstream of the roasters. This change is referred to as High Roast. Off-gas from roasting is treated in a 4-bed single-absorption acid plant, which itself achieves around 98.5% sulphur capture. Tail-gas from the acid plant, containing around 1200 ppm SO_2, is discharged to the smelter main stack.

Electric furnace feed includes the partially roasted calcine from the roasters, silica flux, and petroleum coke, which serves as a reductant. A significant portion of the iron oxidized in roasting is reduced to metal by coke, and collects with the sulphides to form a sulphur-deficient matte. Silica combines with remaining iron oxides to form a fayalite slag. This reductive smelting process ensures high pay metal recoveries. A mixture of carbon monoxide and carbon dioxide is generated by the reduction reactions, and this gas percolates up through the ~0.5 m of unsmelted calcine feed maintained as a covering layer on the surface of the molten bath. Induced draft (ID) fans keep the furnace freeboard under suction to ensure the gases are collected and air is drawn in to combust carbon monoxide. The furnace off-gas stream is treated for particulate removal before joining the smelter main stack. The nature and control of this furnace gas stream is the focus of this paper.

Four batch-converting vessels in the Converter Aisle reduce the iron content of furnace matte from about 30% down to the 2.5% specification of the refinery. Off-gases from all four vessels join the furnace gas just prior to an electrostatic-precipitation cleaning stage ahead of the smelter main stack. The major portion of the Converter Aisle processing does not generate SO_2 due to the sulphur deficiency in the furnace matte. Only in the final stage of blowing does sulphur oxidation occur. Careful control of the fluid-bed roasting stage minimizes the requirement for sulphur removal in the Aisle.

The contributions of the various gas streams to the total amount of SO_2 at the main stack are summarized in Fig. 1. A progression is shown as the smelter implemented first a higher degree of sulphur roasting and second the CFA process combined with

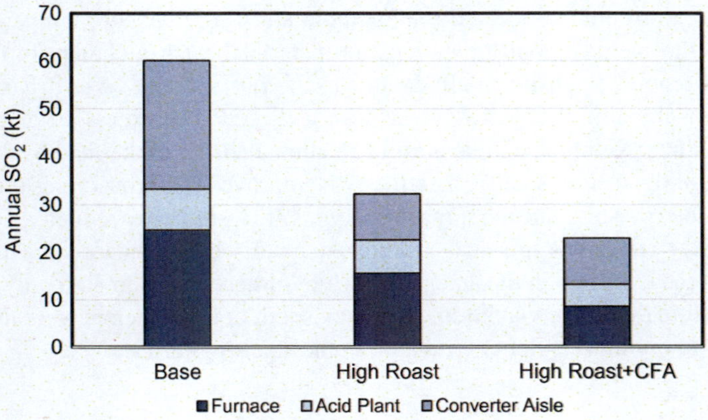

Fig. 1 Changing contributions of gas streams as emissions decrease. (Color figure online)

the higher roasting. The large proportion of sulphur emissions that originate in the furnace, after High Roast reduced Aisle emissions, motivated the development of a new system for the furnace gas stream.

CFA Process Development

A process change to reduce electric furnace SO_2 emissions was first demonstrated with laboratory scale work in 2006. Thermogravimetric analysis (TGA) and crucible tests were completed using industrial roaster calcine samples. It was concluded that a large portion of the emissions arose due to continued roasting of calcine particles as they entered the furnace and contacted ambient air drawn in by the off-gas ID fans. Test work showed that a reduction of the oxygen content in the furnace gas space (the freeboard) reduced this unwanted roasting and could dramatically reduce SO_2 emissions.

This concept was termed "Controlled Furnace Atmosphere" (CFA) as the process change aimed to move from uncontrolled ingress of ambient air to operation with a better-sealed system in which oxygen content was manipulated to minimize SO_2 emissions. This was a departure from conventional operation at the smelter. For many years, the philosophy for freeboard control had been to draw in enough air to ensure full combustion of carbon monoxide, and then additional air to act as coolant, maintaining conditions which did not melt away the feed layer floating on the slag bath. The feed banks acted as an insulating layer, and provided a reaction site for some of the reduction reactions occurring in the furnace.

The requirement to maintain a calcine charge layer in the furnace limited the extent of change that could be adopted with CFA. Although a dramatic change to a completely sealed furnace with a strongly reducing atmosphere was feasible, and has plenty of industrial precedents (for example ilmenite smelting, or ferro-alloy production), this could not be achieved without allowing freeboard temperatures to rise above the melting point of the charge materials. Such a change in temperature would require extensive modifications to the furnace roof and off-gas system. Accordingly, the aim of CFA was to achieve a *reduction* in excess air and oxygen content, but not to eliminate ingress air or to allow melting of the feed banks.

Control of the overall smelter metallurgy is improved if all roasting is completed in the fluid-bed roasters, and the furnace reactions are restricted to reductive smelting. The site SO_2 emissions are also minimized in this case, because the fluid-bed roaster off-gas is treated in the acid plant, while furnace gas is treated only for dust removal before discharge via the smelter stack. Both of these factors were important motivators to minimize sulphur oxidation in the smelting furnace.

Pilot Scale Testing

The laboratory results were further evaluated by two larger-scale pilot campaigns at the electric furnace facilities of Mintek in Randburg, South Africa. Each campaign included smelting 30 tonnes of roasted calcine (normal furnace feed) collected from the Sudbury Smelter, along with suitable additions of silica flux and petroleum coke. The test furnace at Mintek had a 1 m inside refractory diameter. It was fed at about 200 kg/h, and molten slag and matte were tapped into ladles at operating temperatures similar to commercial operation in Sudbury. The pilot furnace freeboard was well sealed, and the composition of the gas atmosphere in the furnace was controlled by injection of metered amounts of air and/or nitrogen. Freeboard gas was continuously sampled to determine the relationship between oxygen content and the occurrence of roasting indicated by SO_2 content.

The first pilot campaign was completed over 8 days in December 2008. The furnace was initially set up with significant amounts of ingress air deliberately injected into the freeboard to generate conditions similar to the commercial operation at that time. Freeboard oxygen content was maintained at about 15%, establishing a baseline condition from which improvement could be measured. Results from the campaign included the following:

- Baseline operating conditions generated freeboard gas composition very similar to then current commercial operation.
- SO_2 generation was reduced from baseline by 20–40% as oxygen content of the freeboard gas decreased from 15% to 5%.
- SO_2 emissions were reduced as much as 57% by the combination of reducing oxygen to 5% and feeding more deeply roasted calcine.

A second campaign was completed in September 21–29, 2009. The pilot furnace feed system was upgraded between campaigns with the installation of a fluoseal in the feed line to prevent air entrainment with the feed. This entrainment had limited the ability to exclude oxygen from the freeboard in the first campaign. Oxygen control

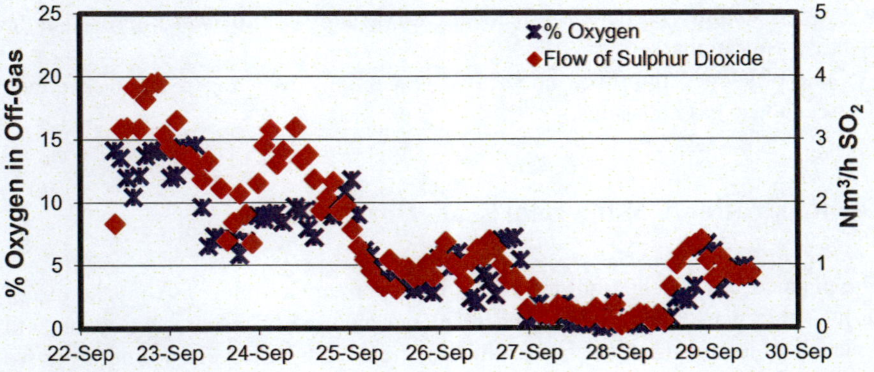

Fig. 2 SO_2 flow from pilot furnace as oxygen content varies over time. (Color figure online)

was improved with this change, and Fig. 2 shows operating levels during the second campaign.

An initial baseline condition was run once again at 15% oxygen to reproduce commercial operation with an uncontrolled atmosphere. Successive operating levels at 10, 5 and 1% oxygen are shown in the figure, along with a return to the 5% level at the end of the campaign. Five percent was the level that was expected to be feasible to implement commercially on the existing electric furnace.

It is important to note that generation of SO_2 decreased as the furnace atmosphere was changed (by excluding air ingress and injecting nitrogen), with reductions above 95% achieved at the most extreme conditions. Reductions at the commercial target of 5% oxygen were on the order of 60%. This validated the laboratory concept of SO_2 abatement through control of smelting furnace atmosphere.

It is also notable that the *variability* in the furnace operation decreased as oxygen levels were reduced. This was reflected in the smoother rate and smaller fluctuations seen in SO_2 generation as the campaign progressed. It was also demonstrated by a measured 15% increase in coke efficiency. The unwanted combustion of this reductant was reduced as the oxygen content of the furnace freeboard decreased. This had the beneficial impact of generating less excess heat from combustion, so the ventilation requirements of the furnace were reduced and stabilized. Control of the freeboard operating temperature was a significant challenge on the existing commercial furnace in Sudbury, ultimately limiting the capacity of the furnace. Any improvement in control due to implementation of the CFA concept would be an important side benefit of the project. Reagent costs for petroleum coke for the commercial furnace could be reduced by this effect, and perhaps more significantly, the greenhouse gas generated by the smelting process would also be reduced.

Results of pilot scale work can be summarized as:

- SO_2 emissions from the electric smelting process were reduced by over 50% for conditions transferable to the commercial operation at the Sudbury smelter.

- Process stability, particularly freeboard temperature control, was improved at the target conditions.
- Coke efficiency improved about 15%, potentially reducing reagent costs and greenhouse gas emissions.

Demonstration/Commercial Scale Development

The planned implementation of CFA on the commercial scale was to make use of an available waste gas stream (acid plant tail gas), with low oxygen content, to replace ambient air used to cool the furnace freeboard. However, in order to prove the concept, it was not feasible to use this gas stream on a temporary basis. In place of this, a supply of bulk nitrogen was installed at the smelter site, using a temporary evaporative system supplied and owned by a commercial gas supply company. Due to logistical constraints with the liquid nitrogen transport to site, this system still could not supply enough nitrogen to suppress furnace oxygen levels to the targeted commercial levels. Instead, the demonstration was planned to operate with furnace oxygen levels reduced from a typical 17% down to 10%. Table 1 gives a comparison of operating conditions for the various plant scale trials completed in 2010.

An intermediate scale between the 1 m diameter pilot furnace and the commercial furnace was considered but ultimately dismissed. This would have required collection of several hundred tonnes of roaster calcine from the production plant, and availability of an appropriately sized furnace was not certain. Construction of a new demonstration plant of appropriate scale would likely have been required, at an expected cost exceeding $10 million. Consequently, development proceeded directly to commercial scale, despite the compromise associated with insufficient nitrogen supply to reach the final target conditions, and despite the risk to commercial production due to the trial. The combination of extensive pilot results at a reasonably realistic

Table 1 Comparison of 2010 Commercial Scale Operating Conditions

	Initial operation	CFA concept	Plant baseline May	Plant demo July	Plant demo Oct.
Calcine feed (tph)	70	70	30	31	60
Ingress Air (Nm^3/hr)	120,000	15,000	33,000	19,000	42,000
Nitrogen/Tail Gas (Nm^3/hr)	–	45,000	–	5,000	7,500
Offgas CO (%)	0.5	0.7	0.3	0.7	0.6
Offgas O_2 (%)	17	5	15	10	11
Offgas Flow rate (Nm^3/hr)	122,000	58,000	45,000	36,000	53,000
Reduction in SO_2 Emissions	Base	55%	Base	37–48%	33%

scale, along with partial implementation at full scale was determined to adequately prove the concept and manage the scale-up risk.

Ingress air was drawn into the furnace, in the baseline operation, through various openings in the freeboard/roof area. For demonstration work, it was necessary to better seal these openings to reduce ingress air, which was not required for cooling in the CFA operating condition. Sealing included the installation of electrode seals to reduce the openings in the roof that the electrodes pass through, and stuffing of similar gaps around the feed pipes and the roof/sidewall interface with refractory blanket.

The full-scale trials were further validated by conducting a baseline trial in May 2010 at full scale. Demonstration conditions were duplicated, except that nitrogen was not introduced—air was drawn into the furnace for cooling as during normal operation. A CFA trial with nitrogen injection was completed in July. Reductions in SO_2 emissions measured during the plant scale work compared favourably with the pilot scale work. For a given oxygen concentration, emission reductions were slightly better at full scale than measurements on the pilot scale.

A second full-scale trial was completed in October 2010. The flow rate of nitrogen was maximized by operating the vapourizers at full capacity, including portable fans to help reduce ice build-up on the system. This flow exceeded the logistical trucking capability, so inventory in the N_2 system was drawn down during the trial, and a maximum test period of 2 days was available. Two conditions were considered in the planning—operation at 30 tph with one roaster or at 60 tph with two roasters. Both results were helpful in confirming performance. One-roaster operation achieved lower oxygen levels than the July test due to increased nitrogen flow rate. Two-roaster operation did not bring oxygen levels as low as the 10% achieved in July, but was more representative of full production levels. Both operating rates were tested as the production demands for the smelter allowed at the time of the trial.

A comparison of reduction achieved versus oxygen in freeboard gas is shown in Fig. 3. The figure shows results at various levels of oxygen during piloting at Mintek, along with a trend line. Ranges, rather than single points, were measured at each oxygen content, reflecting operation at varying freeboard temperature. A target commercial condition was chosen based on this relationship, and on the expected level of 5% oxygen considered feasible for the existing production furnace. Plant scale results are also included in Fig. 3. Measured results at baseline (15%) and test conditions near 10% oxygen confirmed that the pilot plant relationship was followed on the plant scale.

Construction and Implementation

A gas supply system was retrofitted to the electric furnace during a refractory rebuild in 2015. Sixteen inlet slots were installed at the top of the sidewalls just below the roof arch. Tail gas is blown through a 1 m diameter duct from the acid plant by pressure from the existing blower. The flow is throttled by a control valve, which

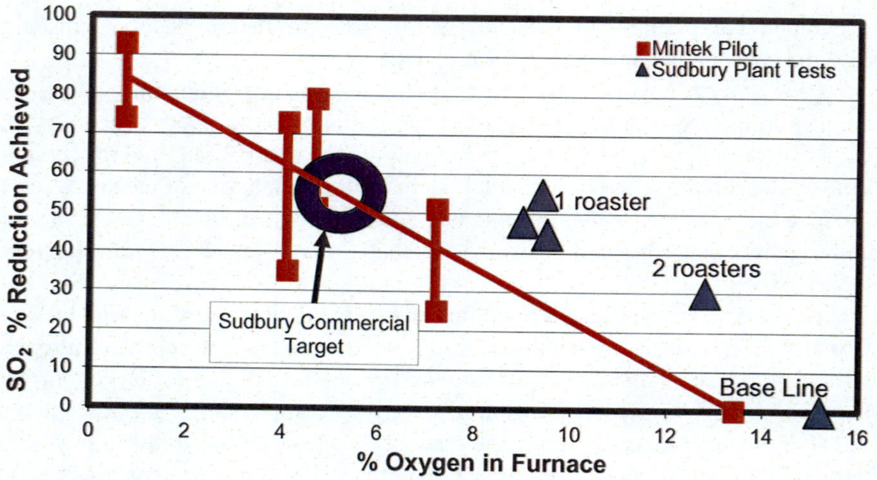

Fig. 3 Reduction in SO_2 generation as oxygen decreases from baseline freeboard concentration. (Color figure online)

responds to freeboard temperature in the furnace, to modulate the cooling effect. The tail gas supply is split amongst the 16 inlet slots, with some ability to bias flow to the hotter areas of the freeboard.

At the acid plant, the tail gas is normally used in a heat exchanger to cool inlet gas to the absorber. This duty does not require the full flow of tail gas, however, and gas is by-passed to the main stack. It is this by-pass stream that supplies the CFA system for the furnace.

Furnace off-gas is monitored for carbon monoxide to ensure adequate oxygen is provided for freeboard combustion. Pilot burners, and a supplementary combustion air supply, ensure that any excess CO burns in a controlled manner in the ductwork upstream of the ID fans. CO is also monitored downstream of the pilot burners to confirm combustion, with an interlock which brings back ambient air to the freeboard if CO exceeds 25% of the lower explosive limit (that is, 3.1% CO).

The equipment installations for the CFA process change were completed during an 8-week planned smelter outage, which included a furnace refractory rebuild. Installation costs for the project comprised about 15% of a larger CDN$130 Million capital project to accommodate High Roast. The major capital items for CFA included:

- Ducting and valves between the acid plant and the furnace freeboard
- Natural gas fired pilot burners (2)
- Air supply fans (2) and ducting for CO combustion at the pilot burners
- Gas analysis equipment for each (2) furnace off-gas duct
- Suction control dampers upstream of the existing furnace ID fans
- Electrode seals and sealing of other furnace roof openings.

Operating costs are minimal, and consist primarily of increased energy consumption by existing and new fan motors, and maintenance costs for the post-combustion and gas analysis equipment.

Industrial Results and Current Operation

A shortcoming of the industrial design was found during the commissioning period. A new furnace pressure control challenge was experienced periodically once the tightly sealed freeboard design was implemented. It was discovered that occasional collapses of feed inventory (calcine banks) result in gas surges. These gas surges were inconsequential prior to the CFA installation, but could cause pressure pulses and force hot gas out of the furnace through remaining openings once the furnace was more tightly sealed. Although the gas surges were not a result of CFA, there was a new requirement to control this phenomenon in order to implement the technology.

An initial response to the problem was to remove the electrode seals, preventing pressure build-up by the gas surges. Improved pressure response was achieved with automation of fans and throttling dampers in the off-gas system. An improved calcine inventory management system was implemented. This combination of improvements eliminated the incidents of hot gas excursions. Unfortunately, the resulting increase in air ingress to the process also reduced the effectiveness of CFA—more sulphur oxidation was possible.

The calcine inventory control, and off-gas automation, along with other process improvements, have given a much-improved ability to control furnace pressure reliably. With this success, the electrode seals are being gradually re-installed. This will permit reductions in freeboard oxygen concentration, and reduced SO_2 emissions. To date, furnace emission reductions up to 30% have been achieved on a monthly basis. This is in line with the development results illustrated in Fig. 3. It is expected that this will be significantly improved once a more fully sealed furnace is re-established, and air ingress is reduced.

The full potential of the new system has not yet been realized, but in combination with increased extent of roasting, it has allowed the smelter to meet lower sulphur dioxide emission limits. Importantly, this new system produces no additional waste, has low operating costs, and provides reduced greenhouse gas emissions.

Carbothermic Reduction Roasting of a Low-Grade Nickel Laterite Ore in the Modified Caron Process

Sadia Ilyas, Hyunjung Kim, and Rajiv Ranjan Srivastava

Abstract Carbothermic reduction-roasting of a low-grade Ni-lateritic was investigated under the modified Caron process. Roasting parameters of reductant (coal), additive (Na_2SO_4), temperature, and time were varied to optimize an efficient reduction of laterite to metallic nickel and cobalt. The degree of phase transformation characterized using XRD revealed a positive effect of additive that improved reduction with respect to increasing temperatures. The laterite mixture (with 10% coal + 9% Na_2SO_4) roasting at 800 °C for a duration of 120 min showed >90% nickel and 68% cobalt in an ammoniacal solution of NH_4OH–$(NH_4)_2CO_3$. The calculated reduction rate indicates the occurrence of a 3-steps reaction, comprising surface-diffusion followed by intermediate-diffusion, and finally the product-diffusion. Subsequently, leaching kinetics of roast-reduced laterite fit to the shrinking core model by following a mixed- and diffusion-controlled mechanism for nickel and cobalt, respectively.

Keywords Extraction and processing · Ni-laterite · Caron process · Carbothermic reduction · Ammoniacal leaching

S. Ilyas (✉) · H. Kim (✉)
Mineral Resources and Energy Engineering, Jeonbuk National University, Jeonju, Jeonbuk 54896, Republic of Korea
e-mail: sadiailyas1@yahoo.com; ilyas.sadia24@gmail.com

H. Kim
e-mail: kshjkim@jbnu.ac.kr

H. Kim
Department of Environment and Energy, Jeonbuk National University, Jeonju, Jeonbuk 54896, Republic of Korea

R. R. Srivastava (✉)
Center for Advanced Chemistry, Institute of Research and Development, Duy Tan University, Da Nang 550000, Vietnam
e-mail: r2.srivastava@gmail.com

Faculty of Natural Sciences, Duy Tan University, Da Nang 550000, Vietnam

C. Anderson et al. (eds.), *Ni-Co 2021: The 5th International Symposium on Nickel and Cobalt*, The Minerals, Metals & Materials Series,
https://doi.org/10.1007/978-3-030-65647-8_27

Introduction

Nickel and cobalt are among the largest used industrial metals. They are widely used as alloying metals to provide the durability and base-strength to alloys like stainless steel (SS) and superalloys (SA), while recent applications in rechargeable batteries (RCBs) is high [1–3]. In the last two decades, more than 90% of primary nickel production is used to fulfill the soaring demands of SS, SA, and RCBs and passed a total of 2.1 million tons. The arduous supply of nickel cannot be accomplished by exploiting the sulfide ores alone [4, 5]. Henceforth, the exploitation of laterite ores that have locked 70% of nickel reserves worldwide is a necessity [6, 7]. The laterites commonly exist as limonite and saprolite which usually differ in thickness and quality of two layers between the orebodies. Iron-based oxides whose surface layer is referred as limonite contains up to 60% goethite (FeO·OH) along with 0.5–1.7% nickel therein with a little of cobalt value [8]. However, nickel can be up to 3.0% with olivine (Mg_2SiO_4) and serpentine ($Mg_3Si_2O_5(OH)_4$) in the saprolitic deposits [9]. A major problem that exists with laterite ore processing is that no significant beneficiation can be achieved [10]. Depending upon the layers and types of lateritic ores, different approaches like high-pressure acid leaching (HPAL) dealing to limonitic laterites and rotary kiln electric furnace (RKEF) dealing with saprolitic orebodies are applied [11, 12] albeit requires high capital and operational costs to get satisfactory leaching of nickel and cobalt [13, 14].

In order to overcome these critical economic points, there has been intensive work aiming at the development of alternative processing routes. Caron process comprises the integrated pyro-hydro process is one of the potential alternatives applied to exploit the limonitic ores [15–17]. In which, the laterite ore is roast-reduced at high temperature in a multi-hearth roaster to convert nickel and iron oxides into ferronickel [18]. Several types of reductants including fuel oil, CO gas, coal, and coke have been applied followed by the ammoniacal leaching of roast-reduced products that selectively leach nickel and cobalt and leaves iron undissolved into the residue [19]. Reduction parameters like reductant dosage, temperature, and time have been studied to obtain the metalized nickeliferous compound [20–22]. With a fair possibility to control iron in ammoniacal leaching, the high temperature roasting in the presence of a suitable reductant and lower recovery of rare metals has been recognized to be non-economical if compared with HPAL or RKEF processing. Additionally, the formation of the ferrite phase at a higher temperature than the required one is identified as a major reason for low recovery in the Caron process. Hence, parametric optimization under the modified process for a maximum metallization of nickel with the minimum iron reduction is imperative along with a minimum loss of nickel and cobalt during leaching [23].

Although a number of studies on reduction-roasting dealing with different types of lateritic ores have been reported that include the addition of additives to enhance the extent of reduction-roasting process [24, 25], the study on Pakistan's laterite reserves is almost none [26]. Consequently, the reduction-roasting behavior of Bajaur's laterite reserves has been studied, wherein, employing locally available coal as the potential reductant. Parameters like temperature and time with/without additive were varied that was followed by ammoniacal leaching in NH_4OH–$(NH_4)_2CO_3$ solution. The extent of phase transformation during reduction-roasting was analyzed using X-ray diffraction (XRD; Bruker, D8 Advance diffractometer) technique. At the optimized condition of reduction-roasting, the reduced product was leached under the varied conditions of temperature and time to determine the leaching kinetics. The present study would be helpful to lay a foundation for the future exploitation of Bajaur's laterite.

Experimental

Laterite sample collected from Bajaur was received from the Pakistan Council of Scientific and Industrial Research (PCSIR, Lahore). Coal sample (fixed carbon, 58.6%; volatile matters, 27.1%; ash 8.6%; and S^0, 7.5%) used as reductant was locally purchased from a Faisalabad based coal-trader. Both samples were separately crushed, homogenized, ground, and sieved to a size below 100 μm. XRD analysis revealed the presence of goethite phase in the sample while the wet-chemical analysis showed the composition as 47.1% iron, 1.43% Ni, 0.16% Co, and 0.65% Mn. A predetermined amount of laterite and coal sample was ball-milled along with 2% of bentonite for 30 min. Additionally, an additive was mixed at the desired amount where it required to examine its influence on metals' reduction. Further, pellets of ~1 mm size were prepared by water spraying to the mixture in a disk-pelletizer. Thus prepared pellets were air-dried for the next 72 h, thereafter, a predetermined amount of dry pellets was taken for reduction-roasting experiments that were conducted in a pot furnace. Notably, roasting time was calculated after reaching the desired temperature of the furnace, while the roast-reduced sample was collected after overnight cooling in the closed condition and collected under an inert atmosphere. The quantitative efficiency of reduction was determined by leaching of roast-reduced pellets. Leaching was conducted in a 250 mL 3-neck flat bottom flask. 5 g reduced sample was charged in 100 mL ammoniacal solution (150 g/L total ammonia) under constant stirring of 400 rpm at room temperature (20 ± 2 °C) and airflow (5 L/h). To determine leaching kinetics, temperature and time were varied as described in the corresponding section. After completion of leaching, the slurry was filtered and the filtrate was analyzed to calculate the leaching efficiency of metals.

Results and Discussion

Effect of Reductant Dosage

For a complete reduction of nickel and cobalt, the reductant dosage was primarily varied in the range from 5 to 15% coal, while the temperature was fixed at 800 °C for 2 h. Figure 1 shows comparative XRD patterns of reduced products in comparison to unroasted laterite. Goethite (FeO.OH) as the main phase was identified in unroasted laterite (Fig. 1a), which was converted to Fe_2O_3 and Fe-Ni alloy phases when roasted with coal (Fig. 1b). A sharp peak of Fe-Ni alloy was observed with a 10% coal mixing dosage (Fig. 1c) which did not change above that dosage (Fig. 1d). Whereas, carbon peaks identified in the feed sample was fully disappeared in all XRD shows the suitability of roasting temperature (at 800 °C).

In order to quantify the reduction efficacy with different coal mixing with laterite, the reduced products were leached in ammoniacal solution (consist of 150 g/L total NH_3 with 165 g/L $(NH_4)_2CO_3$) at a pulp density of 10% for 2 h duration. It was observed that an increase in coal mixing dosage up to 10% improved leaching from 43.4 to 75.2% for nickel and 24.2 to 46.2% for cobalt. Thereafter, increasing the dosage of coal did not improve nickel leaching albeit cobalt leaching was declined.

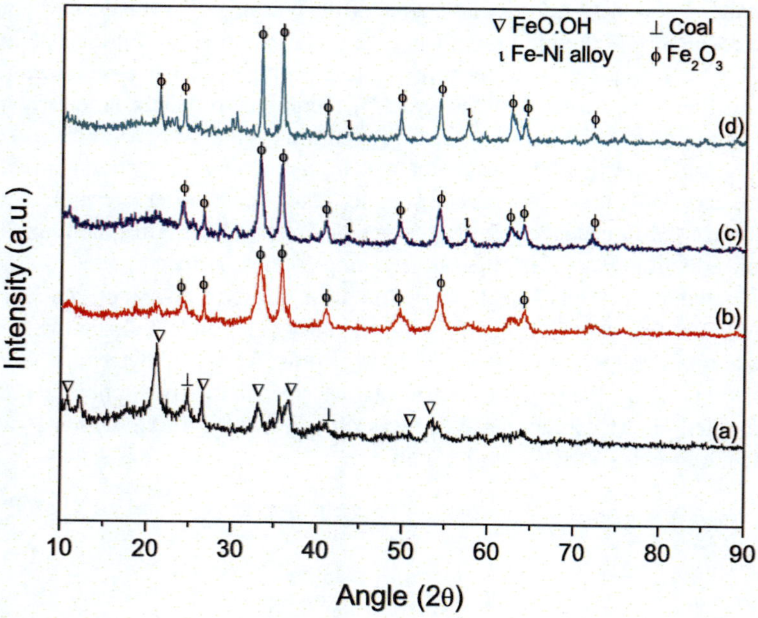

Fig. 1 XRD patterns of feed mixture consist of laterite (**a**), roast-reduced products obtained by mixing of 5% (**b**), 10% (**c**), and 15% (**d**) coal at roasting temperature 800 °C and 2 h duration. (Color figure online)

This phenomenon may be ascribed to the respective increase in iron dissolution possibly due to co-precipitation of cobalt with ferrihydrite [22, 27, 28]. Hence, 10% of coal mixing dosage was optimized.

Effect of Temperature

Temperature has a vital role in reduction and suitable is required to a controlled metallization of nickel and cobalt while keeping iron suppressed in its oxidized form. In this context, temperatures were varied between 600 and 900 °C by maintaining coal mixing dosage (10%) and time (2 h) constant. Figure 2 shows the XRD patterns of roasted products at different temperatures. Goethite phase was found to disappear (except a few at 600 °C) and mostly converted to Fe_2O_3. As the temperature increases from 600 to 900°C, the broader peaks of Fe_2O_3 turned to be sharper, which indicated the increasing crystallinity of the roasted product. However, the phase conversion was visible at 600 °C (Fig. 2a), a significant peak assigned to Fe-Ni alloy has appeared at 700 °C (Fig. 2b). Further, the peak intensity of Fe_2O_3 was enhanced at >800 °C (as shown in Fig. 2c and 2d) albeit peaks for Fe-Ni alloy were unchanged. It can be understood at a lower temperature (600 °C), the nickeliferous phase commences

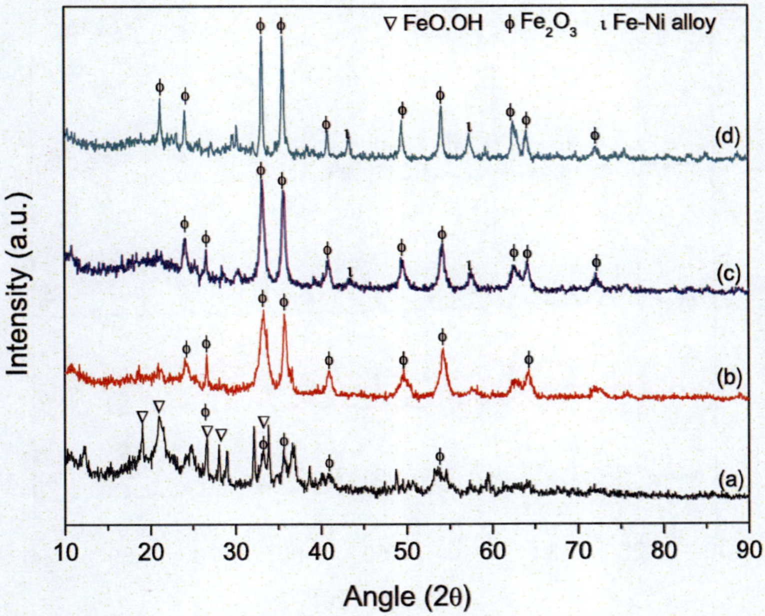

Fig. 2 XRD patterns of roast-reduced products obtained at different temperatures 600 °C (**a**), 700 °C (**b**), 800 °C (**c**), and 900 °C (**d**); while 10% coal mixing and 2 h duration were maintained constant. (Color figure online)

dehydroxylation and gradually decomposes to amorphous phases [29]. On the other hand, reduction at higher temperatures leads to the formation of Fe–Ni alloy [20].

The quantification of reduction efficacy was conducted via ammoniacal leaching of the roast-reduced products obtained at different temperatures. Leachability of metals was found to increase with respect to temperature increased from 600 to 900 °C. Nickel dissolution was relatively higher and increased from 42.8% to 78.6%, whereas, cobalt leaching improved from 26.4 to 48.7% within the temperature variation range. The low recovery of FeNi particles also attributes to be encapsulated by the recrystallization of the amorphous silicate at temperature. Due to a higher dissolution of iron (>20%) at 900 °C, a roasting temperature of 800 °C was optimized.

Effect of Na_2SO_4 Addition

The presence of additive may affect the grain growth of Fe–Ni alloy phase during the reduction-roasting of laterite [26]. Hence, the effect of Na_2SO_4 addition in the range of 3 to 12% with lateritic was studied at 800 °C. Figure 3 reveals that the goethite phase was entirely disappeared by mixing additive regardless of Na_2SO_4 dosage. As the additive mixing increased from 3 to 9%, the phase transformation

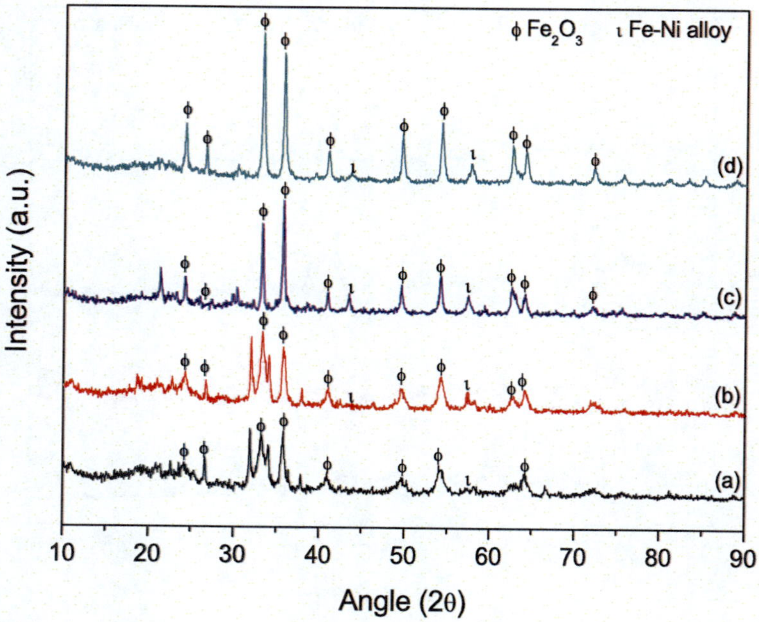

Fig. 3 XRD patterns of roast-reduced products obtained by additive mixing at a dosage of 3% (**a**), 6% (**b**), 9% (**c**), and 12% (**d**) Na_2SO_4, while 10% coal was mixed at 800 °C for 2 h duration. (Color figure online)

from goethite to Fe_2O_3 and Fe–Ni alloy was enhanced (Fig. 3a–c). Further increase of additive by 12% contrarily decreased peak-intensity of Fe–Ni alloy albeit intensity of Fe_2O_3 was increasing (Fig. 3d). It is corroborated to the mixing of close ionic radii of iron (and magnesium) with nickel that increases the occurrence of mix phase [30]. Thus, elevated temperature results in greater deactivation of Ni-oxide and decreases its reduction at higher additive dosage.

To quantify the degree of reduction, roast-reduced pellets were leached in ammoniacal solution at 10% pulp density and 2 h duration. Leaching results showed a major effect on cobalt leaching which could be improved from 48.5 to 67% while mixing $3 \geq 9\%$ Na_2SO_4, receptively (in comparison to 46% cobalt without additive). About 85% nickel could be leached with no significant change by additive mixing. It can be ascribed to the surface modification ability of Na_2SO_4, and the capability of Na^+ ion to liberate the hosted nickel and cobalt from the iron matrix [31, 32]. Furthermore, Na_2SO_4 melts at an elevated temperature which accelerated the reduction and Ostwald ripening processes thereby increasing the Ni and Co recovery (however, it is a matter of future study to observe such phenomenon). To keep a high reduction and leaching efficiency of nickel and cobalt, 9% Na_2SO_4 was optimized.

Effect of Time

The effect of time on reduction-roasting was examined between 15 to 80 min and the reduced products were characterized using XRD (as shown in Fig. 4). As can be seen, the goethite phase started to diminish after 30 min roasting with visible patterns of Fe_2O_3 phases from 60 min albeit with a lower intensity. A reduced phase of Fe–Ni alloy increased after 120 min and did not change further with prolonging the duration of roasting. This behavior can be corroborated to coal conversion into CO and CO_2 which participates in the reduction of laterite by reducing NiO → Ni. However, Fe-phase transformation involves several steps as $FeO{\cdot}OH \rightarrow Fe_2O_3 \rightarrow Fe_3O_4 \rightarrow FeO \rightarrow Fe$ with increasing time [33]. The reactions may commonly be written as shown in Eqs. (1–3) below. No significant change in XRD patterns after 120 min indicated the completion of the reduction process.

$$(Ni, Co)O + CO \rightarrow Ni, Co + CO_2 \tag{1}$$

$$6FeO \cdot OH + CO \rightarrow 2Fe_3O_4 + CO_2 + 3H_2O \tag{2}$$

$$Fe_3O_4 + Ni + 2CO \rightarrow Fe, Ni + 2FeO + 2CO_2 \tag{3}$$

To quantify the reduction efficiency, reduced products were leached in an ammoniacal solution. The results revealed that the metallization of nickel and cobalt was 38.4 and 21.7%, respectively that could reach to 84.3% nickel and 67.6% cobalt

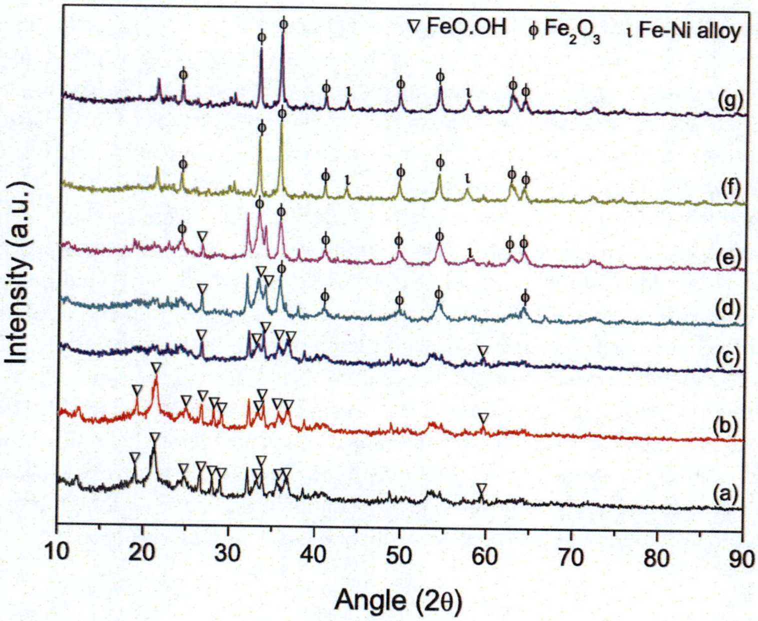

Fig. 4 XRD patterns of roast-reduced products with time 15 min (**a**), 30 min (**b**), 45 min (**c**), 60 min (**d**), 90 min (**e**), 120 min (**f**), and 180 min (**g**); while 10% coal and 9% Na_2SO_4 was mixed for roasting at 800 °C. (Color figure online)

after 120 min of roasting. No change in leaching yield showed that 120 min roasting was optimal. Furthermore, using the leaching data as a function of roasting time was compared to determine the apparent activation energy ($E_{a\text{-}reduction}$) as follows [34]:

$$\ln\frac{t_1}{t_2} = \frac{E_a}{R}\left(\frac{1}{T_1} - \frac{1}{T_2}\right) \quad (4)$$

Here, the roasting at temperatures, T_1 = 600 °C and T_2 = 800 °C for time, t_1 = 120 min and t_2 = 60 min have been considered (the condition on which the leaching yields were approximately found to be similar to each other) for calculating $E_{a\text{-}reduction}$; and R is the Universal gas constant (8.314 kJ/mole). Thus obtained value of $E_{a\text{-}reduction}$ 13.8 kJ/mole indicates that the reduction-roasting process follows a diffusion-controlled mechanism [26, 34]. It can be understood that carbon converted into reducing gases (CO and CO_2) subsequently gets diffused to the laterite surfaces to metalize nickel and cobalt.

Leaching Kinetics and Mechanism

Further, to understand the leaching phenomenon for a solid-liquid heterogeneous reaction occurring onto the surface of unreacted particles, leaching of the roast-reduced products was performed at various temperatures ranging from 20 to 50 °C with respect to time (varied up to 180 min). Pre-assuming that the reaction interface shrinks inward), the shrinking core models with chemical and diffusion control was evaluated as follows [26, 34]:

$$1 - (1 - \alpha)^{1/3} = k_c * t \tag{5}$$

$$1 - 3(1 - \alpha)^{2/3} + 2(1 - \alpha) = k_d * t \tag{6}$$

where α belongs to leaching fraction at any time, t (in min); while, k_c and k_d are the chemical and diffusion rate constant, respectively. Using the kinetic data, graphs plotted for $1–(1–\alpha)^{1/3}$ versus time showed a good fit with straight lines of the regression coefficient (R^2) values $\geq$0.98 for both nickel and cobalt (not shown for the sake of brevity). The obtained rate constant values at different temperatures (the slope values of the straight lines) were used to determine the apparent activation energy of leaching ($E_{a\text{-}leaching}$) through the Arrhenius equation as [34]:

$$k = Ae^{\frac{-E_a}{RT}} \tag{7}$$

where k is the rate constant, A is the pre-exponential factor, R is the universal gas constant (8.314 J/K/mole), and T is the absolute temperature (K). Thus plotted ln$k$ vs. 1/T (shown in Fig. 5) which have a high correlation coefficient ($R^2 > 0.99$). Using the slope values, the activation energy was calculated to be 21.2 and 7.6 kJ/mole for nickel and cobalt, respectively. $E_{a\text{-}leaching}$ value of nickel demonstrates that the process is determined by an intermediate-controlled reaction that shifted from chemical-controlled mechanism at low temperature to the diffusion-controlled mechanism at higher temperature [26, 34]. On the other side, $E_{a\text{-}leaching}$ value of cobalt falls within the range of diffusion-controlled mechanism.

Conclusions

The carbothermic reduction roasting of Bajaur laterite ore for the ammoniacal leaching under the modified Caron process was studied using Na_2SO_4 as an additive and locally available coal as the potential reductant. The parameters viz. coal mixing, temperature, Na_2SO_4 mixing, and time were optimized for the selective reduction of nickel and cobalt. The feed mixture with 10% coal resultant in a fair phase convergence, forming Fe_2O_3 and Fe–Ni alloy at 800 °C. The mixing

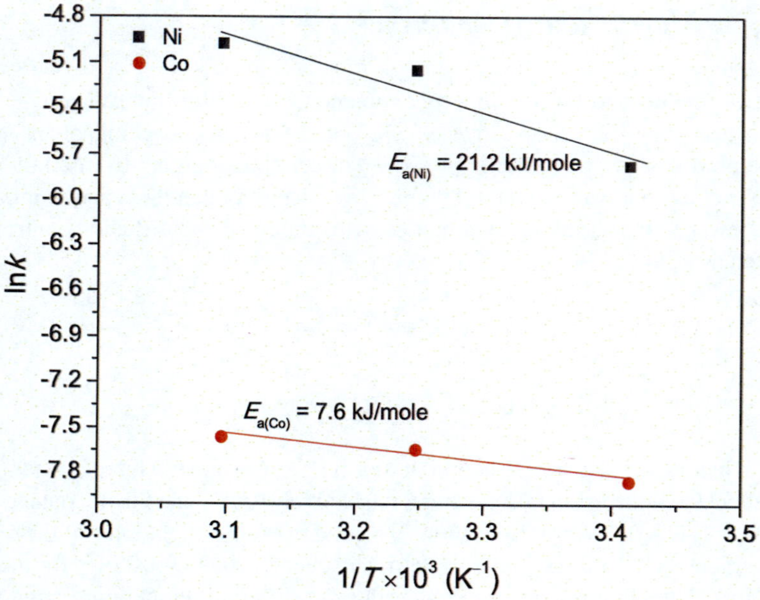

Fig. 5 Arrhenius plots of ln*k* versus 1/*T* for nickel and cobalt. (Color figure online)

of Na_2SO_4 could enhance the leaching yields of nickel and cobalt while leached in $NH_4OH–(NH_4)_2CO_3$ solution. Reduction kinetics with apparent activation energy value of 13.8 kJ/mole indicated to a diffusion-controlled roast-reduction process. Furthermore, the leaching kinetics of reduced laterite was fitted to shrinking core model revealed that nickel and cobalt leaching followed an intermediate-controlled (21.2 kJ/mole) and diffusion-controlled (7.6 kJ/mole) mechanism, respectively.

Acknowledgements This work was supported by the Brain Pool Program through the National Research Foundation of Korea (NRF) funded by the Ministry of Science and ICT (Grant No. 2019H1D3A2A02101993) and Basic Science Research Program through the National Research Foundation of Korea (NRF) funded by the Ministry of Education (Project no. 2020R1I1A1A01074249).

Conflict of Interest The authors declare that they have no conflict of interest.

References

1. Sattar R, Ilyas S, Bhatti HN, Ghaffar A (2019) Resource recovery of critically-rare metals by hydrometallurgical recycling of spent lithium ion batteries. Sep Purif Technol 209:725–733
2. Sattar R, Ilyas S, Kousar S, Khalid A, Sajid M, Bukhari SI (2020) Recycling of end-of-life $LiNi_xCo_yMn_zO_2$ batteries for rare metals recovery. Environ Eng Res 25:88–95
3. Agrawal A, Pathak P, Mishra D, Sahu KK (2012) Solvent mediated interactions for the selective recovery of cadmium from Ni-Cd battery waste. J Mol Liq 173:77–84

4. Ahmad B, Bhatti HN, Ilyas S (2011) Bio-extraction of metal ions from laterite ore by Penicillium chrysogenum. Afr J Biotechnol 10:11196–11205
5. Srivastava RR, Mishra AN, Singh VK (2013) Co-precipitation study of nickel and cobalt from ammoniacal leach liquor of laterite ore. J Metall Mater Sci 55:67–72
6. Farrokhpay S, Fornasiero D, Filippov L (2018) Upgrading nickel in laterite ores by flotation. Miner Eng 121:100–106
7. Xu D, Liu L, Quast K, Addai-Mensah J, Robinson DJ (2013) Effect of nickel laterite agglomerate properties on their leaching performance. Adv Powder Technol 24:750–756
8. Quast K, Connor JN, Skinner W, Robinson DJ, Addai-Mensah J (2015) Preconcentration strategies in the processing of nickel laterite ores Part 1: Literature review. Miner Eng 79:261–268
9. Quast K, Connor JN, Skinner W, Robinson DJ, Li J, Addai-Mensah J (2015) Preconcentration strategies in the processing of nickel laterite ores part 2: Laboratory experiments. Miner Eng 79:269–278
10. Pickles CA, Forster J, Elliott R (2014) Thermodynamic analysis of the carbothermic reduction roasting of a nickeliferous limonitic laterite ore. Miner Eng 65:33–40
11. Whittington BI, Muir D (2000) Pressure acid leaching of nickel laterites: a review. Miner Process Extract Metall Rev Int J 21:527–599
12. Kyle J (2010) Nickel laterite processing technologies—where to next? Paper presented at ALTA 2010 Nickel/Cobalt/Copper Conference, Perth, Western Australia, 24–27 May 2010
13. King MG (2005) Nickel laterite technology—finally a new dawn? JOM 57:35–39
14. Oxley A, Barcza N (2013) Hydro–pyro integration in the processing of nickel laterites. Miner Eng 54:2–13
15. Caron MH (1950) Separation of nickel and cobalt. JOM 188:91–103
16. Oxley A, Barcza N (2016) Why heap leach nickel laterites? Miner Eng 88:53–60
17. Mishra D, Srivastava RR, Sahu KK, Singh TB, Jana RK (2011) Leaching of roast-reduced manganese nodules in NH_3-$(NH_4)_2CO_3$ medium. Hydrometallurgy 109:215–220
18. Rhamdhani MA et al (2009) Advances in research on nickel production through the caron process. Paper presented at the European Metallurgical Conference, Inssbruck, Austria, 28 June–1 July 2009
19. Li B, Ding Z, Wei Y, Wang H, Yang Y, Barati M (2018) Kinetics of reduction of low-grade nickel laterite using carbon monoxide. Metall Trans B 49:3067–3073
20. Utigard T, Bergman RA (1993) Gaseous reduction of laterite ores. Metall Trans B 24:271–275
21. Valix M, Cheung WH (2002) Effect of sulphur on the mineral phases of laterite ores at high temperature reduction. Miner Eng 15:523–530
22. O'Connor F, Cheung WH, Valix M (2006) Reduction roasting of limonite ores: effect of dihydroxylation. Int J Miner Process 80:88–99
23. Han KN, Healy TW, Fuerstenau DW (1973) The mechanism of adsorption of fatty acids and other surfactants at the oxide-water interface. J Colloid Interface Sci 44:407–414
24. Zhou S, Wei Y, Li B, Wang H, Ma B, Wang C (2016) Chloridization and reduction roasting of high-magnesium low-nickel oxide ore followed by magnetic separation to enrich ferronickel concentrate. Metall Trans B 47:145–153
25. Oliveira VdA, Santos CG, Brocchi EdA (2019) Assessing the influence of NaCl on the reduction of a siliceous laterite nickel ore under Caron process conditions. Metall Trans B 50:1309–1321
26. Ilyas S, Srivastava RR, Kim H, Ilyas N, Sattar R (2020) Extraction of nickel and cobalt from a laterite ore using the carbothermic reduction roasting-ammoniacal leaching process. Sep. Purif. Technol. 232:115971. https://doi.org/10.1016/j.seppur.2019.115971
27. Dyer L, Su B, Asselin E (2012) Cobalt loss due to iron precipitation in ammoniacal carbonate solutions. Hydrometallurgy 125–126:144–147
28. McCulloch J, Fittock J (2006) Control of cobalt valence in aerators. Internal Report #50207, Qld, QNI Pty Ltd, Townsville
29. Yang J, Zhang G, Ostrovski O, Jahanshahi S (2013) Changes in an Australian laterite ore in the process of heat treatment. Miner Eng 54:110–115

30. Flavio T, Silva D (1992) Thermodynamic aspects of the roasting processes in the pretreatment of nickelferrous garnierites. Miner Process Extract Metall Rev Int J 9:97–106
31. Rao M, Li G, Zhang X, Luo J, Peng Z, Jiang T (2016) Reductive roasting of nickel laterite ore with sodium sulfate for Fe-Ni production. Part I: Reduction/sulphidation characteristics. Sep Sci Technol 51:1408–1420
32. Rao M, Li G, Zhang X, Luo J, Peng Z, Jiang T (2016) eductive roasting of nickel laterite ore with sodium sulfate for Fe-Ni production. Part II: Phase transformation and grain growth. Sep Sci Technol 51:1727–1735
33. Astuti W, Novita S, Nurjaman F (2017) Reduction mechanism of Indonesian limonite ore by solid reducing agents. J Phys Conf Ser 817:012063. https://doi.org/10.1088/1742-6596/817/1/012063
34. Levenspiel O (1999) Chemical reaction engineering. John Wiley & Sons Inc, New York

Influence of the Cemented Carbides Composition on the Disintegration in Liquid Zinc

Tamara Ebner, Stefan Luidold, Christoph Czettl, and Christian Storf

Abstract The recycling of cemented carbides at the end of their product life cycle is of particular economic, ecological and strategic importance. These composite materials, mainly consisting of tungsten carbide embedded in a Co binder, can be recycled by the so-called zinc process. This research investigates the disintegration of cemented carbides in an excess of liquid zinc in dependence of the WC grain size as well as the Co content of the material, temperature and duration. The data indicate highest disintegration rates at low grain size, average Co content and temperatures as well as a high duration regarding the experimental limits. Therefore, this study provides insight concerning the recyclability of certain cemented carbide grades and discusses the ongoing phenomena.

Keywords Cemented carbide · Recycling · Zinc process · Tungsten · Disintegration

Introduction

Due to their application in machining of metallic and non-metallic materials, chipless forming and wear parts cemented carbides (CC) are an indispensable tool in the manufacturing industry. The required properties for these purposes result from their composition based on the combination of a hard phase embedded in a metallic binder. Generally, tungsten carbide (WC) serves as hard, wear-resistant and high melting phase, whereas, in most cases Co is applied as ductile binder material. Hence, the

T. Ebner (✉) · S. Luidold
Christian Doppler Laboratory for Extractive Metallurgy of Technological Metals, Chair of Nonferrous Metallurgy, Montanuniversitaet Leoben, Franz Josef Str. 18, 8700 Leoben, Austria
e-mail: tamara.ebner@unileoben.ac.at

C. Czettl · C. Storf
CERATIZIT Austria GmbH, Metallwerk-Plansee-Straße 71, AT-6600 Reutte, Austria

C. Anderson et al. (eds.), *Ni-Co 2021: The 5th International Symposium on Nickel and Cobalt*, The Minerals, Metals & Materials Series,
https://doi.org/10.1007/978-3-030-65647-8_28

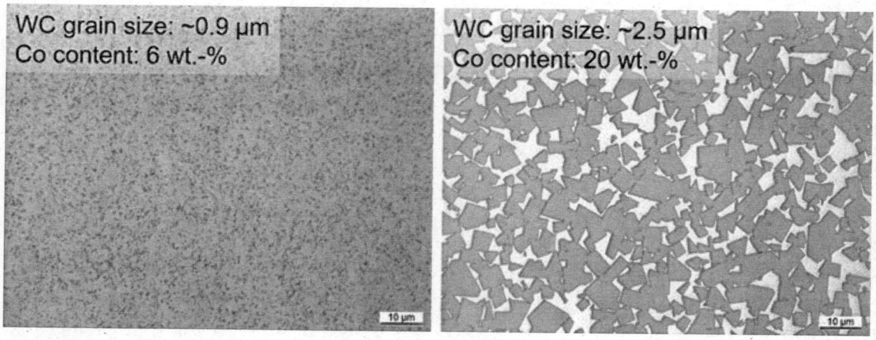

Fig. 1 Microstructure of straight WC-Co grades with different Co contents (Co: white area) and WC grain sizes (WC: gray area) [9, 10]

microstructure of these composite materials depends on the WC grain size and the Co content. Commercially available grades involve grain sizes between submicron to tens of micrometers and 4–16 wt% Co binder. Figure 1 displays cross sections of different cemented carbide grades [1–8].

The two main components of cemented carbides, tungsten and cobalt, are rated as critical raw materials for the European Union, concerning their commercial relevance and possible supply risk. Therefore, the recycling of these commodities at the end of their product lifecycle not only offers a wide range of ecological, economic and strategic advantages but also contributes to the resource efficiency and counteracts future supply shortfalls. Gille and Meier [11] reported cemented carbide recycling rates of 50–70%, for cutting tool inserts even shares of 90% are achieved. For the recovery of these type of scraps direct, semi-direct and indirect methods are available. Direct recycling techniques cover physical-mechanical treatments, which transfer the compact cemented carbides into a powder of the same chemical composition and carbide grain size as the as-supplied material. Typical examples are the zinc process, the coldstream process, the bloating/crushing method or their combinations. In this study, we focused on disintegration of cemented carbides in the zinc process. Thus, a detailed description of this method will be provided [11–16].

In 2010 zinc reclaim powder corresponded to 10% of the raw materials applied in the European cemented carbide industry, which represents a significant rise from less than 5% before 2005, based on the increase in the tungsten price in 2005. Due to the low energy consumption, high tungsten recovery, eco-friendliness and accordingly high amount of treated scrap (12,000 t/a in 2010) the zinc process offers a number of advantages and therefore, constitutes the most important direct recycling method. The feed material for the zinc process includes inserts, drills, compacts from mining, oil and gas industry, wear parts and hot rolls [15, 17, 18].

The first step of the zinc process involves a treatment of the scrap with molten or vaporous zinc at temperatures of 800–1,050 °C in graphite crucibles under Ar or N_2 atmosphere. The mass ratio of cemented carbide to zinc typically accounts for

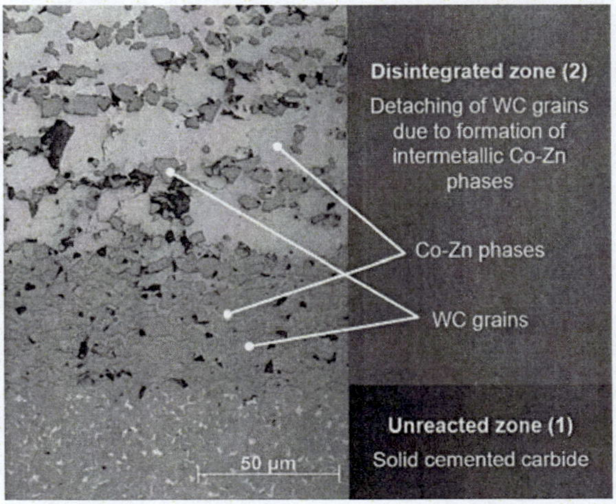

Fig. 2 Chipping of WC grains from the unreacted cemented carbide (**1**) surface due to the formation of intermetallic phases between the liquid zinc and the binder resulting in a disintegrated zone (**2**) consisting of WC grains surrounded by Co-Zn phases

1:1.3. In this stage Zn infiltrates the scrap and forms various intermetallic Co-Zn phases with the Co binder according to the Co-Zn binary phase diagram. These Co-Zn phases exhibit a higher volume than the Co binder. Due to the volume expansion a chipping of WC grains and lamellas occurs as given in Fig. 2, resulting in a swelling and disintegration of the initially compact and solid material. The reaction between the binder and the zinc is a diffusion controlled process, which starts at the cemented carbide surface and proceeds inwards. The zinc typically penetrates the cemented carbide with approximately 1 mm/h, whereas higher rates are stated by the treatment with vaporous zinc [11, 15–17, 19–21].

The second stage of the zinc process involves the removal of zinc from the disintegrated compound. This step involves a distillation at elevated temperatures and/or application of vacuum. After condensing the vaporized zinc, the metal can be reused for a new zincing cycle. The product of the zinc process comprises a porous, spongy and disintegrated cake with zinc contents <10 ppm, which is milled and homogenized to obtain a high quality, so-called zinc reclaim powder. The composition of the produced powder resembles the composition of the treated scrap and serves as feed material in the production of ready-to-press powder for the cemented carbide production. Figure 3 illustrates the zinc process, its feed materials, intermediates and final products with according microsections [11, 15–17].

The exact mechanisms and influence of the microstructure on the disintegration of cemented carbides in the first step of the zinc process is not fully understood yet. As aforementioned, a wide range of cemented carbide grades with varying WC grain size and Co content are available. Additionally, these composite materials can also contain further hard phases such as TaC and TiC, grain growth inhibitors, different binder alloys etc., which affect the microstructure and improve the properties of the product. To generate fundamental knowledge concerning the disintegration of cemented carbides in the zinc process this study takes a step back and focuses on straight WC-Co grades without any further additives and at lower temperatures

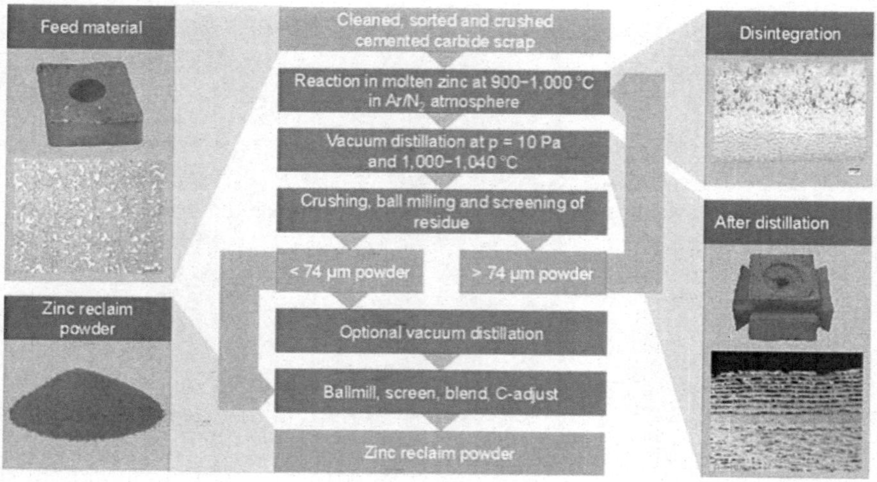

Fig. 3 Flowchart of the zinc process according to Lassner and Schubert [16] and Angerer et al. [15] with intermediates [22] and microsections [9, 23]

of 600–750 °C. The research activities involve the examination of the factors WC grain size (1–5 μm), Co binder content (9–15 wt%), temperature (600–750 °C) and duration (0.5–4 h) influencing the disintegration of cemented carbides with an excess of liquid zinc. Hence, this study provides basic information regarding the disintegration of cemented carbides in the first stage of the zinc process and allows to draw conclusions concerning the recyclability of certain grades.

Experimental Investigations Regarding the Disintegration

To examine the effect of the cemented carbide microstructure on the disintegration behavior in the first stage of the zinc process nine different, straight WC-Co grades with a special geometry and composition, given in Fig. 4, were provided by CERATIZIT Austria GmbH. This allows for the evaluation of the influence of the WC grain size from 1 to 5 μm and Co binder content from 9 to 15 wt% on the decomposition of the compound materials.

The application of the statistical software Modde 12 Pro® assists the creation of a design of experiments by a CCF model (Central Composite Face centered) according to the response surface methodology (RSM) and the generation of reliable data from the experiments. The RSM enables an extensive modelling and optimization by applying quadratic models. This results in a target-oriented performance of the tests as well as in a useful evaluation of the parameters and their mutual interactions. By including the temperature range of interest (600, 675 and 750 °C), the WC grain size (F, M and C) and Co content variation (9, 12 and 15 wt%) variations a Design of Experiments (DoE) was developed according to the CCF model, depicted

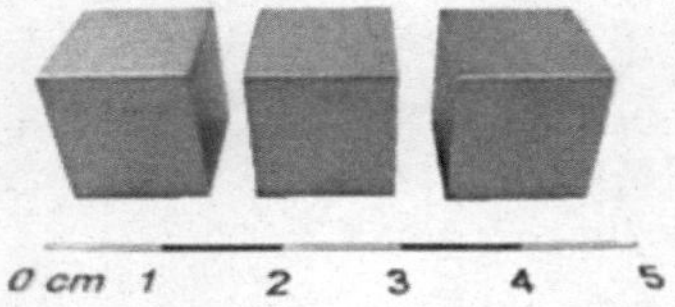

WC-Co grade No.	WC grain size, [µm] and type	Co content, wt.%
1	1.25, F	9.0
2	1.32, F	12.0
3	1.31, F	15.0
4	2.14, M	9.0
5	2.34, M	12.0
6	2.58, M	15.0
7	5.29, C	9.0
8	4.54, C	12.0
9	5.37, C	15.0

Fig. 4 Sample geometry (left) and composition (right) of nine cemented carbide grades, the average WC grain size was determined by electron backscatter diffraction (EBSD) performed with a FEI Quanta FEG 650 with an Oxford Instruments Nordlys Nano EBSD detector

in Fig. 5. These 17 tests involve three repeated center experiments which enable the determination of the reproducibility and further statistical quality factors. Each set of tests was conducted at four different annealing durations (0.5, 1, 2 and 4 h), resulting in a total number of 68 tests.

The experiments were performed according to the developed DoE. Therefore, the WC-Co cubes were placed in a quartz crucible and 99.9999% pure zinc, heated to the target temperature, was casted on each cube. The average Zn:Co ratio accounted for 31.6. This high excess of zinc enables an observation of the disintegration far off the chemical equilibrium, since an excess of zinc is available for the reaction with cobalt. Subsequently, the crucibles remained for certain durations (0.5, 1, 2 and 4 h) in a resistance furnace at the specified temperature (600, 675 or 750 °C) under nitrogen flushing to prevent oxidation. Throughout this treatment a partial disintegration of the cemented carbide occurs, due to the ongoing transactions between the liquid zinc and solid Co binder. The formation of intermetallic Co-Zn phases results in a chipping of WC grains and lamellas. After the annealing a cooling at ambient temperature took place and a metallographic preparation was conducted to enable the determination of the disintegration resp. yield and structure of the material after

Fig. 5 Design of experiments according to the CCF model at a certain duration resulting in 17 tests performed at 0.5, 1, 2 and 4 h (total 68 tests)

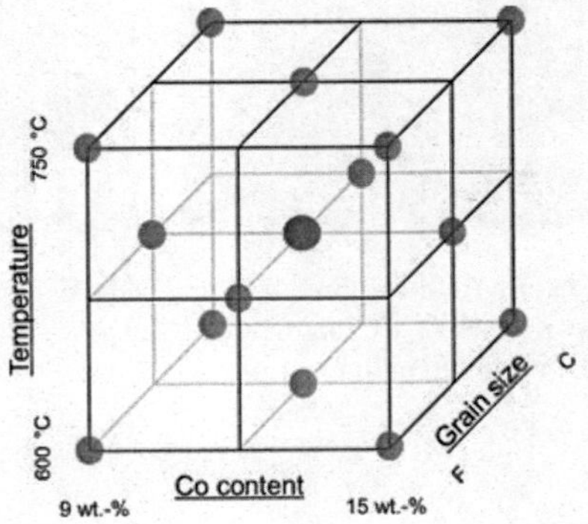

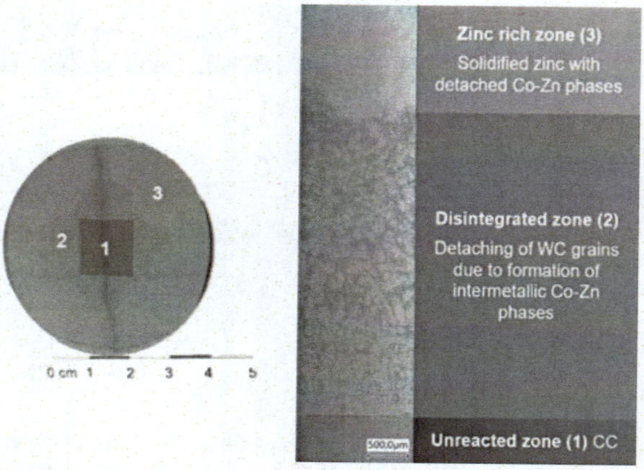

Fig. 6 Sample after zinc treatment (left) and image of a partially disintegrated cemented carbide in solidified zinc with assignation of the different zones (right, taken by the digital microscope Keyence VHX-6000)

the treatment. Figure 6 depicts a sample and an image of the disintegrated structure after zincing. Generally, three different zones can be distinguished:

- Unreacted zone (1): Solid cemented carbide core
- Disintegrated zone (2): Loosened structure with detached WC grains due to the formation of Co-Zn phases
- Zinc rich zone (3): Solidified zinc with detached Co-Zn phases.

The yield for each sample was calculated in percentage by referring the remaining surface of the solid material, measured by stereomicroscopy (Zeiss Stereodiscovery V12) to the cross section of the initial cemented carbide cubes. Figure 7 displays examples of the calculated yields. These results of the measurements were applied for the evaluation by the statistical software MODDE 12 Pro®.

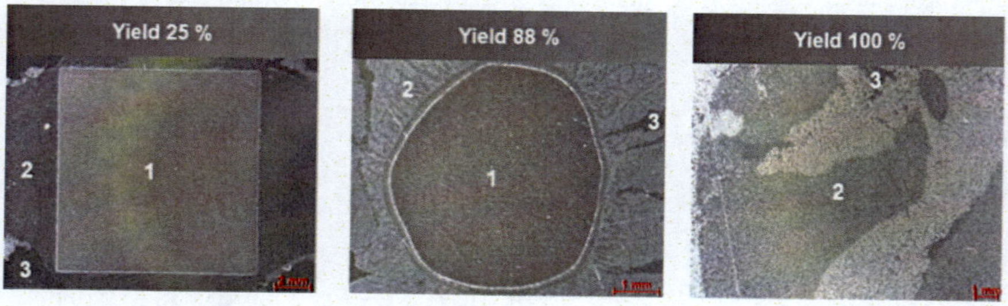

Fig. 7 Yield for three different samples with assignation of the resulting zones (left: GS: C, 15 wt% Co, 600 °C, 2 h; middle: GS: F, 9 wt% Co, 750 °C, 4 h; right: GS: F, 9 wt% Co, 600 °C, 4 h). (Color figure online)

Results of the Zinc Treatment

To evaluate the influence of the investigated factors on the disintegration of cemented carbides and their mutual interactions in the first stage of the zinc process, the calculated yields were analyzed by the statistical software Modde 12 Pro®. Figure 8 depicts the statistical factors R^2, Q^2, model validity (MV) and reproducibility (R) to describe the regression quality of the developed quadratic model equation. The value 1 represents 100%, which means "perfect". R^2 constitutes the model fit, a model with $R^2 = 0.5$ exhibits a low significance. The parameter Q^2 describes the future prediction precision, its value should be >0.5 to ensure a good predictive ability. R^2 overestimates the goodness of fit, whereas Q^2 underestimates it. The difference between these two values is considered to be <20% in most situations. The model displays proper characteristics of the executed regression ($R^2 = 0.652$; $Q^2 = 0.575$). The model validity covers tests of various model problems and should be >0.25. It compares the model error to the pure error. In some cases, MV can be low, as seen in this case, if the pure error exhibits very small values due to extremely good replicates. The reproducibility represents the variation of the responses under constant conditions due to the execution of the tests with regard to the center experiments. This value should be >0.5. The model depicts a high reproducibility of 0.900. Further information concerning the theoretical fundamentals are given in the User guide to Modde, Version 12 [24] by Sartorius Stedim Data Analytics AB [24].

The prediction plots displayed in Fig. 9 show the influence of each parameter (WC grain size (A), Co content (B), temperature (C) and duration (D)) on the yield of the conducted experiments. One of the parameters is varied, whereas the others are held constant at their average values regarding the experimental limits (2.5 µm WC grain size, 12 wt% Co, 675 °C and 2 h). Moreover, the graphs display the lower and upper limits of the 95% confidence interval for the model equation. A rising WC grain size (A) was found to negatively affect the yield. The relationship between the Co content and the disintegration (B) remains positive up to approximately 12 wt% of Co and declines with a further increasing amount of Co. The influence of the temperature

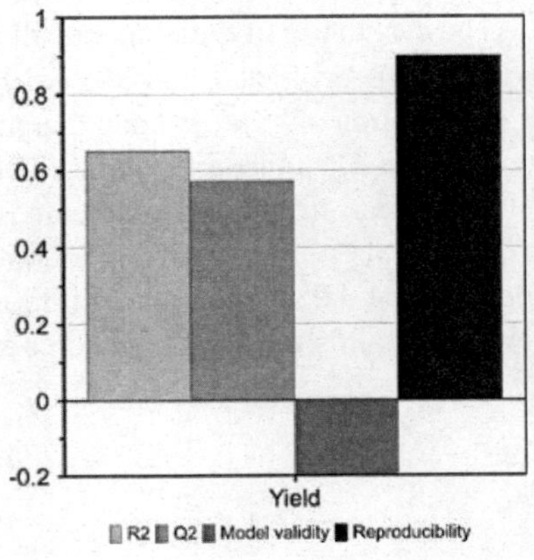

Fig. 8 Summary of fit for the statistical evaluation of the yield generated by Modde 12 Pro®

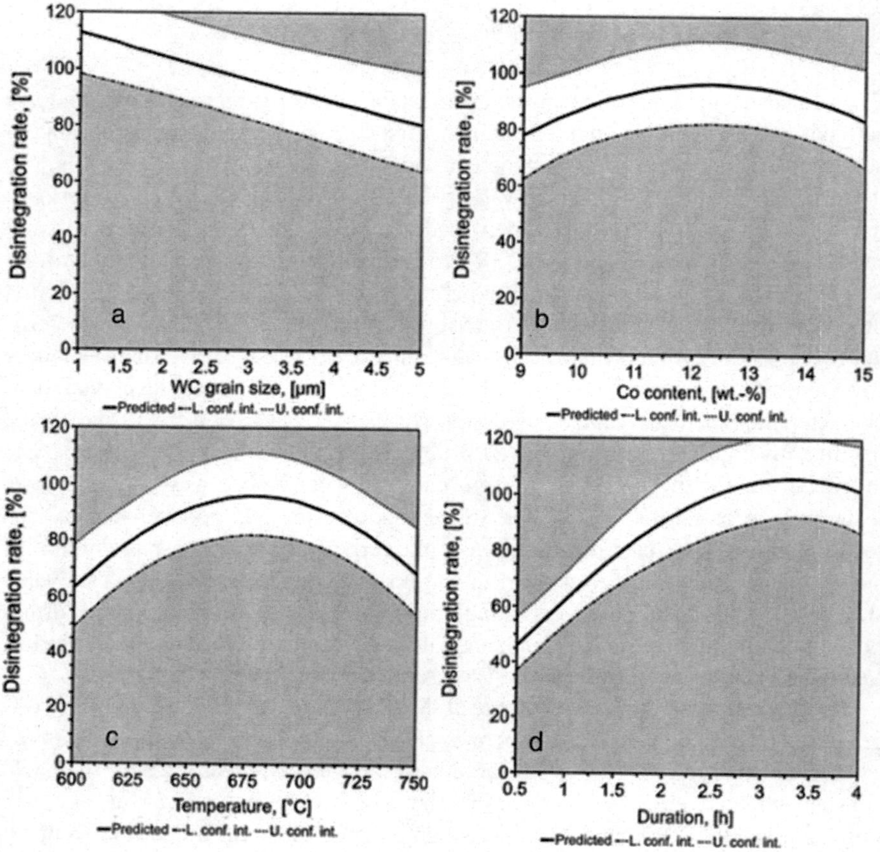

Fig. 9 Prediction plots for the yield of the conducted tests for each parameter, with all other factors held constant at their averages (2.5 µm WC grain size, 12 wt% Co, 675 °C and 2 h)

(C) shows a maximum at ~675 °C. Rising durations result in higher disintegration rates whereas, the curve flattens with increasing annealing durations (D).

The evaluation of these factors allows us to draw conclusions regarding the disintegration of cemented carbides within the investigated limits. Figure 10 depicts a compilation of the examined parameters and therefore, displays the yield in percentage. The inner axes of the contour plots are the factors Co content (9–15 wt%) and WC grain size (1–5 µm), whereas the outer axes represent the temperature (600, 675 and 750 °C) as well as the duration (0.5, 1, 2 and 4 h). These contour plots indicate that the disintegration of the cemented carbides is highest at simultaneously low WC grain size, average Co content and temperatures as well as long duration.

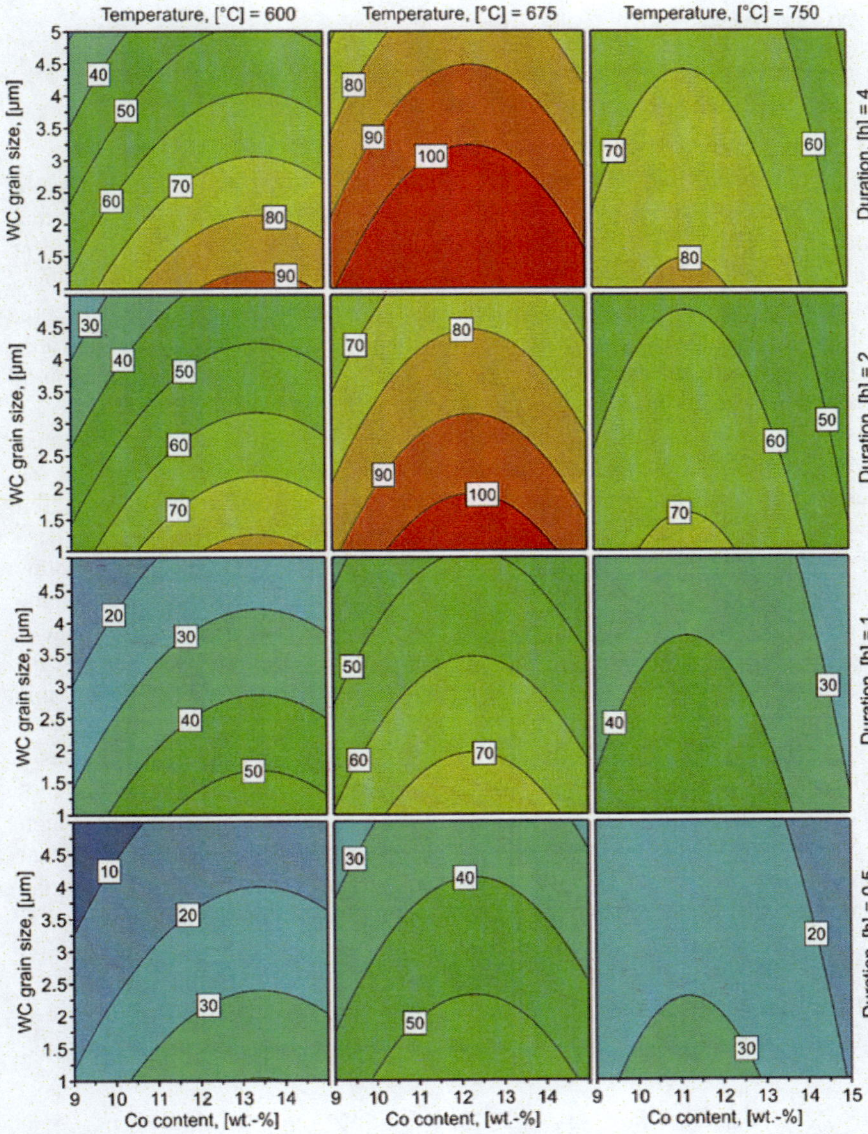

Fig. 10 Yield given in percentage depending on the WC grain size, Co content, temperature and duration generated by Modde 12 Pro®. (Color figure online)

Discussion of the Results

To examine the disintegration behavior of straight WC-Co compound materials the testing material was contacted with an excess of liquid zinc within the stated experimental range:

- WC grain size: 1–5 μm (F, M, C),
- Co content: 9–15 wt%,
- temperature: 600–750 °C and
- duration: 0.5–4 h.

During the treatment of cemented carbide with an excess of zinc a disintegration of the composite occurs due to the formation of Co-Zn phases as a result of mutual diffusion processes of Co and Zn. This leads to the formation of three different zones: an unreacted (1), a disintegrated (2) and a zinc rich zone (3), examples are displayed in Fig. 11. Zone 1 involves the remaining compact cemented carbide core, indicating an incomplete conversion of the cemented carbide. The disintegrated area is characterized by detached WC grains and lamellas as a result of the formation of intermetallic phases. Figure 11b depicts WC grains surrounded by various Co-Zn phases. This article did not state the detailed evaluation of the Co-Zn phases, because it would exceed the scope of this study. Zone 3 covers the solidified zinc and Co-Zn phases, which detached from area 2, depicted in Fig. 11c. The Co-Zn phases exhibit a higher surface expansion compared to the WC grains. At a certain point these intermetallic phases detach from the disintegrated area into the liquid zinc and draw along the WC grains, resulting in a widening of zone 2. It needs to be noted, that the phenomena of detaching Co-Zn phases into the liquid zinc mainly occurs, if zinc is available in excess resulting in a high amount of melt surrounding the sample. By reducing the applied zinc due to economic and ecological interests, the effect of this transaction decreases.

Concerning the impact of the investigated input parameters Fig. 9a depicts a negative trend for the correlation of WC grain size and yield. Therefore, this plot indicates highest disintegration at low grain size of 1 μm. This results from the high specific surface area between the WC grains and the binder, which is concordant with data from the literature. Alkatsev et al. [19] stated that the diffusion of zinc in the cemented carbide mainly proceeds along the phase boundaries between WC and Co. This supports the outcome of our study, which suggests better disintegration behavior of fine grained cemented carbides at the examined operating conditions.

The plot in Fig. 9b shows a maximum indicating highest yields at Co contents of approx. 12 wt%. Based on this data we assume that the formation of intermetallic

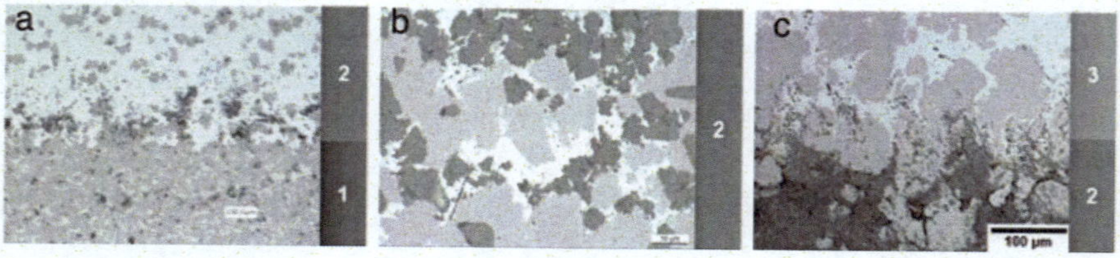

Fig. 11 Resulting zones of different cemented carbides treated with an excess of zinc, **1** unreacted, **2** disintegrated and **3** zinc rich zone

Co-Zn phases is just pronounced weakly at low Co contents (e.g. 9 wt% Co) therefore, the chipping of WC grains from the solid material takes place at a lower rate. Due to the comparably higher supply of Cobalt at higher Co contents (e.g. 15 wt% Co), an increasing amount of solid intermetallic Co-Zn phases arises. Therefore, we hypothesize that the diffusion of Co and Zn atoms, which are required for the phase formation and therefore, the proceeding of the disintegration is impeded in this case. As a consequence, we conclude that average Co binder contents (approx. 12 wt%) represent the ideal balance between formation of intermetallic phases resulting in a chipping of WC grains and the impact of prolonged diffusional paths for Co and Zn in an excess of liquid zinc.

The influence of the duration on the disintegration observed in this investigation, is consistent with our previous studies [10], which focused on the growth of intermetallic phases in Co-Zn diffusion couples. In this preceding examination we evaluated the resulting layer thickness of intermetallic Co-Zn phases while contacting cobalt with liquid zinc at comparable experimental conditions (600–750 °C, 0.5–8 h, liquid zinc in excess). This total layer thickness results from the interaction between the growth of Co-Zn compounds at the interface on the one hand and the detaching of intermetallic phases into the liquid zinc on the other hand. An example for the resulting, adherent intermetallic layer is given in Fig. 12.

The data of the preceding study indicated highest layer thicknesses at 675 °C before the process of detaching intermetallic phases predominates the system with increasing temperatures. As given in Fig. 9c highest disintegration rates are also found at this temperature. This allows the hypothesis that at this temperature the width of the Co-Zn layer is ideal to promote the chipping of WC grains from the solid material before the intermetallic phases detach to the liquid zinc. Therefore, it can be assumed that a temperature of approx. 675 °C favors the loosening of the composite most within the investigated experimental limits.

Fig. 12 Resulting intermetallic Co-Zn layer at the interface of Co with assignation of the different zones [10]

For the impact of the duration on the disintegration we observed a flattening of the curve with increasing time. Generally, typical infiltration speeds of ~1 mm/h for the traditional zinc process are reported [21], whereas higher rates are stated by the treatment with vaporous zinc [19]. Freemantle et al. [25] characterized by-products remaining after the zinc process, which involved solid cemented carbide cores. These by-products occur when the feed materials dimensions are larger than approx. 20 mm and therefore, the zinc is unable to infiltrate the entire cemented carbide which therefore, needs to be treated within the process again [25]. At the expense of increased cost as a result of higher process durations the disintegration can be promoted but nevertheless, in some cases the occurrence of these by-products is inevitable due to the dimension and nature of the inserted scrap [25]. Therefore, Freemantle et al. [25] reported infiltration depths of ~5 mm even for high process durations (24 h at 930 °C for infiltration; vacuum and distillation process at 930 °C for 36 h; gradually cooling). This study supports our findings concerning the flattening of the disintegration rate with rising duration. The progress of the disintegration rests upon the formation of Co-Zn phases as a result of mutual diffusion processes of Co and Zn. These solid intermetallic phases lead to the chipping of WC grains from the cemented carbide surface. These transactions result in the formation of a transition zone between the unreacted, solid cemented carbide and the liquid zinc. The nascent disintegrated zone consists of detached WC grains surrounded by solid intermetallic phases, which widens with increasing durations. Thus, the diffusional paths for Co and Zn atoms are prolonged simultaneously originating from this widening. As a consequence, a slow-down of the disintegration rate occurs with increasing process durations. Therefore, the data indicates that a further raise of the process durations won't improve the disintegration of cemented carbides within the examined experimental conditions at a certain point.

Conclusion and Prospects

The recycling of cemented carbides at the end of their product lifecycle not only offers a wide range of ecological, economic and strategic advantages but also contributes to the resource efficiency and counteracts future supply shortfalls. Especially, because the two main components tungsten and cobalt are rated as critical raw materials for the European Union. Due to the low energy consumption, high tungsten recovery, eco-friendliness and accordingly high amount of treated scrap the zinc process constitutes the most important direct recycling method for cemented carbides. Even though the industry applies this procedure widely on large scale, the exact mechanism of the disintegration progress in the first step of the zinc process is not fully understood yet. To generate fundamental knowledge in this field this study examined the behavior of straight WC-Co grades at lower temperatures of 600–750 °C by contacting these materials with an excess of liquid zinc. The research activities comprise the investigation of the factors WC grain size (1–5 μm), Co binder content (9–15 wt%), temperature (600–750 °C) and duration (0.5–4 h) on the disintegration. While contacting the cemented carbide samples with liquid zinc a disintegration of the compound occurs due to the formation of Co-Zn phases due to mutual diffusion processes of Co and Zn. Thus, three different zones are formed:

- Unreacted zone (1): Solid cemented carbide core,
- disintegrated zone (2): Loosened structure with detached WC grains due to the formation of Co-Zn phases and
- zinc rich zone (3): Solidified zinc with detached Co-Zn phases.

Furthermore, the generated data allows the conclusion that highest yields are achieved under the examined process conditions at simultaneously

- low WC grain size (1 μm),
- average Co binder content (12 wt%),
- average temperatures (675 °C) and
- high durations (4 h).

Based on these findings it can be hypothesized that straight WC-Co grades with low WC grain size and simultaneously average Co content are readily recycled via the zinc process. Whereas, certain challenges in recyclability may arise for grades with high WC grain size and low Co content as well as high grain size and high share of Co binder, both types are indicated in Fig. 13. Within the experimental limits the best disintegration is achieved at 675 °C and 4 h of zincing duration. According to the generated data further raise of the process temperature and duration won't enhance the decomposition of the cemented carbides in an excess of liquid zinc under the examined conditions.

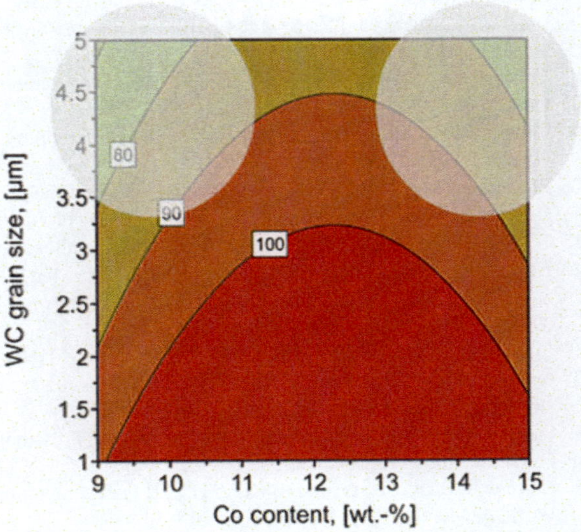

Fig. 13 Yield in percentage depending on WC grain size and Co content at 675 °C treated for 4 h, stressing certain cemented carbide showing worse disintegration behavior. (Color figure online)

Future studies should provide further insight concerning the disintegration behavior of cemented carbides within the zinc process by reducing the amount of applied zinc to the industrially applied zinc to scrap ratio and process temperatures. These investigations should contribute to an improved comprehension of the ongoing transactions and influencing factors in the first stage of the zinc process. This valuably contributes to an enhancement of the process conditions by an efficient resource and energy use and therefore, the optimization of the recycling of the critical raw materials tungsten and cobalt.

Acknowledgements The financial support by the Austrian Federal Ministry for Digital and Economic Affairs, the National Foundation for Research, Technology and Development and the Christian Doppler Research Association is gratefully acknowledged.

References

1. Böhlke W (2002) Hartmetall—Ein moderner Hochleistungswerkstoff. Materialwissenschaft und Werkstofftechnik, 33, 575–580
2. Liedtke M, Schmidt M (2014) Rohstoffrisikobewertung—Wolfram. Deutsche Rohstoffagentur (DERA) in der Bundesanstalt für Geowissenschaften, Berlin
3. Prakash L (2014) Fundamentals and general applications of hardmetals. Compr Hard Mater 1:29–90
4. Schubert WD, Lassner E, Böhlke W (2015) Cemented carbides—a success story. internet. http://www.itia.info/assets/files/Newsletter_2010_06.pdf. Accesses 14 Jan 2015
5. Fang ZZ, Koopman MC, Wang H (2014) Cemented tungsten carbide hardmetal-an introduction. Compr Hard Mater 1:123–137
6. Bose A (2011) A perspective on the earliest commercial PM metal-ceramic composite: cemented tungsten carbide. Int J Powder Metall 47:31–50
7. Ceratizit Austria SA (2016) Rods performs. http://www.ceratizit.com/services/downloads/. Accessed 13 Oct 2016
8. Ceratizit SA (2016) Cutting tools. http://www.ceratizit.com/services/downloads/. Accessed 13 Oct 2016
9. Ebner T (2016) Konzepte zur Optimierung des Hartmetallrecyclings. Dissertation, Montanuniversität Leoben, Leoben
10. Ebner T et al (2020) Studies on the formation of intermetallic compound layers in Co(W)–Zn Diffusion couples. In: Siegmund A et al (ed) 2020—PbZn 2020, pp 661–672
11. Gille G, Meier A (2012) Recycling von Refraktärmetallen. Recycling und Rohstoffe 5:537–560
12. Ad-hoc Working Group: Critical raw materials for the EU. https://nachhaltigwirtschaften.at/resources/e2050_pdf/reports/20100730_critical_raw_materials_for_the_eu.pdf?m=1469660947. Accessed 01 Sep 2016
13. Luidold S et al (2016) Kritische Rohstoffe für die Hochtechnologieanwendung in Österreich. http://www.nachhaltigwirtschaften.at/e2050/e2050_pdf/endbericht_1311_kritische_rohstoffe.pdf. Accessed 09 Aug 2016
14. Oakes J (1999) Recycling of Tungsten Materials. In: Presentation, 12th annual general meeting, Fukuoka. Accessed 12 Oct 1999
15. Angerer T, Luidold S, Antrekowitsch H (2011) Technologien zum Recycling von Hartmetall. World of Metallurgy. ERZMETALL 64:5–14
16. Lassner E, Schubert WD (1999) Tungsten: properties, chemistry, technology of the elements, alloys, and chemical compounds. Springer US

17. Karhumaa T, Kurkela M (2013) Review of the Hardmetal recycling market and the role of the zinc process as a recycling option. In: Proceedings of the 18th Planseeseminar, pp 1.138–1.148
18. Leal-Ayala DR et al (2015) Mapping the global flow of tungsten to identify key material efficiency and supply security opportunities. Res Conser Recycl 103:19–28
19. Alkatsev MI, Svistunov NV, Trotsenko IG (2008) Regeneration of the WC-Co hard alloy with the use of gaseous zinc. Russ J Non-Ferrous Metals 49:156–159
20. Karhumaa T et al (2017) Effect of powder particle size distribution on the properties of submicron hard metal made of WC-Co zinc reclaim powders. In: Proceedings of the 19th Plansee Seminar, RC2/1-RC2-12
21. Zeiler B Recycling von Hartmetallschrott. Hartstoffe, Hartstoffschichten, Werkzeuge, Verschleißschutz, pp 283–303
22. Tikomet Oy (2017) Hartmetallrecycling. http://www.tikomet.fi/de/hartmetall-recycling/. Accessed 28 July 2017
23. Brookes K (2014) Hardmetals recycling and the environment. Metal Powder Report 69:24–30
24. Satorius Stedim Data Analytics AB (Ed) (2017) User guide to MODDE. Satorius stedim data analytics AB, Umeå
25. Freemantle CS et al (2015) PIXE characterization of by-products resulting from the zinc recycling of industrial cemented carbides. Nuclear Instr Methods Phys Res Section B Beam Inter Mater Atoms 363:167–172

Author Index

C. Anderson et al. (eds.), *Ni-Co 2021: The 5th International Symposium on Nickel and Cobalt*, The Minerals, Metals & Materials Series,
https://doi.org/10.1007/978-3-030-65647-8

Subject Index

C. Anderson et al. (eds.), *Ni-Co 2021: The 5th International Symposium on Nickel and Cobalt*, The Minerals, Metals & Materials Series,
https://doi.org/10.1007/978-3-030-65647-8